Smart 3D Nanoprinting

Examining smart 3D printing at the nanoscale, this book discusses various methods of fabrication, the presence of inherent defects and their annihilation, property analysis, and emerging applications across an array of industries. The book serves to bridge the gap between the concept of nanotechnology and the tailorable properties of smart 3D-print products.

FEATURES

- Covers surface and interface analysis and smart technologies in 3D nanoprinting
- Details different materials, such as polymers, metals, semiconductors, glass-ceramics, and composites, as well as their selection criteria, fabrication, and defect analysis at nanoscale
- Describes optimization and modeling and the effect of machine parameters on 3D-printed products
- Discusses critical barriers and opportunities
- Explores emerging applications in manufacturing industries, such as aerospace, healthcare, automotive, energy, construction, and defense

Smart 3D Nanoprinting: Fundamentals, Materials, and Applications is aimed at advanced students, researchers, and industry professionals in materials, manufacturing, chemical, and mechanical engineering. This book offers readers a comprehensive overview of the properties, opportunities, and applications of smart 3D nanoprinting.

Smart 3D Nanoprinting
Fundamentals, Materials, and Applications

Edited by
Ajit Behera
Tuan Anh Nguyen
Ram K. Gupta

CRC Press
Taylor & Francis Group
Boca Raton London New York

CRC Press is an imprint of the
Taylor & Francis Group, an **informa** business

First edition published 2023
by CRC Press
2385 NW Executive Center Drive, Suite 320, Boca Raton FL 33431

and by CRC Press
4 Park Square, Milton Park, Abingdon, Oxon, OX14 4RN

CRC Press is an imprint of Taylor & Francis Group, LLC

ISBN: 978-1-032-03861-2 (hbk)
ISBN: 978-1-032-03862-9 (pbk)
ISBN: 978-1-003-18940-4 (ebk)

DOI: 10.1201/9781003189404

Typeset in Times
by MPS Limited, Dehradun

Visit the Taylor & Francis Web site at
http://www.taylorandfrancis.com

and the CRC Press Web site at
http://www.crcpress.com

Contents

Preface

In this current era, the increased adoption of 3D printing technologies and the incorporation of nanoscale solutions are creating potential growth opportunities for almost all the industries from health care to aero-industries. Nanoscale study on 3D-print products is in an early stage of revolution that is clearly represented in this book. Applying nanotechnology concepts to 3D printing brings many advantages to the engineered materials, such as tailorable property, design flexibility, easy assimilation of materials and devices without a human operator, and many more. Nanomaterials and nanotechnology-based 3D printing techniques can offer many superior qualities over their traditional counterparts, such as higher mechanical-, electrical-, optical-properties, and better corrosion resistance, etc. Concerning these products, the global 3D printing market size was estimated at USD 11.58 billion in 2019 and is expected to reach USD 35.38 billion in 2027 with a compound annual growth rate of 14.6%. This book emphasizes the fundamental concepts associated at surface and interface of any 3D-print products for both nanomaterials and nanotechnology. Various ways of fabrication, presence of inherent defect and their annihilation, property analysis, and emerging application are the main novel focus points of this book. Again, this book will bring a green revolution in the 4.0 Industry. This book can bridge the gap between the concept of nanotechnology and the tailorable properties of 3D-print products, which is a great need for educational and industry society.

Editors

Ajit Behera, PhD, is an Assistant Professor in the Metallurgical and Materials Department at the National Institute of Technology, Rourkela. He earned a PhD at IIT-Kharagpur in 2016. He received the national Yuva Rattan award in 2020 for his contribution to society along with his academic carrier. Also, he received the young faculty award in 2017 and the C.V. Raman Award in 2019. He has published more than 100 publications, including books, book chapters, and journal articles. His research interests include smart materials, additive manufacturing, 3D and 4D printing, NiTi-alloys, plasma surface engineering, nanotechnology, magnetron sputtered thin film, cryo-treatment, and utilization of industrial waste. He has published four patent-related articles on smart materials. He has completed six sponsored research projects, along with several consultancy projects. He is a regular speaker and external examiner on materials science and advanced materials. He is associated with many reputed scientific international organization as a committee member or advisory committee member. More than ten PhD students are from his institute or outside the institute, and four foreign exchange students are working with different projects with him. See https://scholar.google.com/citations?user=R-G7pSoAAAAJ&hl=en.

Tuan Anh Nguyen, PhD, earned a BSc in physics at Hanoi University in 1992 and a PhD in chemistry at the Paris Diderot University (France) in 2003. He was a Visiting Scientist at Seoul National University (South Korea, 2004) and University of Wollongong (Australia, 2005). He then worked as a Post-doctoral Research Associate and Research Scientist at Montana State University (USA) from 2006 to 2009. In 2012, he was appointed Head of the Microanalysis Department at the Institute for Tropical Technology (Vietnam Academy of Science and Technology). His research activities include smart coatings, conducting polymers, corrosion and protection of metals/concrete, antibacterial materials, and advanced nanomaterials.

Ram K. Gupta, PhD, is an Associate Professor of Chemistry at Pittsburg State University, Pittsburg, Kansas, USA. Before joining Pittsburg State University, he was an Assistant Research Professor at Missouri State University, Springfield, Missouri, and then a Senior Research Scientist at North Carolina A&T State University, Greensboro, North Carolina. Dr. Gupta's research focuses on green energy production and storage using conducting polymers and composites, nanomaterials, optoelectronics and photovoltaic devices, organic-inorganic hetero-junctions for sensors, nano-magnetism, bio-based polymers, bio-compatible nanofibers for tissue regeneration, scaffold and antibacterial applications, and bio-degradable metallic implants. See https://scholar.google.com/citations?user=6QwWKBEAAAAJ&hl=en.

1 3D Printing for Hybrid Nanocomposites

Selection Criteria, Fabrication and Defect Analysis

Garima Mittal
Materials Innovation Centre, School of Engineering,
University of Leicester, UK

Shiladitya Paul
Materials Innovation Centre, School of Engineering,
University of Leicester, UK

Materials and Structural Integrity Technology Group, TWI
Ltd, Cambridge, UK

CONTENTS

DOI: 10.1201/9781003189404-1

1.1   INTRODUCTION

Nanomaterials have different properties compared to their bulk counterparts due to the high surface to volume ratio. This, along with the ability to engineer materials at a small scale allows nanotechnology to provide better control over fine-tuning the properties of resultant nanomaterials, including complex composites. 3D printed nanocomposites can have improved properties but the production of such composites can deteriorate the printing quality by increasing the viscosity of the solution, hindering light penetration and anisotropy [1]. A new class of composites, i.e., hybrid nanocomposites, can resolve these issues through hybridisation (interactions at the hybrid interface) [2]. Since the quality and properties of the printed hybrid composite are significantly affected by the production method and characteristics of the material used for printing, this chapter will focus on hybrid composites, their selection criteria for 3D printing, and their 3D printing methods. Despite advancements, the uptake of 3D printing for mass production is very sluggish and limited due to inferior performance and lack of repeatability compared to traditional manufacturing. During printing, different kinds of defects such as porosity, cracking, balling and powder agglomeration occur between separate layers and even within an individual layer. These defects, either process generated or due to environmental factors [3], influence the quality of the final product. Defect detection in 3D printing is a hot topic of discussion as timely detection of defects could save time, cost and resources. Hence, a detailed discussion of types and causes of defects in 3D printed material and their detection techniques is presented in this chapter.

1.2   HYBRID NANOCOMPOSITES

According to IUPAC, "A hybrid material is composed of an intimate mixture of inorganic components, organic components, or both types of components. Note: The components usually interpenetrate on a scale of less than 1 μm" [4]. In other words, hybrid composites are materials formed by the combination of two or more different materials that interact at the submicron level (molecular or nanoscale). Unlike conventional nanocomposites, hybrid nanocomposites exhibit the characteristics that are new or in between the two original components due to the submicron interactions such as chemical bonds and van der Waals attractions, taking place at the hybrid interface rather than just summing up the properties of two individual components [2]. Composites and hybrid composites differ in their functions and properties. Due to interacting at a submicron level, hybrid nanocomposites are more homogeneous. Often the terms 'composites' and 'hybrid composites' are ambiguously used in the scientific literature. Combining two or more kinds of reinforcements in the same matrix or single reinforcement in combining two different types of matrices is also referred to as hybrid composites [5]. These composites could be categorised as hybrids, but by-composition rather than hybrid by-interaction as the hybridisation concept in it is entirely different from hybrid by-interaction composites [6]. Fibre-reinforced nanocomposites are typical hybrid-by-composition, where fibres provide better stiffness and strength to the composite, and nanomaterials overcome the insufficient load-carrying properties by altering the fibre-matrix interface. The final product is often referred to as a hybrid nanocomposite because of its filler composition.

Another class of hybrid nanocomposites is hybrid-by-interaction composites, where a combination of inorganic-organic materials interacts at the submicron level. Hybrid nanocomposites possess a very fine-scale organic-inorganic interface with interconnected molecular co-networks of both parts that potentially synergise both parts as a single phase, providing the same physicochemical behaviour throughout the entire structure. Unlike hybrid-by-composition nanocomposites, inorganic-organic components are non-discrete from each other in hybrid-by-interaction nanocomposites. Based on the type of interactions occurring at the inorganic-organic interface, these hybrid materials could be divided into two classes: Class I and Class II, covering hybrids with weak interactions like hydrogen bonding, van der Waal's force or weak electrostatic interactions and with strong chemical bonding like covalent or ionic-covalent bonding between inorganic-organic components, respectively [7,8]. The modification of an inorganic material using organic molecules and vice-versa are examples of hybridisation [9]. Hybrid nanocomposites could be prepared by various methods, including sol-gel synthesis, hydrothermal synthesis, intercalation and grafting. Sol-gel hybrid nanocomposites and organic clay intercalated polymer composites are typical examples of hybrid nanocomposites and have been used extensively in 3D printing [10,11]. Some of the examples of hybrid nanocomposites used in 3D printing are presented below (Table 1.1):

TABLE 1.1
Examples of Hybrid Nanocomposites Used in 3D Printing

Hybrid Nanocomposites	3D Printing Methods	Applications
Silicone rubber-epoxy/nanoclay	Direct ink writing (DIW)	wearables, healthcare, and soft robotics
Poly L-lactic acid (PLLA) /nano hydroxyapatite (n-HAp)/bioactive glass	Fused filament deposition (FFD)	Bone repair
Polydimethylsiloxane (PDMS)/SiO_2/CNTs	Direct ink writing (DIW)	Piezoresistive sensors
Polyethylene glycol (PEG)/Iron oxide	UV-assisted direct ink writing (DIW)	Bioinspired soft robotic systems
Poly(lactide-co-glycolide) (PLGA)/n-HAp	Fused filament deposition (FFD)	Bone repair
Poly(ethylenglycol)diacrylate, PEGDA/ tetraethoxysilane (TEOS)	Digital light processing (DLP)	Biomimetic applications

1.3 SELECTION CRITERIA OF COMPONENTS IN HYBRID COMPOSITES

Hybrid nanocomposites pave a new pathway to fabricate structures with complex geometries and improved and desired properties such as shape memory, optical transparency and magnetism that can be used in numerous applications. However, 3D-printing inks formulated with hybrid nanocomposites containing organic and inorganic constituents exhibit some issues related to its flowability, mechanical behaviour etc. Ideally, in the case of extrusion-based printing, the material should be dispensed out properly from the nozzle and form a stable structure after deposition. The printability of hybrid feedstock material depends on its viscoelastic behaviour and should satisfy the following four fundamental conditions (Figure 1.1) [12]:

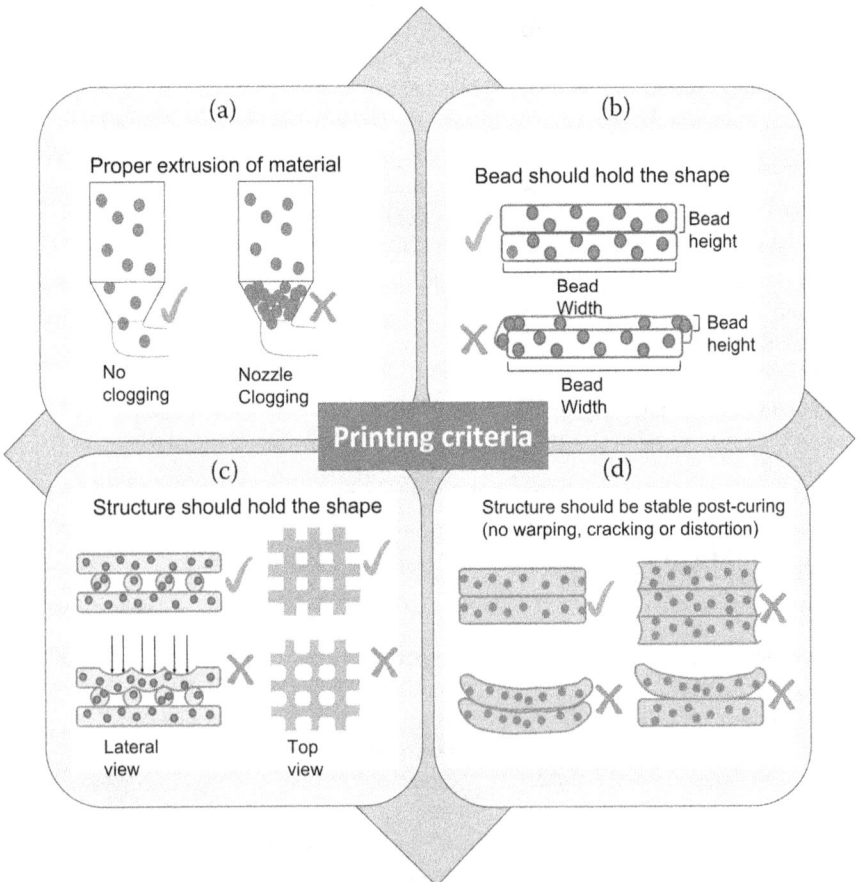

FIGURE 1.1 Four printability criteria for extrusion-based 3D printing feedstock materials: (a) Proper extrusion of material, (b) printed bead should hold the desired shape/height after printing, (c) printed filament should hold its structure over a bridge and (d) printed structure should retain its shape even after curing.

a. A pressure-assisted dispense of the ink must take place out of the printing nozzle with a particular diameter at a specific flow rate.
b. The dispensed material must hold the desired shape after extrusion.
c. The extruded material must be able to hold its structure while forming a bridge over a particular gap.
d. The extruded material must retain its shape during curing into the final structure.

These four criteria help select the feedstock constituents (hybrid composite with better interactions between its components) and optimise various extrusion-based printing conditions like temperature, flow rate, speed, nozzle diameter etc. Other factors that affect the printability (extrusion and non-extrusion-based) of material are discussed below:

1.3.1 FEEDSTOCK COMPOSITION

Nanoscale reinforcements in hybrid nanocomposites provide better control over the manufacturing process, but exhibit challenges in their distribution, dispersion, orientation and interfacial interactions in the matrix, and hence, practical applications do not meet with the theoretical predictions. For better printability, filler related properties such as aspect ratio, loading, orientation and presence of surfactants, and matrix-associated properties such as type, chemical structure, crystallinity, molecular weight and cure kinetics need to be evaluated as they affect the processing parameters, rheological properties of the ink and thermomechanical properties of the printed structure.

By tuning the ratio of organic and inorganic components in the hybrid systems, the flowability and print stability of the ink could be controlled. For example, with increased epoxy content in the epoxy/silicone system, the print stability decreases [10]. Nanomaterials could help achieve the desired flowability of the ink and desired solid-like behaviour after deposition, depending on the viscosity of the hybrid resin. The filler content also affects the surface finish as low filler content might not solidify the feedstock material, and printed filaments fuse and lose individual definition [13]. With increased filler content, features related to individual filament deposition become prominent in the final structure. High filler content could also cause agglomeration and inferior printability and surface finishing. The filler aspect ratio is another factor that influences the warpage and shrinkage of the printed component. Sometimes, isotropic fillers are more useful than the high aspect ratio fillers (e.g., Fibers) because of introduction of isotropic properties in the printed material [14]. Although 3D printing helps in the homogenous distribution of nanofillers in the printed composite structure, filler alignment affects ink properties and the printability of the final product. For proper orientation, techniques like electrohydrodynamic 3D printing (e-jet), magnetic-field-assisted 3D printing, dual-nozzle strategies and acoustic focusing are used. Along with aligning fillers into the matrix, these techniques help deposit material with increased filler volume concentration [15].

1.3.2 THERMOPHYSICAL PROPERTIES

Printing temperature (and distribution) strongly influence the thermophysical and rheological properties of the feedstock material [16]. Ideally, the feedstock material should have desired thermophysical properties such as a low coefficient of thermal expansion (CTE), low shrinkage/retention, no phase changes after solidification, little or no volatile compounds and high heat resistance. For thermosets, printing takes place mostly at room temperature, while for thermoplastics, this is not fixed. After extrusion, due to the transition of material from its glass transition temperature (Tg) to chamber temperature, defects (cracks, inter or intra-layer delamination) might occur, affecting the strength and size of the printed material. The elevated (~Tg of the material) printing chamber temperature helps avoid these deformations by enhancing the interlayer diffusion and bond strength [17,18]. Print speed is another parameter that affects the structure and properties of the printed material. If the printing speed is too low, an array of spherical voids with spacing approximately equal to the nozzle diameter could be observed in the cross-section of the material [13]. Similarly, more residence time (material staying in the hot zone) with slower printing speed results in a final structure with many voids due to the degradation of polymer components [19]. If loaded in optimal volume, thermally conductive nanofillers could improve the thermal conductivity and reduce the warpage of the printed structure. Too low filler content might result in low thermal conductivity, and agglomeration could be observed for too high filler content [20]. Good interfacial interaction between filler and matrix enhances the thermal conductivity by reducing interfacial thermal resistance [20]. Besides, nanofillers help improve the printability of filamentous hybrid material by improving its buckling resistance. Ideally, the critical buckling pressure of the filament must be higher than its extrusion pressure and the filament must be flexible to allow printing without fracture.

1.3.3 RHEOLOGICAL PROPERTIES

Rheological properties such as viscosity, elastic modulus, yield stress also affect the printability of a material and shape fidelity and structural stability of the printed part. Ideally, during printing, the ink should pass through the nozzle optimally (high shear-thinning behaviour), and post-printing, it should retain its predefined shape (high yield stress). Generally, smaller nozzles are preferred for less viscous inks, while for more viscous inks, bigger nozzles with rigid struts are suitable. Apart from its inorganic-organic content, the rheological property of a hybrid ink varies with its filler content. If the filler content is too little, the printed structure might lose its integrity. Hence, for proper shear thinning behaviour and shape-retaining, an optimal filler content should be incorporated [21]. Longer residence time in the nozzle is also responsible for poor printing quality due to heat-induced viscosity reduction [22]. This information, along with temperature, helps attain the best possible flow by varying screw speed or nozzle geometry.

1.4 FABRICATION METHOD IN 3D PRINTING USING HYBRID NANOCOMPOSITES

3D printing processes can be categorised into liquid-based and solid-based methods. There are different sub-categories of these methods based on the layer deposition approach. Material extrusion is the basic idea behind liquid-based methods, where molten or semi-molten material deposits and solidifies in a layer after passing through a heated nozzle. This ink is generally a polymeric material with or without reinforcing materials like fibres, metal or ceramic nanoparticles, etc. While solid-based techniques work on the sintering or melting principle, where a heat source such as a laser or electron beam is used to melt the powder bed of the intentional material (could be metal, ceramic or plastic), which later solidifies in the form of a layer. Different synthesis routes can be used to form ink containing organic-inorganic components. In class I hybrid nanocomposites, the polymer component is entrapped with an inorganic (e.g., Silica) network during condensation. The in-organic network entangles with the polymeric chains by mechanical means and hydrogen bonding [23]. For class II hybrid nanocomposites, inorganic and organic components are linked with the help of a coupling agent or a polymer with silane bonds (such as polydimethoxysilane (PDMS)). In both cases, paste-like sol could be directly printed on a substrate before gelation. The detailed discussion of 3D printing techniques is presented in previous chapters so, this chapter briefly dis-cusses some of the commonly used 3D printing techniques for hybrid nanocomposites:

1.4.1 FUSED FILAMENT FABRICATION (FFF)

3D components are printed by depositing liquid or paste material on a selected path in a layer-by-layer assembly. FFF is one of the extensively employed techniques for printing hybrid nanocomposites. Besides, due to the cost-efficient nature and easy scalability, this method is suitable for industrial production. The printing proces-sability and properties of the printed component depend on various process para-meters such as feed rate, print speed, material properties, inorganic-organic ratio and filler content. In FFF, hybrid nanocomposite filaments or pellets are auto-matically fed into a printing system with a predefined feeding rate, heated and extruded from the printing nozzle. Mainly thermoplastics such as acrylonitrile butadiene styrene (ABS), polystyrene (PS), polylactic acid (PLA), polycarbonate (PC) and nylon are printed using FFF. In a study, hybrid nanocomposite filaments of PLLA/n-HAp/bioactive glass were formed in two steps. At first, inorganic components (n-HAp/bioactive glass) were dispersed into an aqueous solution, followed by mixing with PLLA granules for homogeneous mixing of both parts. The obtained solution was dried for 12 h at 50 °C. Then, a twin-screw extruder was used to form hybrid filaments, where the composite granules were extruded, fol-lowed by palletisation and second-time extrusion for homogenous dispersion of all phases. These filaments showed excellent printability for the FFF 3D printing method [11].

1.4.2 DIRECT INK WRITING (DIW) OR ROBOCASTING

DIW is a selective extrusion-based layer-by-layer assembly printing method used to print various polymeric and ceramic inks. Multiple parameters such as printing pressure and speed, nozzle size, printing medium and ink composition significantly influence the DIW process [24]. The difference between FFF and DIW is that in FFF, the filament/pellet is melted, extruded with the desired flow, followed by deposition and solidification after cooling. While DIW depends on pastes with specific rheology (no melting or cooling); good shear thinning for the desired throughput and good yield stress with quick elastic recovery for structure stability becomes crucial. This could be achieved by chemical or physical cross-linking, solvent evaporation and rheological behaviour tuning. DIW provides the ability to print various materials with a wide range of viscosity.

UV-assisted DIW is a variation of this technique, where UV light is used to solidify photopolymers printed in the form of monomers. For UV-DIW, apart from sol-gel-based inks, colloidal solutions of inorganic nanoparticles (or sometimes aqueous suspension of inorganic particles) encapsulated into monomer droplets of UV-cured polymers are also used to form a stable 3D printed structure [25]. The colloidal solutions of inorganic-organic components offer low viscosity and better dispersion of inorganic components at sub-micron length scales within the polymeric matrix. Inhomogeneous distribution and agglomeration of nanomaterials could limit the potential of DIW printing, but the surface treatment of nanomaterials before incorporation into the organic matrix could overcome the problem [10]. Another approach to mix thermoset resins with photocurable resins is that during printing, photocurable material restores the shape through rapid curing while thermoset part of the material helps to cross-link [26]. Freeze extrusion fabrication (FEF) is another variation of DIW. The inorganic-organic paste is extruded and printed in a layer by layer pattern under cryogenic conditions, helping stable structure formation after printing [27].

1.4.3   DIGITAL LIGHT PROCESS (DLP)

DLP is a variation of stereolithography (SLA) in which resin in a photocurable resin tank polymerises in the presence of UV-light at the predefined points in the x-y plane. Layer thickness is controlled by altering the distance between the plate and the bottom of the tank. DLP differs from SLA in building orientation of the final structure and illumination method. In DLP, the light is projected from the bottom of the tank, while the printing substrate is immersed into the tank from above. The illumination source contains an LED lamp and a digital mirror that helps in polymerising the resin in a predefined UV pattern, unlike a point-by-point pattern as in SLA, allowing faster printing. Because of the targeted and fast UV curing, produced structures show high resolution and dimensional accuracy. Factors like the intensity of light, exposure, scanning speed, and post-printing thermal or UV treatments influence the quality of the printed component [28]. Because of light scattering, nanomaterials reduce the UV penetration depth, but the surface treatment of nanomaterials or in-situ generation of nanoparticles after printing can resolve the issue. For example, silica precursor mixed with an oligomer (PEGDA) and suitable photoinitiators and an organic-inorganic coupling monomer to fabricate PEG/SiO_2 3D hybrid nanocomposite structures using DLP showed homogenous dispersion of nanomaterials into the matrix [1].

A comparison among three widely used conventional techniques for 3D printing of hybrid nanocomposites is presented in Figure 1.2.

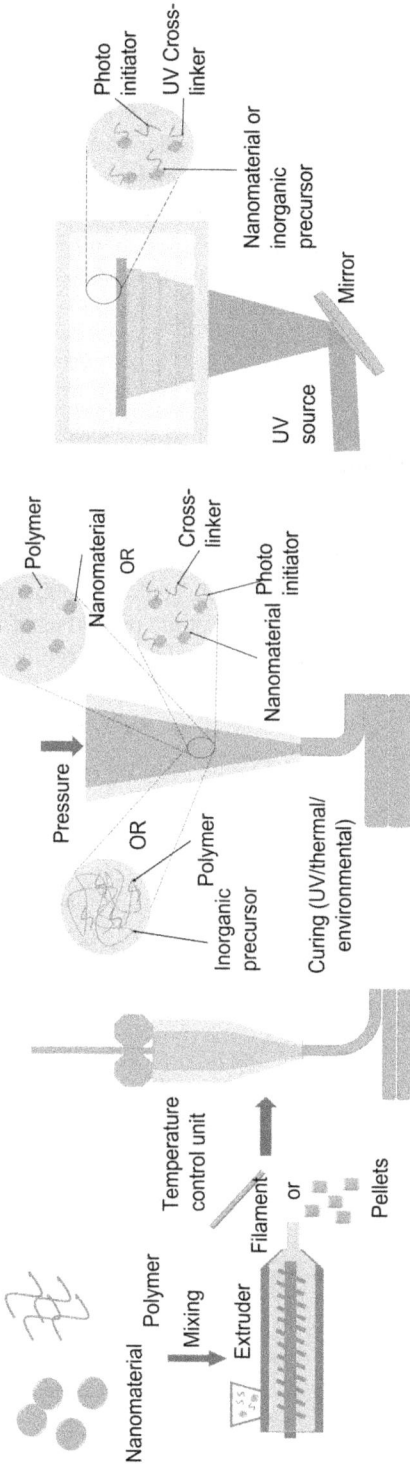

FIGURE 1.2 Comparison among conventional 3D printing techniques, fused filament fabrication (FFF), direct ink writing (DIW) and digital light processing (DLP) used for printing hybrid nanocomposites structures.

	Fused Filament Fabrication (FFF)	Direct ink writing (DIW)	Digital light processing (DLP)
Compatible Ink	Mostly thermoplastic polymers	Wide range of viscosity materials	Photopolymerisable polymers
Advantages	Cost-efficient, simple, suitable for large-scale printing, multi-material printing possible	Cost-efficient, simple, suitable for wide range of materials, large-scale printing, multi-material printing possible	Fine resolution, good surface finishing, quick, suitable to produce complex geometries
Disadvantages	Poor surface finishing, nozzle clogging and degradation issues, poor filament interaction	Required precise control over rheology, low resolution, poor stability of final structure (easy to collapse)	Limited printing materials, UV penetration issues, poor dispersion fillers (in the absence of applied electrical, magnetic, or acoustic field)

Other fabrication methods are powder-based processes, where selective regions of a powder bed are fused or sintered using thermal energy from a laser (selective laser sintering; SLS) or electron beam (electron beam melting; EBM) or bonded layer-by-layer using plastic binders (such as polyvinyl alcohol (PVA)). Multiple jet fusion (MJF) and binder jetting are variations of the powder-based process. Recently, functionally graded 3D printed components with tailored properties have also been produced by printing either single material with graded geometry or multi-materials (or materials with varied elements) across the geometry [29]. Multiple printing nozzles are required for printing multiple materials, and the current most suitable technique for that is polyjet printing.

1.5 DEFECT ANALYSIS

3D printed structures suffer from different kinds of defects that influence the precision and repeatability of the process and properties of the printed material. For defect monitoring, various sensors, automated systems and constant in-line monitoring are integrated into the printing system that helps in improving the process quality and reducing material and time loss. Sometimes post-printing treatments also enhance the quality of the printed material. These defects could be equipment, process or material generated. Different parameters that affect the properties of 3D printed hybrid nanocomposites are presented in Figure 1.3. For defect reduction and process quality improvement, an understanding of defect types and causation is essential.

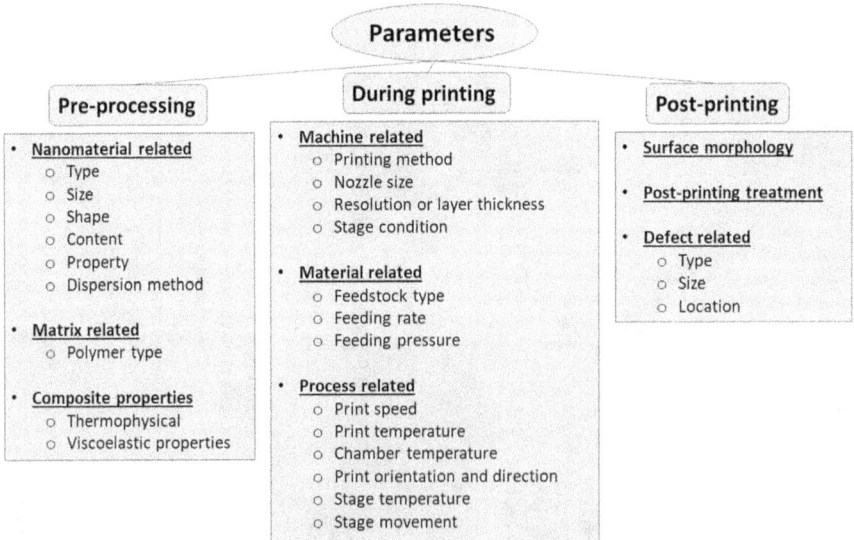

FIGURE 1.3   Common pre-processing, during printing and post-processing parameters that affect the 3D printed structure properties.

1.5.1 DEFECT TYPES

1.5.1.1 Porosity

Porosity is a prevalent defect type found in additive manufactured (AM) components, divided into two main categories: interlayer and Intra-beads defects. The common reason for intra-bead voids in nanocomposites is the poor filler-matrix interaction. Although nanomaterials help reduce the porosity of the matrix, increased nanofiller content due to agglomeration and poor interactions with matrix causes intra-bead porosity. For example, a 3D printed structure of PLA nanocomposite with 5 wt.% of cellulose nanofibers (CNF) showed more agglomeration and porosity than the nanocomposite with 1 wt.% of CNF [30]. Other sources of intra-bead porosity are entrapped air in the feed section of the screw or filament/pellets, polymer degradation, off-gassing or material shrinkage after printing [31]. During printing process, deposition pattern, contact area and thermal fusion are the factors that influence both inter-layer and inter-bead porosity. When a thermoplastic filament passes through a circular nozzle, it does not fill the volume of the CAD design, giving rise to macro-size pores. The size of these pores mainly varies with the nozzle size, layer thickness and extrusion temperature. For example, PLA filaments printed at 250 °C (printing temperature) give reduced inter-layer pore volume in nanocomposites compared to filaments printed at 210 °C because of the increased cohesion surface between the printed composite layers [32]. Printing patterns also affect the inter-layer pore geometry. For example, hexagonal printing arrangement forms six pores at the corners of hexagon while square printing arrangement forms four pores at the corner of the arrangement [33].

1.5.1.2 Anisotropic Properties

Most 3D printed components display distinctive mechanical or thermal properties along the print direction (x-axis) and through successively deposited layers (z-axis). Hence, printing orientation (filament arrangement and orientation) needs to be optimised to minimise the anisotropy. The tensile strength of a printed component is usually highest for the longitudinal orientation of the filament and lowest in the transverse direction [34]. The reason behind it is the insufficient interactions (air gaps or voids formation) between adjacent printed filaments and the inherent anisotropy, affecting the load-bearing capacity of the printed material [13,35]. The interfacial diffusion between printed filaments and layers can be improved through sintering/heating of the printed filament above its critical sintering/glass transition temperature [36]. For good adhesion among printed filaments, methods like exposure to ionising radiation, local heating through microwaves, bimodal blends with low-molecular-weight additives, "z-pinning" approach and infrared heating are used [37]. Reinforced nanomaterial is also responsible for the anisotropic behaviour of the printed material. Anisotropic behaviour is more prevalent in systems reinforced with high aspect ratio materials such as nanofibers, nanotubes and lamellar materials, depending on filler volume and arrangement [13,38]. During extrusion, high-aspect-ratio fillers stably orient in the axial direction of the flow due to the dominant shear flow at the proximity of nozzle walls. With moving towards the bead center, the shear flow decreases gradually. Hence, more disoriented filler arrangement is observed [38].

Different techniques such as acoustic focusing, external magnetic field (for magnetic materials), external electrical field (for conductive materials) and rotating print head are widely used for the spatial control over filler orientation [15,39,40].

1.5.1.3 Distortion and Cracking

Cracking, distortion or warpage, mainly due to thermal stress, are common defects found in 3D printed nanocomposites. For thermoplastic polymers, cooling of extruded material from Tg to chamber temperature give rise to residual stress, leading to inter-layer or intra-bead deformation of the printed component in the form of cracking, de-lamination or warpage. Thermoset polymers experience shrinkage or warpage upon curing while printing or post-printing curing [41]. Crack propagation in nanocomposites is reduced by nanoscale fillers reinforcement that acts as bridging material, limiting the crack propagation through requiring more energy to pull it out or break it [42]. If the nanofiller content is too high, it could lead to crack formation in the composite due to the in-homogeneous distribution and agglomeration of nanomaterials. Sometimes, corners of the printed material start lifting and detaching from the substrate (out-of-plane distortion) depending on material properties like chemical structure, composition, and curing temperature, leading to mechanical failure due to residual stress [43].

1.5.1.4 Stringing

These are surface-related defects, often observed in thermoplastics, that could be detected visually and removed through post-processing. It introduces additional time and cost, especially for scale-up production cases. Misconfiguration in withdrawal settings during designing printing parameters could be a reason for string formation [44].

The pictorial representation of defects is depicted in Figure 1.4 and the generation and solution of different defects are summarised in Table 1.2.

1.5.2 DETECTION METHODS

With computerisation and digitisation of the 3D printing processes, defect analysis of 3D printed components has also evolved, providing information about external and internal defects. The printed components could be analysed ex-situ (off-line; post-printing) or in-situ (online; during printing). Ex-situ techniques are well-established and cost-effective compared to in-situ techniques, but in-situ techniques provide the freedom to analyse process-induced internal structures formation mechanisms. In in-situ defect detection techniques, the probe is placed under the printing bed, and the camera/receptor is mounted on the printing head, scanning each printed layer. The data acquired from these detectors is fed back to the system to modify printing parameters. Defect characterisation techniques such as Infrared imaging, ultrasonic detection and computed tomography (CT) scanning are commonly used. Machine learning-based defect detection technologies are comparatively new methods for anomaly detection in 3D printing.

1.5.2.1 X-Ray Microcomputed Tomography (μ-XCT or MicroCT)

MicroCT is an image-based non-destructive evaluation (NDE) technique that uses x-rays to provide 3D information about internal defects, geometry, porosity, surface

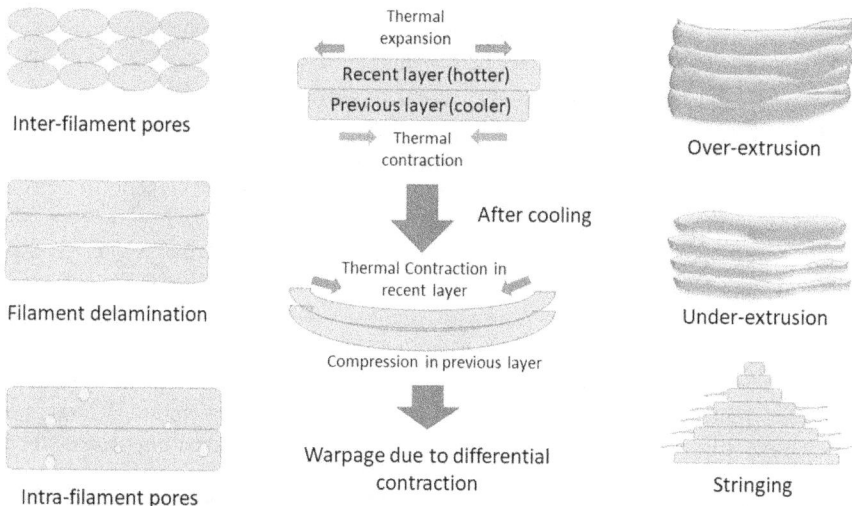

FIGURE 1.4 A pictorial representation of different kinds of defects (inter-layer/filament pores, intra-bead/filament pores, filament delamination, warpage, over-extrusion, under-extrusion and stringing) often found in 3D printed components.

TABLE 1.2
Types of Defects/Discontinuities Observed in Hybrid Nanocomposites

Defects Type	Generation Cause	Solution
Micro-size pores (intra-bead)	Entrapped air in system, vaporization of volatile material, poor interaction between nanomaterials and matrix due to agglomeration of nanomaterials, improper melting of material, poor process parameters	Optimal nanomaterial content and proper dispersion into the matrix, in-situ monitoring (to close the printing loop in the presence of defect), optimization of printing parameters such as print speed and temperature
Macro-size pores (inter-layer)	Poor fusion between printed filaments as well as layers	Post-printing annealing, Z-pinning using nanotubes
Warping	Poor adhesion of first printed layer to the printing bed, thermal stress trapped between layers	Selection of suitable printing bed and optimization of bed temperature
Anisotropy	Printing orientation and direction	Selection of suitable printing orientation
Cracking	Poor interlayer fusion, thermal shrinkage	Optimization of printing process, in-situ monitoring
Stringing	Poor retraction distance and speed	Optimization of printing process, post-printing surface finishing
Curling	Excessive heating	Post-printing surface finishing

morphology and structural quantification. X-rays, projected on the specimen at different rotating angles with small increments, measure the absorption and form a sequence of 2D images. Longer scan time generates high-resolution images, providing information about the density and composition of the specimen. Computational stacking of these images generates a volumetric view comprising volumetric pixels (voxels) of the sample, where each voxel represents the x-ray density of the material. Resolution of generated scan varies with detector type, the x-ray source and hardware. For higher resolution scans, an advanced version of microCT, i.e., off-axis CT, where rotation around a predefined point in the specimen is used to provide information for small size samples. MicroCT could be used to detect various defects (open or closed pores, filament overlapping, post-printing swelling behaviour, printing direction and anisotropy) in composites [34,45]. Figure 1.5 represents the 2D slices and reconstructed volumetric microCT images of 3D printed structures of pure PLA and PLA/CNF (1 wt.%) nanocomposites, where

FIGURE 1.5 X-ray microtomographic images: 2D scan of fractured cross-sections of 3D printed (a) polylactic acid (PLA) and (b) PLA/CNF (1wt.%). Reconstructed x-ray microtomographic 3D images of 3D printed (a) polylactic acid (PLA) and (b) PLA/CNF (1wt.%). (Reproduced with permission from reference [30]. © 2020 Society of Plastics Engineers).

Figure 1.5 (a), (c), (b) and (d) are the 2D scans of the fractured surface and re-constructed volumetric scan of pure PLA and PLA/CNF (1 wt.%), respectively [30]. MicroCT is suitable for small 3D printed components. Limited resolution used for larger components might fail to detect small defects. The minimum detectable defect size should be at least three voxels wide. The scan resolution is limited up to the one-2000th time of the largest part of the specimen. Also, due to inefficient penetration of X-rays in large and high-density specimens, image quality is compromised. MicroCT is a costly detection technique, and it does not have relevant industrial standards. Despite limitations, microCT is extensively employed to detect various defects, improving the knowledge of 3D printing process parameters.

1.5.2.2 Infrared (IR) Detection

IR or thermography is an NDE technique, providing defect information through thermal radiation difference between irregularity and rest of the material, visualised through infrared images. Different components of the printed material exhibit different heat flow due to varied heating and cooling rates. This contrast in emitted radiation is depicted by infrared camera, revealing the anomaly based on its radiation intensity [46]. The temperature rise at an interfacial gap or void will be slower than the 'non-defect' material, or a thinner layer would heat more quickly than a thicker layer. In some studies, thermal imaging is used in conjunction with x-ray scattering to provide information on crystalline structure progress and related thermomechanical properties [47]. Before scanning, thermal calibration of the camera is required and it needs to be assured that the temperature range is suitable for the detection camera. Different heat sources could be used to generated heat stimulation depending on the type of defect of interest. This technique is also suitable for in-situ monitoring of the printing process. It has already been integrated into 3D printing techniques like powder bed fusion and selective laser melting for solidification and quality monitoring. Figure 1.6 represents the thermal images of a powder bed fusion 3D printed bar with intentional defects (b), which were corrected through re-melting (c) by re-scanning performed using a build-file (a) [48].

1.5.2.3 Ultrasonic Evaluation

In this NDE technique, ultrasonic waves are used to examine internal defects of 3D printed materials. Ultrasonic pulses (ranging from 1 to 10 MHz) generated from a transducer pass through the specimen when placed upon it (contact or immersion mode). Different signals are reflected from different interfaces, and variation between transmitted and reflected signals reveals the material's integrity, including defect size, shape and location. For the size and shape detection of internal defects, ultrasonic C-scans are the most suitable, consisting of amplitude and depth signal views. Calibration of acoustic impedance, propagation velocity, and attenuation using the same test material need to be done before detecting differentiation of anomalies in the specimen. The sensitivity of this technique depends on the testing frequency. With wavelength increment, detection capability decreases. It is inefficient in providing information about subsurface irregularities due to the surface roughness of the printed material. Laser ultrasonic detection is an advanced version

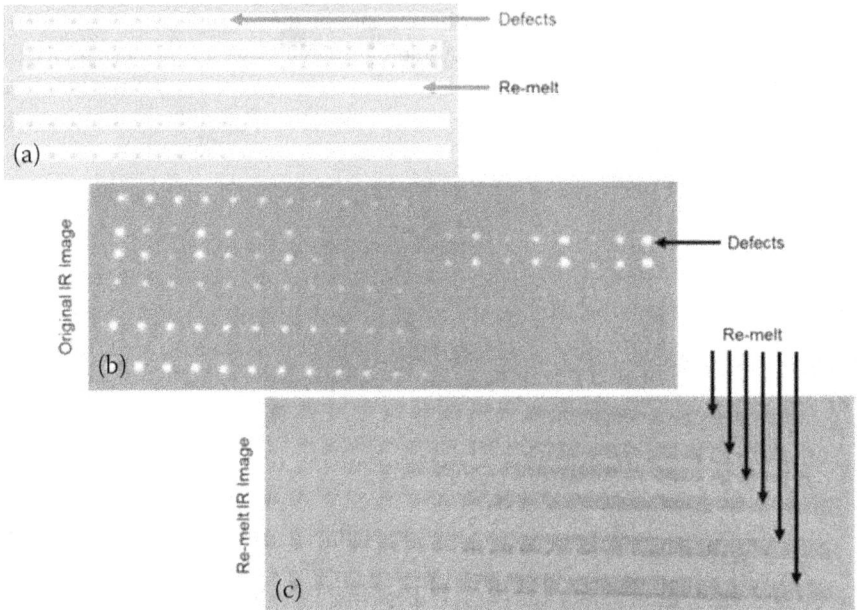

FIGURE 1.6 Thermographic images of a powder bed fusion 3D printed bar. Correction of un-melted powder through layer re-melt shown by (a) the build file with the re-melt area shown by the red arrow, (b) showing the defect present in the IR image by the black arrow and (c) showing the defect not shown on the left column due to the re-melt. (Reproduced with permission from reference [48]. © IOP Publishing. All rights reserved).

of ultrasonic detection, where a laser beam is used to generate ultrasonic waves. The defect detection in hybrid nanocomposites is different from bulk materials due to inorganic-organic interface boundary scattering and orientation-dependent wave speeds. However, it helps estimate curing and nanomaterials dispersion related irregularities. Researchers have used ultrasonic elastic waves to evaluate internal defects and characterise elastic properties of materials [49]. Although comparatively less work has been done to establish standards for ultrasonic defect detection in 3D printed hybrid nanocomposites than metal components, detection goals and methods are similar [50].

There are other techniques to detect irregularities, but they are limited to a particular kind of material. For instance, eddy current detection is limited to conductive materials and magnetic particle testing is limited to ferromagnetic materials [51]. Liquid penetration is the conventional method that detects surface defects in nonporous hard 3D printed components, where the specimen is treated with a simple or fluorescent dye that penetrates surface pores through capillary force [52]. Pores up to 5-microns could be detected through simple dye, while fluorescent dye allows visualising pores up to 1–2-microns. Table 1.3 compares some of the commonly used NDE defect detection techniques.

TABLE 1.3

Commonly Used Conventional NDE Defect Detection Methods

Detection Method	Advantages	Disadvantages	Defects
Micro-computed tomography (MicroCT)	• 2D and 3D information of features • High resolution • No detection medium required • Unaffected by surface morphology of the specimen	• Lack of standards • Expensive • Size and density dependent	Suitable for all kinds of (even microscopic) defects
Ultrasonic method	• Easy & cost-efficient • Suitable for all materials • Detect internal defects • Accurate	• Inefficient for sub-surface defects and complex geometries • Surface finishing required • Cannot detect defects parallel to the ultrasonic wave	Cracks, anisotropy, porosity and internal defects (excluding surface and sub-surface)
Infra-red detection method	• Quick, easy & low-cost • Less sensitive to surface finishing	• Limited penetration	Suitable for surface and sub-surface cracks, porosity, layer defects
Penetration method	• Easy & low-cost • Suitable for all materials, specimen sizes/shapes • Suitable for pin hole defects	• Detect only surface defects • Non-effective for porous specimen • Viscosity of penetration liquid might vary with dyes and temperature	Only surface porosity
Eddy current method	• Easy & quick • Suitable for detection in high temperature	• Limited to electrically conductive specimens • Poor detection accuracy • No information about deeply situated defects	Suitable for surface and sub-surface cracks and defects

1.5.2.4 Machine Learning-Based Detection Methods

Lack of real-time monitoring and feedback pushes the need for converging artificial intelligence (AI) with 3D printing. The algorithms used in AI help in learning and making decisions with the least human interference. With machine learning (ML), predictions and approximations can be made based on the information derived from the existing data sets, which helps in automatic process improvements, defect detection and quality assessment. Image formation and quality in ML are influenced by various aspects like capturing angle, lighting situation, environmental conditions and defects analysis, and precision vary with image processing algorithms. Establishing a correlation between printing parameters and 3D printed structures is required, which is nonlinear, complex and time-consuming. ML helps in creating a definite relation between input and output parameters, optimising the process, and some examples are presented in Table 1.4.

TABLE 1.4

Applications of an ANN in the Field of 3D Printing

3D Printing Process	ANN Input Parameters	ANN Output Parameter
Selective laser sintering	Laser power, scanning speed, hatch spacing, powder layer thickness	Density Geometrical dimensions
	Laser power, scanning speed, hatch spacing, powder layer thickness	Manufacturing time Shrinkage percentage
	Vertical height, deposited volume, bounding box	Part porosity Tensile Strength
	Laser power, scanning speed, hatch distance, powder layer thickness, scanning mode, temperature distribution, the processing time	Density
	Powder layer thickness, laser power, scanning speed	
	Laser power, scanning speed, hatch distance, powder layer thickness, temperature distribution	
	Laser power, scanning speed, hatch distance, powder layer thickness, scanning mode, temperature distribution, the processing time	
Stereolithography	Powder layer thickness, curation time, hatch distance, filling cure depth, filling spacing depth	Geometrical dimensions (precision)
Laser-melting deposition	Laser power, scanning speed, powder feed rate	Geometrical dimensions (precision)
Fused deposition modeling	Layer thickness, positioning, raster angle and width, air gap	Compressive strength Wear
	Layer thickness, positioning, raster angle and width, air gap	Deposition error in volume Dimensional precision
	Positioning, slice width	
	Layer thickness, positioning, raster angle and width, air gap	
Binder jetting	Layer thickness, printing saturation, heater power ration, drying time	Surface roughness, shrinkage

Different artificial neural networks (ANNs) used for defect and quality assessment in 3D printed components are convolutional neural networks (CNNs), deep residual neural network method, recurrent neural network method and autoencoder network method. In a study, CNN was used to detect warping during liquid deposition modeling (LDM) printing, where the real-time layer-by-layer print was captured with extracted corners of the components passed through CNN to check the possible warping. If identified, a pause signal was sent to create a closed-loop in-situ detection system [54]. Similarly, a deep convolutional neural network method was used to monitor stringing during printing, capable of terminating the printing or optimising printing parameters after identification of stringing [44].

1.6 CONCLUSION

Hybrid nanocomposites provide convergence of the properties of both inorganic-organic components, but their use in 3D printed commercial functional parts is not yet ready due to some pre-printing, printing and post-printing issues. Issues like development of suitable hybrid nanocomposites with good flowability and good structural and functional performance could be resolved by selecting suitable nanomaterials and their stable and homogenous dispersion in various polymers. During printing, the effect of multiple parameters such as print speed, flowrate, printing window and post-printing treatments on the structure and properties of the printed material needs to be understood. Advancements in printing instruments such as multi-material printing, hybrid printing and printing methods with better resolution would also be helpful in utilizing the full potential of hybrid nanocomposites. Defects in printed parts are inevitable, but their timely and accurate detection through in-situ methods can save time, material and cost, significantly. Through these innovative solutions, better 3D printed structures could be achieved.

REFERENCES

1. Chiappone, A., et al., *3D printed peg-based hybrid nanocomposites obtained by sol–gel technique.* ACS Applied Materials & Interfaces, 2016. **8**(8): pp. 5627–5633.
2. Singh, A., N. Verma, and K. Kumar, *Chapter 2 – Hybrid composites: A revolutionary trend in biomedical engineering,* in *Materials for Biomedical Engineering,* V. Grumezescu and A.M. Grumezescu, Editors. 2019, Elsevier. pp. 33–46.
3. Wu, M., Phoha, V.V., Moon, Y.B. and Belman, A.K., Detecting Malicious Defectsin 3D Printing Process Using Machine Learning a nd Image Classification. *Proceedings of the ASME 2016 International Mechanical Engineering Congress and Exposition. Volume 14:Emerging Technologies; Materials: Genetics to Structures; Safety Engineering and Risk Analysis.* Phoenix, Arizona, USA. November 11–17, 2016. V014T07A004. ASME. https://doi.org/10.1115/IMECE2016-67641
4. Alemán, J.V., et al., *Definitions of terms relating to the structure and processing of sols, gels, networks, and inorganic-organic hybrid materials (IUPAC recommendations 2007). Pure and Applied Chemistry,* 2007. **79**(10): pp. 1801–1829.
5. Nimbagal, V., Banapurmath, N.R. , Sajjan, A.M. et al., *Studies on hybrid bio-nanocomposites for structural applications. Journal of Materials Engineering and Performance,* 2021. 30: pp. 6461–6480. https://doi.org/10.1007/s11665-021-05843-9

6. Nanko, M., *Definitions and categories of hybrid materials. Advances in Technology of Materials and Materials Processing*, 2009. **11**: pp. 1–8.

7. Gómez-Romero, P. and C. Sanchez, *Hybrid materials, functional applications. An introduction. Functional Hybrid Materials*, 2003: pp. 1–14. https://onlinelibrary. wiley.com/doi/10.1002/3527602372.ch1

8. Judeinstein, P. and C. Sanchez, *Hybrid organic–inorganic materials: A land of multidisciplinarity. Journal of Materials Chemistry*, 1996. **6**(4): pp. 511–525.

9. Saveleva, M.S., et al., *Hierarchy of hybrid materials—The place of inorganics-in-organics in it, their composition and applications. Frontiers in Chemistry*, 2019. **7**(179): 1–21.

10. Joseph, V.S., et al., *Silicone/epoxy hybrid resins with tunable mechanical and interfacial properties for additive manufacture of soft robots. Applied Materials Today*, 2021. **22**: pp. 100979.

11. Melo, P., et al., *Processing of Sr2+ containing poly l-lactic acid-based hybrid composites for additive manufacturing of bone scaffolds. Frontiers in Materials*, 2020. **7**(413): pp. 1–12.

12. Duty, C., et al., *What makes a material printable? A viscoelastic model for extrusion-based 3D printing of polymers. Journal of Manufacturing Processes*, 2018. **35**: pp. 526–537.

13. Hmeidat, N.S., J.W. Kemp, and B.G. Compton, *High-strength epoxy nanocomposites for 3D printing. Composites Science and Technology*, 2018. **160**: pp. 9–20.

14. Spoerk, M., et al., *Optimization of mechanical properties of glass-spheres-filled polypropylene composites for extrusion-based additive manufacturing. Polymer Composites*, 2019. **40**(2): pp. 638–651.

15. Collino, R.R., et al., *Deposition of ordered two-phase materials using microfluidic print nozzles with acoustic focusing. Extreme Mechanics Letters*, 2016. **8**: pp. 96–106.

16. Zhong, W., et al., *Short fiber reinforced composites for fused deposition modeling. Materials Science and Engineering: A*, 2001. **301**(2): pp. 125–130.

17. Sood, A.K., R.K. Ohdar, and S.S. Mahapatra, *Parametric appraisal of mechanical property of fused deposition modelling processed parts. Materials & Design*, 2010. **31**(1): pp. 287–295.

18. Spoerk, M., et al., *Polypropylene filled with glass spheres in extrusion-based additive manufacturing: Effect of filler size and printing chamber temperature. Macromolecular Materials and Engineering*, 2018. **303**(7): pp. 1800179.

19. Samperi, F., et al., *Thermal degradation of poly(butylene terephthalate) at the processing temperature. Polymer Degradation and Stability*, 2004. **83**(1): pp. 11–17.

20. Halim, N.A., et al., *A review on 3D printed polymer-based composite for thermal applications. IOP Conference Series: Materials Science and Engineering*, 2021. **1078**(1): p. 012029.

21. Heggset, E.B., et al., *Viscoelastic properties of nanocellulose based inks for 3D printing and mechanical properties of CNF/alginate biocomposite gels. Cellulose*, 2019. **26**(1): pp. 581–595.

22. Serafin, A., et al., *Printable alginate/gelatin hydrogel reinforced with carbon nanofibers as electrically conductive scaffolds for tissue engineering. Materials Science and Engineering: C*, 2021. **122**: pp. 111927.

23. Gao, C., et al., *Robotic deposition and in vitro characterization of 3D gelatin–bioactive glass hybrid scaffolds for biomedical applications. Journal of Biomedical Materials Research Part A*, 2013. **101A**(7): pp. 2027–2037.

24. Rocha, V.G., et al., *Direct ink writing advances in multi-material structures for a sustainable future. Journal of Materials Chemistry A*, 2020. **8**(31): pp. 15646–15657.

25. Scott, P.J., et al., *Polymer-inorganic hybrid colloids for ultraviolet-assisted direct ink write of polymer nanocomposites. Additive Manufacturing*, 2020. **35**: pp. 101393.

26. Chen, Q., et al., *A dual approach in direct ink writing of thermally cured shape memory rubber toughened epoxy. ACS Applied Polymer Materials*, 2020. **2**(12): p. 5492–5500.
27. Shao, G., et al., *Freeze casting: From low-dimensional building blocks to aligned porous structures—A review of novel materials, methods, and applications. Advanced Materials*, 2020. **32**(17): pp. 1907176.
28. Quan, H., et al., *Photo-curing 3D printing technique and its challenges. Bioactive Materials*, 2020. **5**(1): pp. 110–115.
29. Dermanaki Farahani, R. and M. Dubé, *Printing polymer nanocomposites and composites in three dimensions. Advanced Engineering Materials*, 2018. **20**(2): p. 1700539.
30. Ambone, T., A. Torris, and K. Shanmuganathan, *Enhancing the mechanical properties of 3D printed polylactic acid using nanocellulose. Polymer Engineering & Science*, 2020. **60**(8): pp. 1842–1855.
31. Keleş, Ö., et al., *Stochastic fracture of additively manufactured porous composites. Scientific Reports*, 2018. **8**(1): p. 15437.
32. Kuznetsov, V.E., et al., *Increasing strength of FFF three-dimensional printed parts by influencing on temperature-related parameters of the process. Rapid Prototyping Journal*, 2020. **26**(1): pp. 107–121.
33. El Moumen, A., M. Tarfaoui, and K. Lafdi, *Additive manufacturing of polymer composites: Processing and modeling approaches. Composites Part B: Engineering*, 2019. **171**: pp. 166–182.
34. Guessasma, S., et al., *Anisotropic damage inferred to 3D printed polymers using fused deposition modelling and subject to severe compression. European Polymer Journal*, 2016. **85**: pp. 324–340.
35. Ma, G., et al., *Mechanical anisotropy of aligned fiber reinforced composite for extrusion-based 3D printing. Construction and Building Materials*, 2019. **202**: pp. 770–783.
36. Torrado, A.R., et al., *Characterizing the effect of additives to ABS on the mechanical property anisotropy of specimens fabricated by material extrusion 3D printing. Additive Manufacturing*, 2015. **6**: pp. 16–29.
37. Sweeney, C.B., et al., *Welding of 3D-printed carbon nanotube–polymer composites by locally induced microwave heating. Science Advances*, 2017. **3**(6): p. e1700262.
38. Hmeidat, N.S., et al., *Mechanical anisotropy in polymer composites produced by material extrusion additive manufacturing. Additive Manufacturing*, 2020. **34**: p. 101385.
39. Raney, J.R., et al., *Rotational 3D printing of damage-tolerant composites with programmable mechanics. Proceedings of the National Academy of Sciences*, 2018. **115**(6): p. 1198.
40. Kokkinis, D., M. Schaffner, and A.R. Studart, *Multimaterial magnetically assisted 3D printing of composite materials. Nature Communications*, 2015. **6**(1): p. 8643.
41. Karalekas, D. and A. Aggelopoulos, *Study of shrinkage strains in a stereolithography cured acrylic photopolymer resin. Journal of Materials Processing Technology*, 2003. **136**(1): pp. 146–150.
42. Yang, Y., et al., *Biomimetic anisotropic reinforcement architectures by electrically assisted nanocomposite 3D printing. Advanced Materials*, 2017. **29**(11): p. 1605750.
43. Alsoufi, M. and A. El-Sayed, *Warping deformation of desktop 3D printed parts manufactured by open source fused deposition modeling (fdm) system. International Journal of Mechanical & Mechatronics Engineering*, 2017. **17**: pp. 7–16.
44. Paraskevoudis, K., P. Karayannis, and E.P. Koumoulos, *Real-time 3d printing remote defect detection (stringing) with computer vision and artificial intelligence. Processes*, 2020. **8**(11): pp. 1–15.
45. du Plessis, A., et al., *X-ray microcomputed tomography in additive manufacturing: A review of the current technology and applications. 3D Printing and Additive Manufacturing*, 2018. **5**(3): pp. 227–247.

46. Lu, Q.Y. and C.H. Wong, *Additive manufacturing process monitoring and control by non-destructive testing techniques: Challenges and in-process monitoring. Virtual and Physical Prototyping*, 2018. **13**(2): pp. 39–48.
47. Shmueli, Y., et al., *In-situ X-ray scattering study of isotactic polypropylene/graphene nanocomposites under shear during fused deposition modeling 3D printing. Composites Science and Technology*, 2020. **196**: p. 108227.
48. Mireles, J., et al., *Analysis and correction of defects within parts fabricated using powder bed fusion technology. Surface Topography: Metrology and Properties*, 2015. **3**(3): p. 034002.
49. Yuan, S. and X. Yu, *Ultrasonic non-destructive evaluation of selectively laser-sintered polymeric nanocomposites. Polymer Testing*, 2020. **90**: p. 106705.
50. Honarvar, F. and A. Varvani-Farahani, *A review of ultrasonic testing applications in additive manufacturing: Defect evaluation, material characterization, and process control. Ultrasonics*, 2020. **108**: p. 106227.
51. Lu, Q.Y. and C.H. Wong, *Applications of non-destructive testing techniques for post-process control of additively manufactured parts. Virtual and Physical Prototyping*, 2017. **12**(4): pp. 301–321.
52. Chen, Y., et al., *Defect inspection technologies for additive manufacturing. International Journal of Extreme Manufacturing*, 2021. **3**(2): p. 022002.
53. Mahmood, M.A., et al., *Artificial neural network algorithms for 3D printing. Materials*, 2021. **14**(1): pp. 1–27.
54. Saluja, A., J. Xie, and K. Fayazbakhsh, *A closed-loop in-process warping detection system for fused filament fabrication using convolutional neural networks. Journal of Manufacturing Processes*, 2020. **58**: pp. 407–415.

2 3D Nanoprinting in the Aero-Industries

Alperen Doğru
Aviation Higher Vocational School, Ege University,
Bornova, Turkey

M. Batıkan Kandemir
Department of Materials Science and Engineering,
İzmir Katip Çelebi University, Izmir, Turkey

M. Özgür Seydibeyoğlu
Department of Metallurgical and Materials Engineering,
Muğla Sıtkı Koçman University, Muğla, Turkey

CONTENTS

DOI: 10.1201/9781003189404-2

2.1 INTRODUCTION

2.1.1 Additive Manufacturing

Additive manufacturing, known as three-dimensional printing, is a production system in which the material is combined layer by layer using computer-aided design (CAD) data of physical models; this system is unlike the material reduction processes from solid raw materials seen in traditional production methods, such as machining. Additive manufacturing is used in a variety of industries to quickly produce a system or part prototype before the final product or commercialization. Additive manufacturing reduces the production cost in terms of shortening the lead time and using a small number of parts. A product, for which three-dimensional CAD data is created with additive manufacturing methods, can be produced directly without the need for tools, molds, and apparatus. In addition, additive manufacturing simplifies the production processes of objects with complex geometries and provides design freedom. Additive manufacturing methods enable lightweight designs, assembly-free parts, on-site production, direct production of directional materials, in-building supports, and personalized products.

Using these technologies, highly complex parts with complex geometry can be produced. These parts, which are produced with minimum material waste and post-processing requirements, can be used in a wide variety of areas. Manufacturers can use additive manufacturing to create parts in low volume and at a low cost without the need for assembly. It can also be called a tool that increases design freedom. Another driving force behind this technology is that it is eco-friendly and environmentally friendly, thanks to its ability to produce lightweight components with high structural integrity [1].

In addition to being utilized alone, additive manufacturing methods can also be utilized in combination with subtractive manufacturing to improve the qualities of finished parts. Conventional subtractive and additive manufacturing processes can be easily combined [2].

Additive manufacturing technologies are used in various industries, such as aero-industries, defense, space, automotive, transportation, and medical. The aviation industry, in particular, has recently shown interest in additive manufacturing technology due to a huge number of lightweight components with unusual designs [3]. Additive manufacturing technologies help manufacturers in the aero-industry eliminate production steps other than process planning for downstream processes, shorten delivery times, provide cost and fuel efficiency for active aircraft, reduce energy consumption, and reduce waste of raw materials and environmental (carbon) footprints [3]–[5]. Therefore, the adoption of additive manufacturing technologies is manifested by continued growth in the global aero-industries, defense, and space market. Problems such as the lack of globally acknowledged certification and standardization of additive manufacturing methods, high material costs, limited component size, and relatively sluggish production speed, on the other hand, need to be addressed [4], [6]. The capabilities and limitations of additive manufacturing technologies given in Table 2.1 are factors that affect the popularity of this technology in this industry.

TABLE 2.1

An Overview of Additive Manufacturing

Advantages	Limitations
• Low cost of production • Nearly net shape production • Production of unique and complex structures (design freedom) • Reduced part assembly necessary • Minimum materials waste • Short time to market (reduced lead time) • Green manufacturing capability • Lightweight production possibility • Tooling and fixturing elimination • Reduced scrap	• High first-time buy cost of AM equipment, materials, and software • Low reliability regarding mass production • Lack of global certifications and standardizations • Limited component size and building volume • Low production speed compared to the subtractive manufacturing processes • Costly and high volume production • Limited materials option • Metallurgical defects such as porosity, hot cracking • Unsatisfactory dimensional accuracy

Additive manufacturing has many advantages that make it preferable when compared to traditional methods. With additive manufacturing, lighter and functional parts can be produced due to the high potential of producing lattice structures and optimized designs. Some structures cannot be produced with traditional production methods or can be produced with several different sub-processes and require post-production assembly processes (complex designs such as cooling channels, closed volumes); these can be integrated into a single operation with additive manufacturing. Contrary to traditional methods, the result of the production of the part layer by layer and the recyclability of the raw material that is not used during production provides material savings. For example, the recyclability rate for metal powders in some processes is in the range of 95%–98%. In addition to the advantages it provides, additive manufacturing technology also has some disadvantages and limitations. These can be listed as follows: Initial investment costs and system prices are high. Depending on the material and method, additive manufacturing parts require post-production treatments to improve surface quality, improve mechanical properties, and remove support structures. Large parts may require additional procedures, such as post-production assembly due to the limited size of parts that may be manufactured based on machine specifications.

2.1.2 Additive Manufacturing Methods

The production of objects drawn in the CAD program with additive manufacturing was first carried out in the 1980s. These first models, produced for prototype purposes, enabled the ideas developed by engineers to become reality. With this developed method, time and cost savings were achieved, and human-induced problems were minimized [7]. In addition, all kinds of shapes that are difficult to process with traditional methods can be produced with additive manufacturing.

Today, although additive manufacturing technologies are developing for the production sector, they are frequently used by scientists, doctors, and artists. With rapid prototyping, scientists transform their theoretical studies into models and analyze them, doctors can design and examine human body parts, and artists can easily produce their works. With rapid prototyping, many more models can be produced in a short time. Today, with the developments in the plastic material sector, not only draft models but also final products can be produced with rapid prototyping [8], [9]. The ancestor of the method called 3D printing today is rapid prototyping [10]. The advancement of additive manufacturing technology has been hastened by advances in CAD, computer-aided manufacturing (CAM), software, and part processing methods [9], [10]. However, the limited production volumes of additive manufacturing devices still cannot be an alternative to the traditional CAM-assisted manufacturing method to produce large-sized parts due to precision problems [1]. Furthermore, while the number of materials that may be utilized in additive manufacturing technologies is growing every day, it still has a limited selection compared to other methods [11]. Additive manufacturing methods are divided into three: solid, liquid, and powder-based. These methods are summarized in Figure 2.1.

2.1.2.1 Laminated Object Manufacturing (LOM)

In this method, as seen in Figure 2.2, the material is spread by a roller and is in the form of a sheet. The laser cuts the material layer by layer following the CAD model. Paper, metal, and composite can be used as the layer material. Each layer is bonded together by heat-resistant adhesive, pressure, and heat. The advantage of the LOM method is that it is low in cost, does not require support material, does not need extra post-processing, does not cause deformation in the layer material, and allows the production of large parts. The disadvantage is that it produces excessive waste

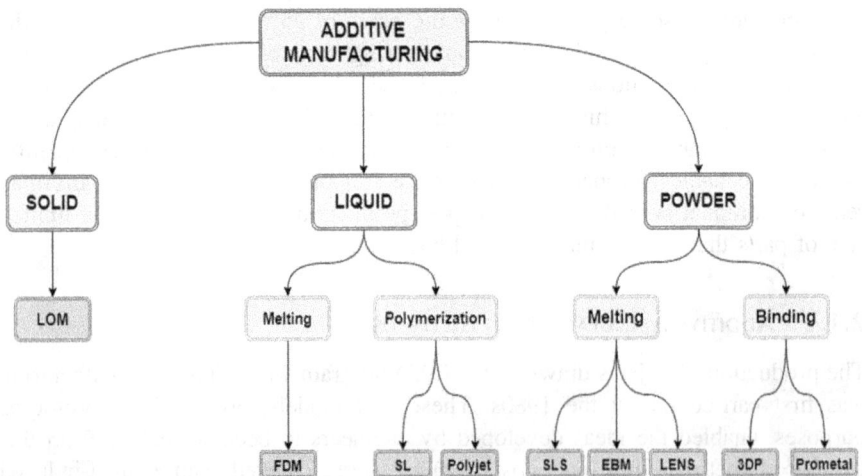

FIGURE 2.1 Additive manufacturing methods [8].

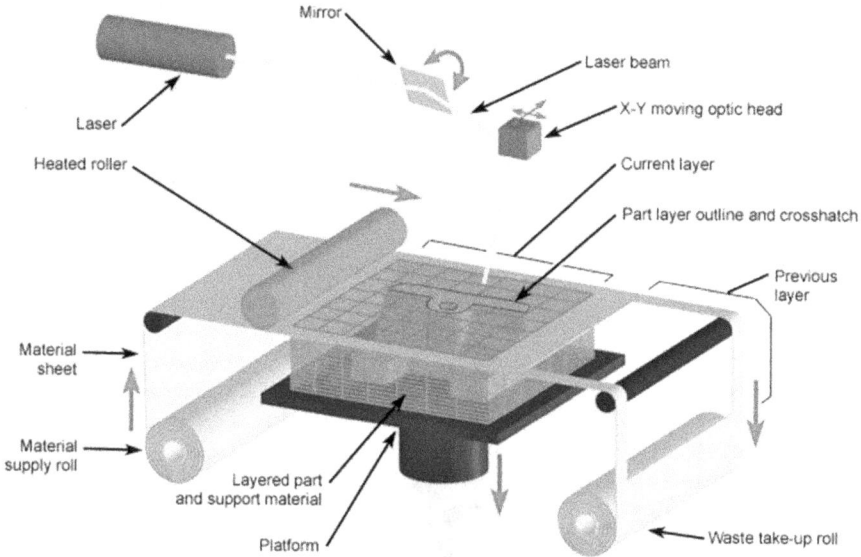

FIGURE 2.2 Illustration of laminated object manufacturing technique [12].

material, and, depending on the direction, the final product's mechanical properties change. The limitations of the process is the difficulty in creating a hollow structure in the model, and the limited material choice [10].

2.1.2.2 Fused Filament Fabrication (FFF)

In this method, as seen in Figure 2.3, a thin polymer filament passes through a heating nozzle. The nozzle, which moves in the Cartesian system, extrudes the molten polymer layer by layer at certain diameters. This extrusion process continues until all sliced layers are finished. Acrylonitrile butadiene styrene (ABS), polycarbonate (PC), polyamide (PA), polyetheretherketone (PEEK), polylactic acid (PLA), etc., materials are used as filament material. The advantage of this method is that it does not require resin for curing and chemical treatment, and it has a low cost [9]. The z-axis resolution is low, and the mechanical properties are weak on this axis. Additional processing, such as sanding, may be required in the FFF method to remove defects on the surface after printing. This additional processing leads to a waste of time for large parts. When the filling ratio is decreased and the printing speed is increased, the mechanical strength decreases [3]. Another common name for the FFF method in the industry is FDM.

2.1.2.3 Stereolithography (SL)

The process of creating layer-by-layer models by curing photosensitive polymers with ultraviolet (UV) light, which acts as a catalyst, is known as stereolithography, or SL. The SL method shown in Figure 2.4 is among the most popular rapid prototyping techniques. Today, a layer sensitivity of 10 μm can be achieved with the micro stereolithography method [2]. SL printing requires a delicate process.

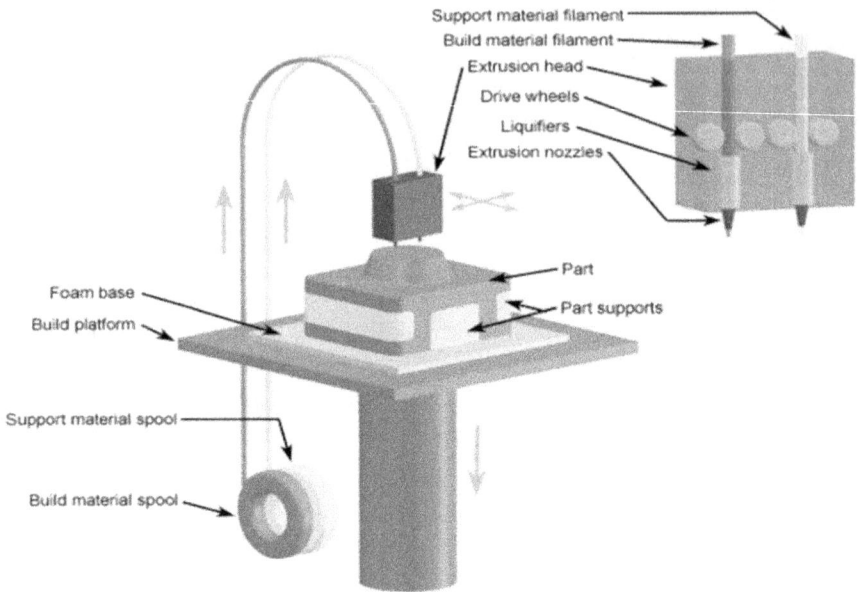

FIGURE 2.3 Illustration of fused filament fabrication technique [13].

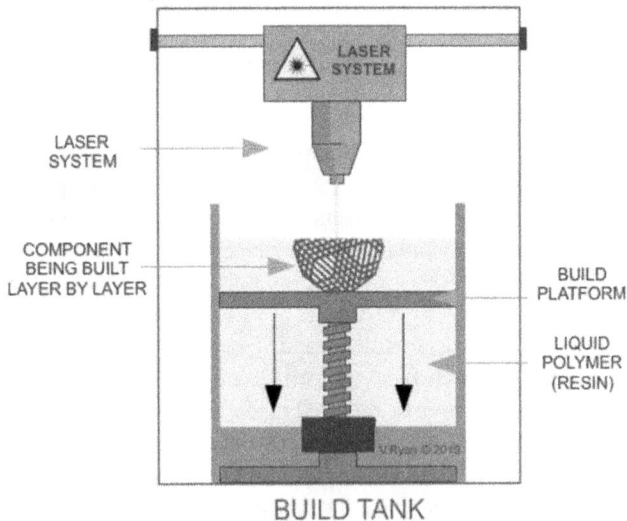

FIGURE 2.4 Illustration of stereolithography technique [14].

Overcuring can cause a break between the substrate and the top printed layer. In addition, since the resin is highly viscous (scanned line shape), the layer thickness may vary. The type of stereolithography that uses more than one material is called "multiple material stereolithography."

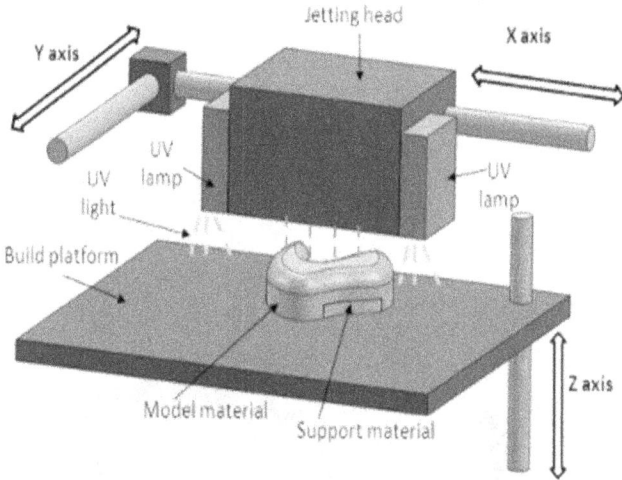

FIGURE 2.5   Illustration of polyjet technique [15].

2.1.2.4   Polyjet

In this method, which uses inkjet technology, the photopolymer resin, which is sprayed layer by layer from the ink nozzle, is cured under UV light. The material is sprayed with the special jetting head in Figure 2.5. The layer thickness of the produced model is in the order of 16 μm, and the resolution is high. The strength of the produced parts is lower compared to the SL and SLS methods. A gel polymer is used as the support material. After printing, the final piece is cleaned by spraying water. The polyjet method is suitable for the production of multicolored parts [4], [5].

2.1.2.5   Selective Laser Sintering (SLS)

In this method, the powdered material is sintered with the laser directed by the laser source in Figure 2.6. Metal, alloy powders, polymer (polystyrene, polyamide), and metal/ceramic combinations are used as materials. In addition, composite and fiber-reinforced polymers (for example, glass fiber reinforced polyamide) are suitable for producing parts in the SLS method [6], [16]. In this method, the temperature of the printing zone is very close to the melting temperature. The laser melts at the points determined by the drawing program. The part is supported during printing by the non-fused powder, which eliminates the requirement for a separate support structure. The build chamber is then lowered one layer by the mobile platform, and the procedure is repeated. The advantage of the SLS method is the high material variety and the ease of powder recycling. The downsides of the SLS method include that print quality is dependent on particle size, the process requires an inert gas environment to prevent oxidation, and the process requires a temperature so close to the melting temperature. Another name for the SLS method in the industry is "direct metal laser" [8].

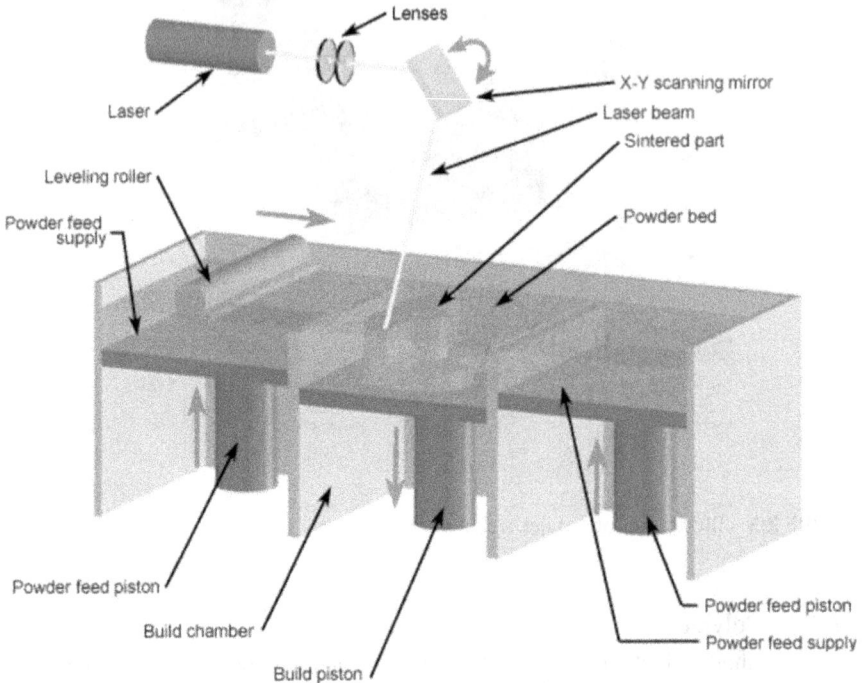

FIGURE 2.6 Illustration of selective laser sintering technique [17].

2.1.2.6 Electron Beam Melting (EBM)

The EBM method is similar to the SLS method in many ways. The difference from the SLS method is that the powder is melted with a high voltage (60 kV) electron laser, and the melting temperature is exceeded during the process. In addition, a high vacuum chamber is used, as seen in Figure 2.7, to prevent oxidation of metal parts during the process. The use of "pre-alloyed metal" powder is one of the advantages of the EBM method. Thanks to the temperature reaching 1000°C, the formation of residual stress is prevented. Although it is not widely used today, this additive manufacturing method has a great potential for aero-industries production and the aviation industry due to its high vacuum [8], [18], [19].

2.1.2.7 Laser Engineered Net Shaping (LENS)

A high-power laser beam injects molten metal powder into a precise location in this process. The material solidifies by spontaneous cooling. A closed environment containing argon gas is used to prevent oxidation during the process. The demonstration of the method is in Figure 2.8. Stainless steel, nickel-based alloys, TĪ6AL4V, tool steels, copper alloys, and alumina are materials suitable for the LENS method. This process is used in the repair of difficult and expensive parts due to its high precision. However, for sensitive applications such as turbine blade repair, attention should be paid to the formation of residual stress caused by uneven heating and cooling, which adversely affects the mechanical properties [21].

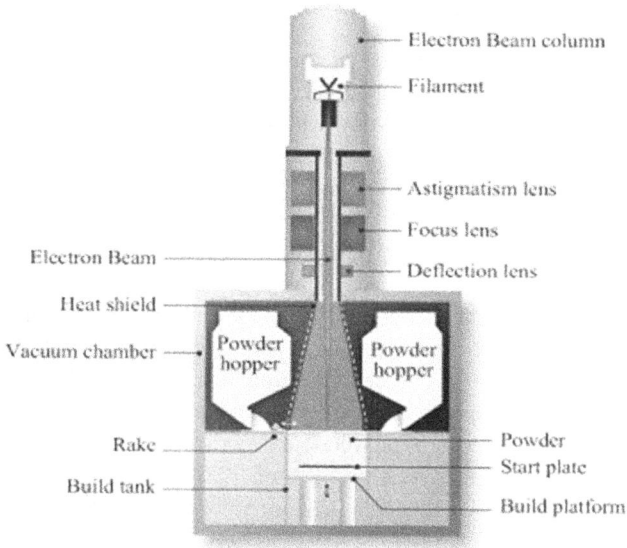

FIGURE 2.7 Illustration of electron beam melting technique [20].

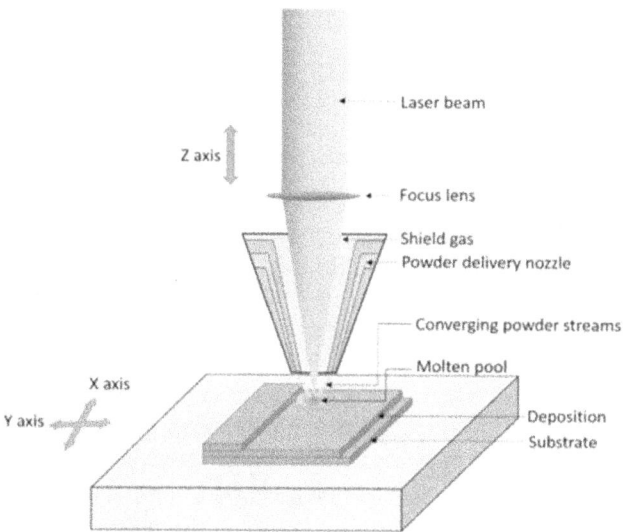

FIGURE 2.8 Illustration of laser engineered net shaping technique [22].

2.1.2.8 Three-Dimensional Printing (3DP)

In this method, which was patented by MIT in 1993, a water-based binder liquid is sprayed onto a starch-based powder. The binder allows the layers to stick together. Because of its resemblance to inkjet printing, which is 2D printing on paper, this process was given the moniker 3DP [9].

2.1.2.9 Prometal

In this method, which is preferred for injection mold/part production, stainless steel, bronze, and tungsten can be used as powder material. After the process has started, a binder liquid is sprayed onto the powder from the jets, and heat is applied. The layer then descends and a piston feeds powder material for each new layer. This process is repeated until the product is finished. After the process, the part is cured, and the residual dust is cleaned with air. There is no need for an extra process to produce molded parts, but sintering, infiltration, and finishing processes are required for the production of functional parts [9], [11]. For the infiltration process, the bronze powder is added to the metal powder in the ceramic chamber and a diffusion connection is established between the aluminum oxide and the part. After the heat treatment, the runner part is cut and the process is finished. Depending on the material type, sintering parameters such as temperature and time may vary. The mechanical strength of the parts produced by this method is higher than the parts produced by traditional manufacturing methods [23], [24].

The International Standards Organization/American Society for Testing and Materials Standards (ISO/ASTM 52900:2015) classifies additive manufacturing methods according to on their working principle, materials, and energy types [25]. All commercially accessible additive manufacturing methods are divided into seven categories by the ISO/ASTM 52900:2015 standard. These categories are directed energy deposition (DED), boat photopolymerization (VP), powder bed fusion (PBF), binder sputtering (BJ), material sputtering (MJ), sheet lamination (SL), and material extrusion (ME) [26].

2.1.3 Importance of Additive Manufacturing for Aero Industries

The manufacturing industry has created new processes and technologies for the low-volume manufacture of creative and sustainable parts with complicated geometries and unique technical needs over the last few decades. The most striking of these developing technologies is additive manufacturing technology. Additive manufacturing methods, which have been used to produce prototypes for many years, have found a wider application area with research on the properties of materials, developments in new production methods, and development of existing methods. Additive manufacturing technology, which started with the use of polymer raw materials, has turned into a system in which metal and ceramic materials are used, composite structures can be produced, laser technology is integrated, and sensitivities are increased. In addition, thanks to the developments, device alternatives have increased, and a competitive environment has been created. Additive manufacturing paved the way to produce in one piece designs that are limited in traditional production methods. The innovative aspects of additive manufacturing have been used to redesign many existing applications. Aircraft cooling ducts, turbine blades, fuel injectors, hearing aids, and prostheses used in the medical sector are examples of these. The standards developed in the process of additive manufacturing methods paved the way for the parts produced by these methods to be used directly in practice. Additive manufacturing methods have changed the manufacturing and design paradigms in the aviation industry. Today, with the possible standards and certifications, additive manufacturing is now being used in many applications in the aircraft, defense, and space industries. In aircraft parts, more efficient and performance components can be created with additive manufacturing technologies, from structural parts to engine turbines. Thus, manufacturers save on raw materials, time, and cost. In the aviation industry, where weight reduction is an important focus, additive manufacturing methods have brought a new dimension to part design and production.

Additive manufacturing creates opportunities to improve the performance of parts and to remove design constraints. With additive manufacturing, parts with almost half the original weight can be produced. The benefits of additive manufacturing technologies for the aero-industry are shown in Figure 2.9. Thanks to topology optimization in the aero-industries, the performance and function of the part can be improved along with weight reduction. However, it is necessary to pay attention to the formation of complex geometry, which makes production difficult during topology optimization. Thanks to additive manufacturing, it is possible to eliminate all these restrictions. In addition, weight reduction is an application that directly reduces flight costs, increases payload carrying capacity, and reduces carbon emissions.

Additive manufacturing provides significant freedom in design, and parts are produced with the closest accuracy to digital design. The most important factor that ensures the success of all these processes is the developments in material technology. The increase in the variety of materials that can be used in additive manufacturing methods has expanded the application areas. Aero-industries, defense, and space applications are the most variable application areas of additive manufacturing technologies. Aircraft, satellites, rockets, and spacecraft parts are exposed to extreme temperatures, high vibration, acoustic load, high pressure, and speed

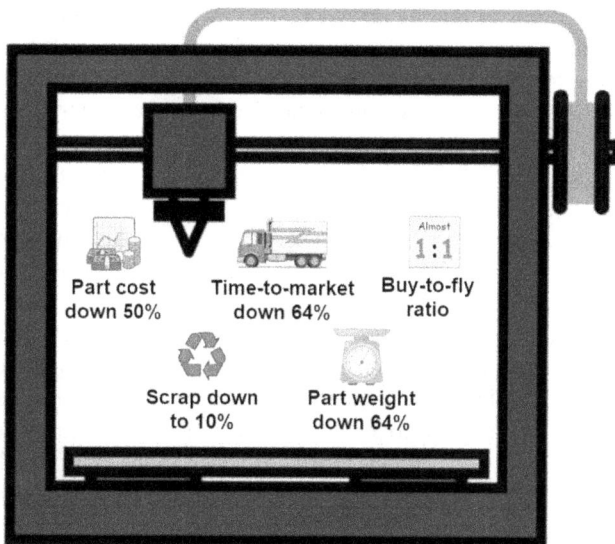

FIGURE 2.9 Benefits of additive manufacturing technology to the aero-industry.

during use. Similarly, launch vehicles are exposed to high structural, vibration, and acoustic loads. Aircraft and satellites, in particular, must withstand extreme temperatures, from liquid cryogens to rocket engine combustion and exhaust. Combustion chambers, turbopump assemblies, and jet turbines all have high pressures and speeds. Longer-term aging and fatigue difficulties should be considered for aero-industries applications, even though some of these impacts are of shorter duration. In addition, aero-industries manufacturers should consider the repair, repair, and renewal activities of the parts they produce more than aerospace-oriented applications. All these factors have increased the demand for alloy materials that are more performant, functional, and capable of meeting a wide variety of simultaneous needs.

The possibility of using different types of materials, such as shape memory alloys, nanocomposites, and carbon nanomaterials in additive manufacturing technologies, expands the scope of applicability of these methods. The development of nanotechnology causes continuous innovation in materials science. Today, the use of nanomaterials in informatics [27] paint [28], energy [27], beauty products [29], pharmaceutical [30], automotive, aero-industries [31] is increasing. New methods and approaches are being developed to integrate additive manufacturing technologies and nanotechnology. Nanomaterials are used to improve mechanical, optical, electrical, magnetic, and biological properties. To improve the mechanical, electrical, magnetic, thermal, and flow properties of parts used in the aero-industries, studies and applications are carried out on the use of nanomaterials in additive manufacturing methods.

2.2 MATERIALS USED IN ADDITIVE MANUFACTURING

When it comes to additive manufacturing, it's critical to select materials that have the necessary qualities for the job. Powder, sheet, wire, and liquid are some of the materials that can be employed. Additive manufacturing parts often require additional processing to reduce porosity, roughness, geometric tolerance, and improve microstructure. Polymer structures are divided into two as amorphous and semi-crystalline according to their crystal structures. Amorphous polymers are suitable for extrusion due to their large viscous softening temperature. The wide viscous softening temperature allows the extruder to operate in a wider temperature range. This causes minimal distortion in the material. Since the viscous softening temperature is narrow in semi-crystals, sensitive changes in extrusion temperature complicate the process. Therefore, "powder bed fusion" methods such as SLS are more suitable for this type of polymer. Methods based on "Vat polymerization," such as SLA, can only use photosensitive thermosets as feeder material. In these methods, attention should be paid not to precipitation of the resin, viscosity, and unwanted laser diffraction due to nanoparticles [32]. Binder jetting is a suitable method for photosensitive thermosets. Particles added to the resin should not cause an increase in viscosity to a degree that prevents printing [33]. In methods such as "binder jetting" and "sheet lamination," any feeder material that can be turned into powder or sheet can be used. As a rule, any metal that can be cast and welded is suitable for metal powder bed fusion. Distortion can be minimized by changes in the support structure, cooling rate, and laser design. Metallic powder or wire is used as feeder material in directed energy deposition processes involving laser or electron beams [34].

2.2.1 NANOSCALE MATERIALS USED IN ADDITIVE MANUFACTURING

The combination of nanomaterials and additive manufacturing opens new opportunities in nanocomposites. The use of different material types such as shape memory alloys, composite materials, nano-sized material additives, and nanoparticles in additive manufacturing technologies has expanded the applicability of these methods. The reinforcement of carbon nanotubes, carbon nanofibers, graphene, nanocellulose, and metal nanoparticles to various polymer matrices in productions using additive manufacturing technology improves properties such as tensile strength, toughness, and shear modulus. [35]. The integration of nanomaterials and additive manufacturing technologies enables the improvement of material properties of parts containing nanomaterials with customized geometries, rapid design updates, monolithic production, and increased part integration.

Multi-walled carbon nanotubes and carbon nanofibers are known to be added to the polymer matrix to improve mechanical characteristics, according to literature research. For use in additive manufacturing processes such as FFF and SL, nanoscale carbon fibers have been added to matrix materials, such as ABS [36], PLA [37], and epoxy [38]. The addition of nano additives caused an increase in tensile strength between 30% and 50%. Nanocarbon fiber reinforcement increased the hardness values. However, in additive manufacturing technologies, the bonding between the interface is important. In the use of composite structures, the junctions at the fiber/matrix interfaces are expected to be ideal. The large surface area and high reactivity of nanomaterials provide advantages in interfacial bonding.

Carbon-based nanomaterials, including nanotubes, nanofibers, graphene, and carbon black, are common nanocomposites with high mechanical strength. It is also possible to change the optical properties of materials with nanomaterial additives. Nanomaterial additions can alter the reflection and transmission of light in components significantly. The color change is created in materials with gold nanoparticles and lanthanoid nanoparticle additives. Nano additives show unique optical properties that create different colors during the reflection and transmission of light in the material. By using such additives on aircraft windows, harmful rays can be absorbed, and invisible in optical radars can be achieved. In addition, it is important to use certified materials in the aviation industry because products coming from outside the quality system negatively affect flight safety. Optical properties gained by nanomaterial additives can develop new approaches to prove the originality of parts and eliminate counterfeiting [39].

The discovery of nanographene revolutionized electrical conductivity. With the use of graphene in electronic card printing and additive manufacturing technologies, efficiency increases in data transfer speed and cooling performance are achieved [40].

One of the aerospace applications is antenna construction employing graphene nano conductive ink-based materials and the additive manufacturing process. Due to high mechanical properties benefits, graphene-based antennas are being replaced with traditional copper-based antennas. Using graphene nano reinforcement and additive manufacturing technology, it is possible to produce antennas that are reduced by one-third [41].

In metal additive manufacturing technologies, mostly micro-sized particles are used. Porosity and surface quality are two major drawbacks of metal additive fabrication. The use of nano-sized particles creates opportunities to overcome these constraints. In the production of cobalt (Co) and tungsten carbide (WC) materials at the nanoscale by the selective laser melting (SLM) method, an increase in wear resistance and hardness has been achieved. Because nano cobalt powder has a larger surface area than micron powder, it has a lower melting point than micron powder. This allows for the creation of a liquid melt pool in the WC with a uniform distribution of unmelted micron particles [42]. In the SLM method, in which titanium carbide nanoparticles are used, it has been observed that TiC reinforcement improves tribological properties, affects the friction coefficient, and reduces wear resistance [43].

The ability to increase mechanical qualities, produce complex-shaped goods, and build functional products is enabled by the unique material combinations of a wide range of nano powder materials [44].

Nanomaterial additions improve mechanical characteristics, boost thermal and electrical conductivity, lower sintering temperatures, and impact dimensional correctness, according to studies in the literature. These are the variables that have resulted in the aerospace, defense, and space sectors adopting additive manufacturing technologies in large numbers. This indicates that additive manufacturing technology will continue to gain traction in the worldwide aerospace, defense, and space industries.

2.3 CHALLENGES

Nanomaterial production involves costly processes. Adding nanomaterial by making composite as reinforcement material, obtaining nano-sized particles with homogeneous distribution, showing similar properties of these particles (fixed surface areas, etc.) involve special processes. Therefore, unit prices of nano-doped materials are higher than their pure forms. The fact that additive manufacturing methods eliminate waste of material to a large extent creates significant opportunities in this sense.

Lightness is a concept that affects the technical and economic performance of structural parts of aircraft. The strengths under load expected from structural parts are limited to a certain safety factor. These factors, which are crucial for aircraft safety, also have a significant impact on performance. The detrimental consequences of mass reduction on mechanical performance can be mitigated thanks to nanomaterial additives in additive manufacturing technology. Reducing the mass of the system provides a reduction in fuel costs and emission rates while increasing payload capacity and flight range.

Fuel consumption is foremost among aircraft operational expenses, accounting for 33% of total expenditures. By lowering the weight of each aircraft by 1 pound

(0.45 kg), an airline with a fleet of more than 600 commercial aircraft may save around 11,000 gallons (41,639 L) of fuel each year [45].

Past research has shown promising results in incorporating nanomaterials into additive manufacturing technologies. Additive manufacturing methods create significant opportunities in the aerospace and defense industries where the number of productions is low, parts with complex geometry and topology optimization are intense. Additive manufacturing technologies, one of the focal points of today's literature studies, have also taken their place in industrial applications in the aerospace, defense, and space industries.

Another alternative to meet the requirements of additive manufacturing technologies is the use of nano-sized building materials. Nano-sized construction materials provide excellent strength, hardness, and wear resistance while maintaining appropriate ductility [46].

2.4 APPLICATIONS AND TRENDS

Aviation, defense, and space industries are high-strategy technologies that show the economic power and industrial level of a country. Additive manufacturing technologies create many opportunities in these industries, such as on-site production, narrowing the supply chain, and reducing labor costs by reducing assembly. In addition, thanks to the production of complex geometries and the freedom of design, many parts can be redesigned quickly, and repair processes are facilitated. In this way, products with increased functionality and performance can be obtained. Flow characteristics are an important performance criterion, especially in aircraft fuel systems. With additive manufacturing technologies, unique internal flow configurations that were nearly impossible for conventional manufacturing techniques could be produced.

Jet engine manufacturer GE Aviation has produced the 30,000th fuel nozzle tip (Figure 2.10 (a)) for LEAP jet engines using additive manufacturing technology. The fuel nozzle, consisting of 18 separate parts, could be produced in one piece with the additive

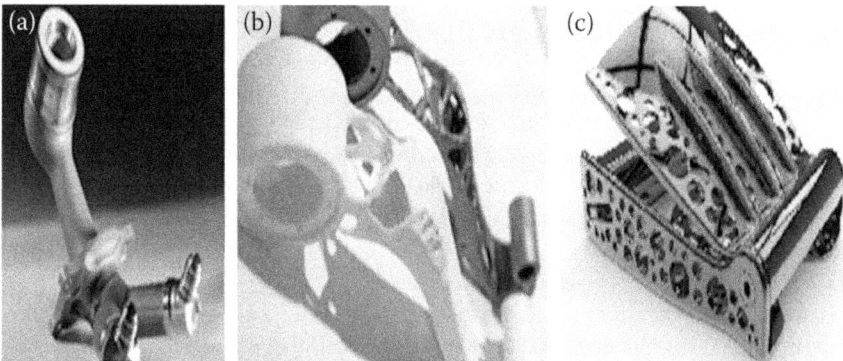

FIGURE 2.10 (a) GE Aviation LEAP turbine engine fuel nozzle tip [47], (b) Virgin Atlantic Airlines TV monitor arm [48], (c) lightened seat belt buckle [49].

manufacturing method. With the new design, the fuel nozzle has become 25% lighter than the old design. In addition, the new design protects the nozzle against the high temperature caused by the combustion of fuel, thanks to its cooling channels, and increases fuel efficiency. GE Aviation expects 50% of jet engine parts to be manufactured using additive manufacturing methods during their current life [47].

In a project funded by the Technology Strategy Board (TSB) for Virgin Atlantic airline, the part (Figure 2.10 (b)) that holds the TV monitor in the airline's first-class seats was rebuilt to be manufactured using the additive manufacturing method. With the lattice structure, the weight of the arm has been reduced by 50%. It is stated that the aircraft will save 20,000 liters of Jet A1 fuel during its 30-year flight life [50].

The weight of an airline seat belt buckle produced by conventional manufacturing methods varies between 155 g and 70 g (Figure 2.10 (c)). When seat belt buckles are produced by additive manufacturing from titanium alloy material without sacrificing durability, the weight is reduced to 68 g. The standard steel buckle used on the Airbus A380 passenger airplane (853 economy class passengers) may be replaced with a titanium buckle made via additive manufacturing, resulting in a total weight reduction of 74 kg. Over the life of the airplane, this saves 3,300,000 liters of fuel and reduces CO_2 emissions by 0.74 million tons [51].

Today, flaws in globally accepted certification and standardization for the use of additive manufacturing technologies in the aerospace, defense, and space sectors are being addressed. With the establishment of ASTM International Additive Manufacturing Technologies Committee F42 and the activities carried out by the committee, standardization studies have accelerated. Continuing to develop and oversee new standards for growing additive manufacturing technologies, the committee is leading the expansion of its application areas. The active work of regulatory institutions on additive manufacturing will contribute to the spread of these technologies in the future.

To understand the additive manufacturing technologies well and to know the advantages they provide in design, it is necessary to increase the training on this subject. It is very important to train qualified manpower on these technologies, which change the perspective of designers, to catch up with the pace of technological development. It is necessary to carry out tests that show that additive manufacturing methods meet constantly renewed safety expectations and to conduct studies on the literature.

In the future, it is predicted that the advantages of nanomaterials will be discussed more specifically in additive manufacturing technology, and the combination of these two technologies will create a wide variety of opportunities in subjects such as lightening, mechanical, optical, electrical, and thermal conductivity.

2.5 CONCLUSION

It is well known that applications in the aviation sector hasten the development of additive manufacturing technology. Many uses of additive manufacturing technology in the aviation sector will be the topic of research and development studies. Although several studies on the integration of nanomaterials with additive manufacturing technologies have been undertaken, the research done thus far has been restricted. To successfully overcome the restrictions of additive manufacturing and nanomaterials, investigations should be done to collect new data and create solutions to include nanostructures in additive manufacturing technologies. There aren't enough studies on the use of existing nanoparticles in additive manufacturing methods. The interaction of nanocomposites with additive manufacturing is poorly understood. Standardizing process parameters and synthesis methodologies for various nanomaterials and processes require research. It is known that nanomaterials can offer effective solutions in subjects such as lightness, thermal conductivity, electrical efficiency, and mechanical performance in the aviation industry and that integrating nanomaterials with additive manufacturing technologies will create opportunities in the supply chain and complex geometry part production.

REFERENCES

1. T. Wohlers, "Additive manufacturing advances," *Manuf. Eng.*, vol. 148, pp. 55–56, 2012.
2. A.D. Taylor, E.Y. Kim, V.P. Humes, J. Kizuka, and L.T. Thompson, "Inkjet printing of carbon supported platinum 3-D catalyst layers for use in fuel cells," *J. Power Sources*, vol. 171, no. 1, pp. 101–106, Sep. 2007.
3. S. Morvan, R. Hochsmann, and M. Sakamoto, "ProMetal RCT(TM) process for fabrication of complex sand molds and sand cores," *Rapid Prototyp.*, vol. 11, pp. 1–7, 2005.
4. T. Wohlers, "Wohlers report," 2010. https://www.wohlersassociates.com/2010report.htm
5. V. Petrovic, J. Vicente Haro Gonzalez, O. Jordá Ferrando, J. Delgado Gordillo, J. Ramon Blasco Puchades, and L. Portoles Grinan, "Additive layered manufacturing: Sectors of industrial application shown through case studies," *Int. J. Prod. Res.*, vol. 49, no. 4, pp. 1061–1079, Feb. 2011.
6. T. Hwa-Hsing, C. Ming-Lu, and Y. Hsiao-Chuan, "Slurrybased selective laser sintering of polymer-coated ceramic powders to fabricate high strength alumina parts," *J. Eur. Ceram. Soc.*, vol. 31, pp. 1383–1388, 2011.
7. S., Ashley, "Rapid prototyping systems," *Mech. Eng.*, vol. 113, p. 34, 1991.
8. K.V. Wong and A. Hernandez, "A review of additive manufacturing," *ISRN Mech. Eng.*, vol. 2012, pp. 1–10, 2012.
9. K. Cooper, "Rapid prototyping technology: Selection and application," *Assem. Autom.*, vol. 21, no. 4, pp. 358–359, Dec. 2001.
10. R. Noorani, "Rapid prototyping, principles and applications," *Assem. Autom.*, vol. 30, no. 4, p. 377, 2010.
11. J.P. Kruth, "Material incress manufacturing by rapid prototyping techniques," *CIRP Ann. –Manuf. Technol.*, vol. 40, no. 2, pp. 603–614, Jan. 1991.
12. X. Cui, S. Ouyang, Z. Yu, C. Wang, and Y. Huang, "A study on green tapes for LOM with water-based tape casting processing," *Mater. Lett.*, vol. 57, no. 7, pp. 1300–1304, Jan. 2003.
13. K. Stetz, "Fused deposition modeling (FDM) | Kyle Stetz /// Rapid Prototyping Study," 2009. [Online]. Available: https://kylestetzrp.wordpress.com/2009/05/20/fused-deposition-modeling-fdm/
14. J., Sandoval and R.B., Wicker, "Functionalizing stereolithography resins: effects of dispersed multi-walled carbon nanotubes on physical properties", Rapid Prototyping Journal, vol. 12, no. 5, pp. 292–303, 2019. https://doi.org/10.1108/13552540610707059
15. R. Udroiu and I.C. Braga, "Polyjet technology applications for rapid tooling," in *MATEC Web of Conferences*, 2017, vol. 112. https://www.matec-conferences.org/component/makeref/?task=show&type=html&doi=10.1051/matecconf/201711203011

16. G.V. Salmoria, R.A. Paggi, A. Lago, and V.E. Beal, "Microstructural and mechanical characterization of PA12/MWCNTs nanocomposite manufactured by selective laser sintering," *Polym. Test.*, vol. 30, no. 6, pp. 611–615, Sep. 2011.

17. CustomPartNet, "Rapid prototyping – Selective laser sintering (SLS)," *CustomPartNet*, 2017. [Online]. Available: https://www.custompartnet.com/wu/selective-laser-sintering %0A http://www.custompartnet.com/wu/selective-laser-sintering.

18. C. Semetay, "Laser engineered net shaping (LENS) modeling using welding simulation concepts," Lehigh University, 2007.

19. L.E. Murr *et al.*, "Metal fabrication by additive manufacturing using laser and electron beam melting technologies," *Journal of Materials Science and Technology*, vol. 28, no. 1. Elsevier, pp. 1–14, Jan. 2012.

20. M. Galati and L. Iuliano, "A literature review of powder-based electron beam melting focusing on numerical simulations," *Additive Manufacturing*, vol. 19. Elsevier, pp. 1–20, Jan. 2018.

21. Y. Xiong, *"Investigation of the laser engineered net shaping process for nanostructured cermets,"* University of California, 2009.

22. J. Long, A. Nand, and S. Ray, "Application of spectroscopy in additive manufacturing," *Materials*, vol. 14, no. 1. Multidisciplinary Digital Publishing Institute, pp. 1–29, Jan. 2021.

23. R.C. Prometal, "ProMetal RCT rapid prototyping and digital sand casting services," 2010. https://www.youtube.com/watch?v=Z8MaVaqNr3U&ab_channel=PrometalRCT

24. C. Lafayette, "Additive manufacturing: ProMetal three dimensional printing (ExONE R1)," 2013. https://www.youtube.com/watch?v=cXbFSg96wV0&ab_channel=LafayetteChBE

25. I. Gibson, D. Rosen, and B. Stucker, "Introduction and Basic Principles," in *Additive Manufacturing Technologies*, Springer New York, 2015, pp. 1–18.

26. 52921-13 ASTM International ISO/ASTM, "Standard terminology for additive manufacturing – Coordinate systems and test methodologies," 2013.

27. R.S. Aga *et al.*, "Laser-defined graphene strain sensor directly fabricated on 3D-printed structure," *Flex. Print. Electron.*, vol. 6, no. 3, p. 032001, Apr. 2021.

28. A. Al-Kattan *et al.*, "Release of TiO2 from paints containing pigment-TiO2 or nano-TiO2 by weathering," *Environ. Sci. Process. Impacts*, vol. 15, no. 12, pp. 2186–2193, 2013.

29. T. Smijs and S. Pavel, "A case study: Nano-sized titanium dioxide in sunscreens," in Dolez, P.I. (Ed.), *Nanoengineering: Global Approaches to Health and Safety Issues.* Elsevier, 2015, pp. 375–423.

30. S. D'Souza, "A review of in vitro drug release test methods for nano-sized dosage forms," *Adv. Pharm.*, vol. 2014, pp. 1–12, Nov. 2014.

31. J.C. Najmon, S. Raeisi, and A. Tovar, *Review of additive manufacturing technologies and applications in the aerospace industry.* Elsevier, 2019.

32. J. Ledesma-Fernandez, C. Tuck, and R. Hague, "High viscosity jetting of conductive and dielectric pastes for printed electronics," in *Proceedings –26th Annual International Solid Freeform Fabrication Symposium – An Additive Manufacturing Conference, SFF 2015*, 2020, pp. 40–55.

33. T. Wu and T. Das, "Theoretical modeling and experimental characterization of stress development in parts manufactured through large area maskless photopolymerization," 2011, pp. 748–760. http://hdl.handle.net/1853/54274

34. D. Bourell *et al.*, "Materials for additive manufacturing," *CIRP Ann. – Manuf. Technol.*, vol. 66, no. 2, pp. 659–681, 2017.

35. R. Redón, L. Ruiz-Huerta, Y.C. Almanza-Arjona, Y. Rojas-Aguirre, and A. Caballero-Ruiz, "Nanocomposites for additive manufacturing," *Am. J. Chem. Res.*, vol. 1, no. 5, pp. 1–14, 2017.

36. M.L. Shofner, K. Lozano, F.J. Rodríguez-Macías, and E.V. Barrera, "Nanofiber-reinforced polymers prepared by fused deposition modeling," *J. Appl. Polym. Sci.*, vol. 89, no. 11, pp. 3081–3090, Sep. 2003.

37. E.A. Papon and A. Haque, "Tensile properties, void contents, dispersion and fracture behaviour of 3D printed carbon nanofiber reinforced composites," *J. Reinf. Plast. Compos.*, vol. 37, no. 6, pp. 381–395, Feb. 2018.

38. C.B. Sweeney *et al.*, "Welding of 3D-printed carbon nanotube–polymer composites by locally induced microwave heating," *Sci. Adv.*, vol. 3, no. 6, p. e1700262, Jun. 2017.

39. L. Kool, A. Bunschoten, A.H. Velders, and V. Saggiomo, "Gold nanoparticles embedded in a polymer as a 3D-printable dichroic nanocomposite material," *Beilstein J. Nanotechnol. 1043*, vol. 10, no. 1, pp. 442–447, Feb. 2019.

40. L. Jiao *et al.*, "Laser-induced graphene on additive manufacturing parts," *Nanomaterials*, vol. 9, no. 1, p. 90, Jan. 2019.

41. P. Ram, R.J.L. Rajakumaran, R. Chithoor Santharam, J. Nancheri, and M.G. Ogirala, "Feasibility analysis of additive manufacturing method for graphene based super solar body mounted patch antenna for satellite applications," *SN Appl. Sci.*, vol. 2, no. 4, p. 93, Aug. 2020.

42. S. Grigoriev, T. Tarasova, A. Gusarov, R. Khmyrov, and S. Egorov, "Possibilities of manufacturing products from cermet compositions using nanoscale powders by additive manufacturing methods," *Materials (Basel).*, vol. 12, no. 20, pp. 1–16, 2019.

43. D. Gu, H. Zhang, D. Dai, M. Xia, C. Hong, and R. Poprawe, "Laser additive manufacturing of nano-TiC reinforced Ni-based nanocomposites with tailored microstructure and performance," *Compos. Part B Eng.*, vol. 163, no. April 2018, pp. 585–597, 2019.

44. B. Zheng, J.E. Smugeresky, Y. Zhou, D. Baker, and E.J. Lavernia, "Microstructure and properties of laser-deposited Ti6Al4V metal matrix composites using Ni-coated powder," *Metall. Mater. Trans. A Phys. Metall. Mater. Sci.*, vol. 39 A, no. 5, pp. 1196–1205, May, 2008.

45. IATA, "Montreal-March 6, 2013 Maintenance Cost Task Force (MCTF) 1 Airline Cost Management Group (ACMG)," *MCTF*, 2015.

46. I.V. Gorynin, A.S. Oryshchenko, V.A. Malyshevskii, B.V. Farmakovskii, and P.A. Kuznetsov, "III international scientific and engineering conference 'nanotechnologies of functionalmaterials': Additive technologies based on composite powder nanomaterials," *Met. Sci. Heat Treat.*, vol. 56, no. 9–10, pp. 519–524, Feb. 2015.

47. GE Aviation, "New manufacturing milestone: 30,000 additive fuel nozzles," *GE*, 2018. https://www.ge.com/additive/stories/new-manufacturing-milestone-30000-additive-fuel-nozzles

48. Ç. Gürbüz, "Additive manufacturing for lightweight aviation parts," in Karakoc, T.H., Ozerdem, M.B., Sogut, M.Z., Colpan, C.O., Altuntas, O. and Açıkkalp, E. (Eds.), *Sustainable Aviation: Energy and Environmental Issues*, Springer, Cham, 2016, pp. 333–339.

49. Crucibledesign, "3D printing with metal: Product design that delivers new technology," 2018. [Online]. Available: https://www.crucibledesign.co.uk/blog/design-that-delivers-new-technology.php.

50. Royal Academy of Engineering, "Additive manufacturing: Opportunities and constraints," 2013, no. November 2013, pp. 1–34. https://www.raeng.org.uk/publications/reports/additive-manufacturing

51. R. Huang *et al.*, "Energy and emissions saving potential of additive manufacturing: the case of lightweight aircraft components," *J. Clean. Prod.*, vol. 135, pp. 1559–1570, Nov. 2016.

3 Smart 3D Nano-Printing in Automobile Industry

Lokanath Barik
Department of Mechanical Engineering, National Institute of
Technology, Rourkela, India

Ajit Behera
Department of Metallurgical and Materials Engineering,
National Institute of Technology, Rourkela, India

CONTENTS

DOI: 10.1201/9781003189404-3

3.1 INTRODUCTION

During the product development phase of an automobile, a designer initially creates a scaled model of the actual part (prototype) and, after successful testing, a full-scaled model enters the production line. Before production, a series of fabrication steps occurs, and each subpart of the final assembly is fabricated individually through various machining processes. Instead of these cumbersome procedures, 3D printing generates a fully functional prototype that can be designed digitally through commercial software like CATIA, Solidworks, etc. Saving time and capital, this technology can produce any desired part as per customers' demand. It can manufacture any replacement parts that normally take days or weeks to produce. Collectors of antique cars generally find it challenging to reproduce parts of those cars since in-line assembly is no longer active. Parts of such models often come at a greater price due to antiquity, and their availability index is low. However, 3D printing can develop parts of these vintage car models or rare spares, which are difficult to obtain, with better mechanical properties and closer dimensional tolerance [1]. Several companies like Porsche, Rolls-Royce, BMW have switched to 3D printing to develop parts or sub-assemblies that are limited in the reach of the buyers. These parts could be of any limited-run vehicle or customized imported one whose productivity is limited in numbers. Conventionally speaking, if the productivity and storage limit is 60,000 and the warehouse runs out of these standard parts, then specific production planning and tooling is required to generate more of them in time, as well as to keep customers' satisfaction in check. Generally, when high demand occurs in a shorter period, it is eventually observed that the quality of the product is compromised or there is production lag (delay). The option of 3D printing is much better than reworking or manufacturing that specific part.

Additionally, it is worth noting that inventory storage space can be significantly reduced since it can produce the desired components in minimum hours or days. The benefit of it could be reduced indirect inventory cost or storage cost, whichs benefits the customers as well as the manufacturer, making it more cost-effective [2].

Nevertheless, interior design customization, like color, seat designs, and add-ons, has been limited to average car owners; on the road, all these cars look alike. Buyers of elite class cars, those with deep pockets, demand fully customized cars as per their choice; that means it is difficult to deliver cars at the stipulated time for car owners of, let's say Rolls-Royce (popularly known for customized cars). As per a survey, 30% to 40% of the Rolls-Royce owners prefer to buy a stock model from showrooms rather than order a customized model. Hence, Rolls-Royce has adopted this novel technology, giving its customers an option to decorate their car model from interior to exterior as per their luxury, compatibility, and choice. It was a beneficial step taken by the car company to satisfy its customers and make a profit out of the business [3]. Commercial car manufacturers or sports car manufacturers have always competed for lightweight and energy-efficient cars. For decades, many types of research have been conducted for improving thermal efficiency and possibilities of developing waste energy recovery methods. For decades, the automobile industry has come a long way, from old vehicles that ran fewer miles per gallon to more fuel-efficient cars; but, lightweight-ness is always of prime concern in aerospace, as well as the automotive field. Therefore, 3D printing could be the right direction for further enhancements since it generates latticed structures from aluminum alloys. These components are made equally strong and precise as their counterparts, while reducing the weight to the extent of 80%. Additionally, lightweight cars would generate less pollution and the frictional loss could be reduced considerably. This option can also counteract global working and raise the standard norms like BH4, BH6 engines specified by the government of India [4].

As mentioned earlier, while a cost is associated with materials and tooling, this method guarantees almost zero wastage. Since 3D printing is an additive manufacturing process, no subtractive operations, like cutting or drilling, are performed on the parts. The advantage of this could be reduced overall cost of the car, making it accessible to average drivers and an appropriate use of natural resources in this era of scarcity. In short, with this technology, cars ranging from 12–20 lakhs can be made available for between 7–10 lakhs with improved performance and functionality. 3D printed parts are successfully manufactured, but the whole car model is yet to strike the market since research work continues. Productivity of large parts generally requires more complex analysis. However, one automobile manufacturer in Hong-Kong XEV has developed a 3D printed electric vehicle named LSEV that can go up to 90 miles per hour and has a top speed of about 43 miles per hour. It is worth noting that, right now, the time taken to produce this car is somewhat higher than expected; that leaves the young researchers a research direction: to improve its efficiency on a larger scale [5].

3.2 MATERIAL AND TECHNIQUES USED IN THE AUTOMOBILE INDUSTRY

3.2.1 MATERIALS

A wide variety of materials is available for commercial 3D printing applications. Materials may be present in liquid form, such as resin; powder bed form, such as metals, composites, plastics, etc.; or in fibrous form, such as nylon [6]. The following materials, discussed below, are generally used in automobile sectors to develop parts and prototypes.

3.2.1.1 Plastics

Among all the materials available for 3D printing, plastics are most used to manufacture auto parts, especially the frame and accessories of the car. A wide variety of different plastic materials are available for 3D printing, such as polyethylene terephthalate (PET), ASA, etc. Due to their low cost, water repellent properties, and ease of manufacturing, plastics are popular in multiple sectors. Manufacturing processes associated with them, in general, are SLS, FDM, SLA [7].

3.2.1.2 ABS

ABS (acrylonitrile butadiene styrene) is one of the most frequently used thermoplastics used in the automobile industry, mainly to print the vehicle framework or body parts. It is resistant to shocks and is quite flexible, having elastomeric polybutadiene as its base material. Additionally, it can be easily welded and, most importantly, is recyclable. It can resist a wide range of temperatures, i.e., from −20 °C to 80 °C, making it suitable to be implemented in rockets or aircraft. Generally, the FDM method is used to 3D print ABS materials [8].

3.2.1.3 PLA

When it comes to biodegradable materials, PLA (polylactic acid), derived from renewable raw materials, is the best option. It is easier to 3D print without the involvement of a heated platform, unlike ABS, which is neither biodegradable nor able to be printed at low temperature. However, because of its low-temperature manufacturing, it is more difficult to manipulate to the desired shape than any other material. Here also, the FDM technique is used to 3D print PLA material, and it can be manufactured in a variety of colors [9].

3.2.1.4 Polycarbonate

Polycarbonates are not an ideal choice when the part is subjected to high-temperature operations, but not so high so that it will melt. They are 3D printed above 150 °C, the temperature at which they deform. They possess high strength temperature resistance and can provide strong resistance against deformation. However, these materials cannot be used where the component is subjected to wet regions and low temperatures; they tend to absorb moisture from surrounding and at low temperatures, so the 3D printed layers begin to separate. Hence, polycarbonates are not widely accepted in the automobile industry. They can be produced using the FDM process [10].

3.2.1.5 Polyamides

Polyamides commonly known as nylon semi-crystal polymers, the most used and widely accepted. They offer a high degree of strength, flexibility, rigidity, and, most importantly, they are shockproof at various impact conditions, such as ground disturbances. The FDM process is involved in synthesizing polyamide parts, and they are mostly used in the fibrous form (nylon), instead of powdered granular form. They are found to be useful in a variety of applications, such as gears, engine covers, manifolds, etc. [11]

3.2.1.6 Polypropylene

Polypropylene is a thermoplastic polymer, exclusively used in the automobile industry as compared to other sectors. It possesses high rigidity and flexibility, high resistance against shock loads, and resistance to abrasion and wear, etc. Polypropylene in the purest form is quite sensitive to temperature and UV rays, which an automobile body is normally subjected to. Under these conditions, an advanced modified alternative of it is used, one having more advanced physical and mechanical properties, making it more suitable for making auto parts [12].

3.2.1.7 Metals

Metal powder-based 3D printing is an emerging newer technology in the automotive sector, and it is gaining acceptance. Metals are manufactured by processes such as EBM, wire-feed EBP (electron beam printing), powder bed, and magneto-hydrodynamic printhead. Granular powder of pure metals, such as aluminium, titanium, etc., and alloys of steel, such as stainless steel, can be used. These metals must be melted above their melting point, most preferably by powerful laser beams or electron beams. The liquified metal is then deposited layer by layer and is allowed to solidify, forming the final product. Due to the lack of technological advancements, only small parts and prototypes can be generated. However, as per manufacturers, there is a vast possibility that metal-based 3D printing at a large magnitude can be implemented shortly, providing an alternative to the conventional subtractive manufacturing process [13].

3.2.1.8 Composites

Composite materials such as carbon and glass fiber are generally used for 3D printing of vehicle body parts. The printed body parts are found to have much higher strength, and these composite material fibers can be manipulated or optimized at each subsequent layer. This provides design flexibility of properties and appearance. Carbon fibers can impart very high strength, and, additionally, when mixed with different polymers (mentioned above), they can make the component flexible, heat resistant, lightweight, etc. Carbon fibers with resins can be commonly seen in manufacturing dashboards or any vibration-resistant prototypes. Moreover, some plastic-based composites mixed with wood filaments can provide a more organic-looking texture and can improve their reusability, making them biodegradable [14].

NANOPARTICLE POWDER FOR 3D PRINTING
AUTOMOBILE PARTS

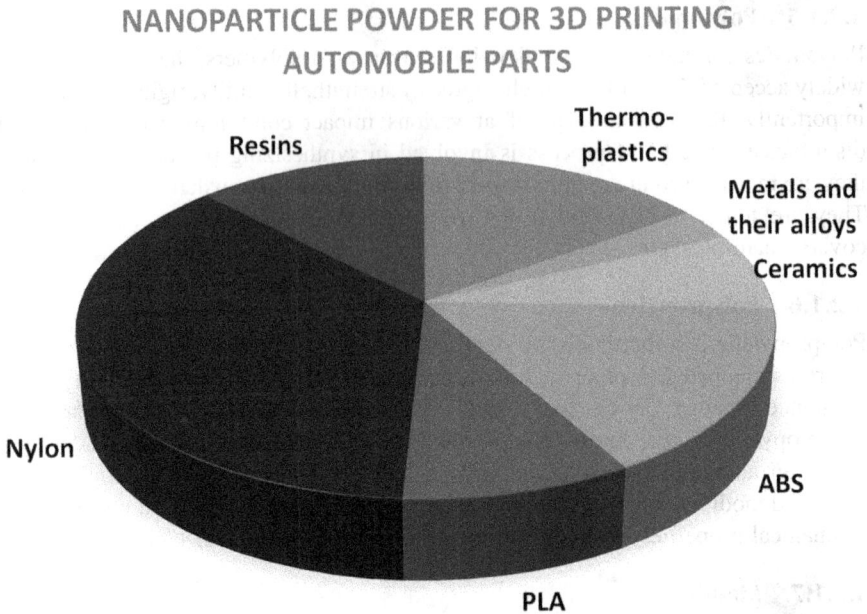

FIGURE 3.1 Usage value of different nanoparticles in the automobile industry.

Figure 3.1 denotes the usage value of different powder materials for 3D printing applications where nylon is the most used fiber material and metals are used less because of their latent heat of fusion. Mostly, the plastic groups are easily 3D printed and have flexibility in designing. Substitutes of metallic parts can be developed by combining these powders to an optimum ratio [15].

3.2.2 VARIOUS TECHNIQUES

3.2.2.1 SDL

Selective deposition lamination, or SDL, is a paper-based 3D printing technology that uses paper material at subsequent layers and bonding adhesives to generate the desired 3D shape. Initially, a few layers of paper are attached to the base plate, and then adhesives are applied by the 3D printer to bond each subsequent layer of paper to be fed. The adhesives are optimized such that more is present at the support sections, where strength is desired. Similarly, hundreds and thousands of such layers are formed and finally cut by a sharp blade to the desired shape of the digitally designed model. Mainly accessories of cars, such as logos, decorative parts, etc., are manufactured [16].

3.2.2.2 EBM

Electron beam melting (EBM) is a technique that utilizes a high-energy beam of electrons to melt the metallic powders and join them to form the exact replica of the desired CAD model. The procedure is prone to contaminations, such as impurities or oxidation; hence, the laser forming setup is present inside a vacuum chamber and is guided by magnetic fields to the desired angle and position. It is more accurate and precise since the laser can be concentrated at narrow zones and the energy dissipated by the beam of energized electrons can easily melt the materials. It is used in various sectors where metal components or tolling are required [17].

3.2.2.3 DLP

Digital light processing (DLP) utilizes photopolymers to 3D print parts by using a light source of an arc lamp with the liquid crystal display panel. The printer prints the entire surface in one pass of the vat of photopolymer resin. A highly accurate and much faster printing experience is observed. A relatively shallow vat of resin is required for synthesis, resulting in a lower cost and minimal wastage [18].

3.2.2.4 MULTIJET FUSION

Multijet fusion is indeed an economic 3D printing technology that can create working nylon models and market-ready parts in just one day. When compared to methods like SLS, the final products have a better surface finish, finer features resolution, and better mechanical qualities. Engineering-grade materials with excellent overall qualities are available from MJF. MJF also has a better surface texture, sharper features, uniform mechanical qualities, and quicker construction times [19].

3.2.2.5 POLYJET

PolyJet is a cutting-edge 3D printing process that creates smooth, precise parts, tooling, and prototypes. It adopts a unique microscopic layer-resolution technique that can provide accuracy up to 0.014 mm; hence, it can create thin and complicated topologies with the largest variety of powder materials available with any other technology. It is generally used to produce tools, such as jigs and fixtures, with higher accuracy and precision [20].

3.2.2.6 SLM/DSLM

SLM and DMLS are relatively comparable techniques, similar in process and utilizing metal powder bed-fusion mechanisms. The energized beam of light-assisted laser melts the metal powders layer by layer until the final CAD profile is obtained. Fusion of metal powders even occurs at the molecular level, providing uniform strength and mechanical properties. Pure metals such as titanium and aluminium can also be 3D printed using this technology. The sintered pieces have a slightly rough finish that might require additional post-processing operations, such as grinding and polishing. This technique offers several advantages, such as the formation of complex contours, reduced lead times, multiple prototyping, etc. [21].

3.2.2.7 SLA

The automotive market uses SLA-based 3D printing extensively for manufacturing vehicle components and their prototyping, even for parts that are directly introduced in the product. BMW, Lamborghini, and Jaguar Land Rover all have embraced this new trend, generating design concepts, functioning prototypes, and finished parts in-house without outsourcing them to vendors. This approach is not only beneficial to designers and the research and development wings, but it may also significantly decrease costs and shorten the time it takes for new innovative designs to reach the market [22]. Stereolithography technology can produce concept models, rapid prototypes, and complicated products with complicated topologies in a shorter time. The parts can be made from a variety of materials having higher resolutions and good surface textures. SL uses an ultraviolet laser-assisted process directed on a liquid thermostat resin. Each subsequent layer is imaged on the resin surface, after which it is given the desired 3D shape as programmed in the CAD software.

3.2.2.8   SLS

Selective laser sintering 3D prints nylon powder by using laser light to fuse these nylon fibers, mapping the exact geometry layer by layer starting from the bottom and moving to the top, as developed in the CAD profile. It can create accurate designs and operational production parts within a day. There are a variety of nylon-based polymers available, all of which produce extremely durable finished items. Each layer of part shape is sintered into a heated bed consisting of nylon-based powdered material on the SLS equipment. A roller rolls all over the bed until each layer is merged to disperse the subsequent layer of powder layer by layer. The process is continued until the design is complete [23].

3.2.2.9   FDM

FDM 3D printing is the most widely used method among all. It works in a horizontal- and vertical-oriented method in which an extrusion nozzle traverses across a build platform. The procedure uses elastomeric thermoplastic material that achieves its melting temperature and then is driven out in layers to make a three-dimensional object. During manufacturing, each layer comes out to like a horizontal cross-section. This unique technology is used by many organizations; it enables the manufacturing of intricate things. As a result, engineers are employing it to evaluate parts for quality and durability [24].

Figure 3.2 shows 2018 market statistics of automobile manufacturing 3D printing processes for manufacturing auto-parts [25]. It can be concluded that the FDM method has gained a lot of popularity and is expected to grow in the near future. Methods like FDM, SLS, and SLA use plastic materials and their derivatives; hence, they are easier to manufacture. EMB mainly works on metal powder beds by heating the powder metals to their fusion temperature. However, large component formation is difficult and time-consuming. With subsequent development in heating efficiency, soon enough, this technique will be considered among the widely used techniques.

3.3   RECENT DEVELOPMENTS

The imprints of 3D printing are growing in every sector. Manufacturers and shareholders are staking their shares in this emerging technology. Considering its added benefits, all the concerned sectors are making a profit beyond the contribution

3D PRINTING PROCESSES IN AUTOMOBILE INDUSTRY

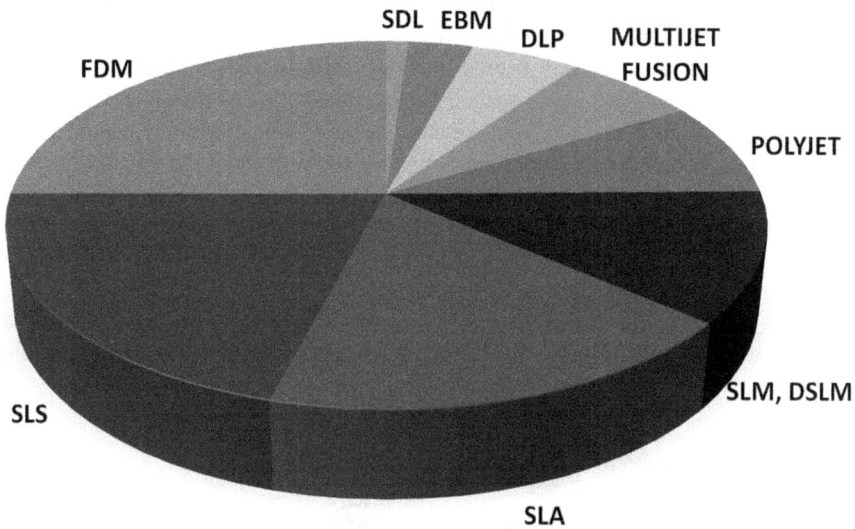

FIGURE 3.2   Application of different methodologies in the automobile industry.

margin. It can be successfully forecasted that 3D printing will soon take over many markets and production facilities [26].

As per a recent study conducted by [27], the growing demand for 3D printed parts is shown in Figure 3.3, where it can be seen that the stock price and shares of this technology are growing at an exponential rate. There was a time when the net

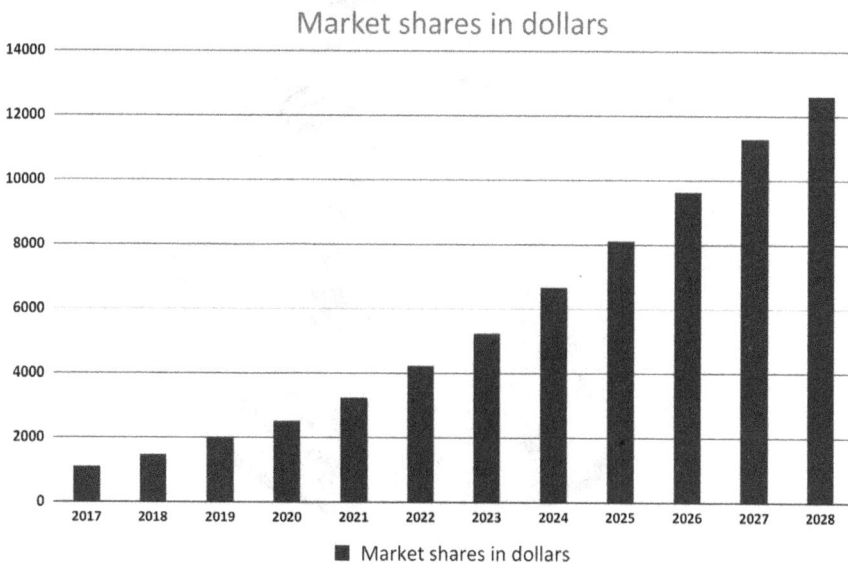

Market shares in dollars

FIGURE 3.3   Recent trends in automobile 3D printing stock market growth and prediction.

worth of shares was a mere $1,086, and now in 2021, it has touched $3,225. Considering its demand and applicability, we can predict that the shares will touch $12,614 dollars by the end of 2028.

Various research work has been conducted recently; researchers successfully developed material-based and method-based auto-parts. Mohanavel et al. [28] discussed 3D printing of the front knuckle (Figure 3.4), which is a part of the suspension system; steering component, manufactured through the forging process; and ball bearings, which support axial load by providing relative motion between two rotating components and are manufactured through powder metallurgy. 3D printing of these parts (Figure 3.4 and Figure 3.5) provides better accuracy in dimensions than their conventional counterparts.

[29] Wang et al. developed a printing strategy in which the modeled functional part, which could be printed, was created separately with appropriate printing

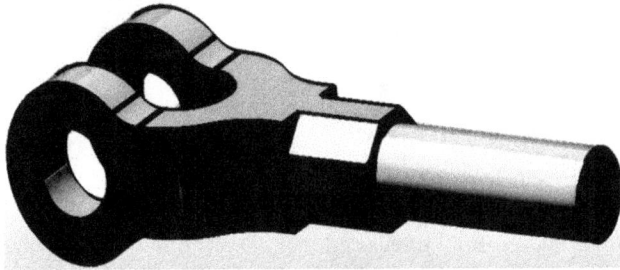

FIGURE 3.4 Front knuckle.

FIGURE 3.5 Ball bearing meshing.

materials, whereas the unbounded part was provided with a non-printing powder, such as photosensitive liquid resin. Upon 3D printing, the functional part model was printed and the unbounded region either did not print or was removed. This 3D printing technique was tested in the fabrication of spring washer, vehicle chassis dampers, and certain electronic accessories. [30] Cen et al. developed a connecting rod through laser printing technology using Fe powder subjected to the desired process parameters. A connecting rod is a part of the engine assembly that converts the reciprocating motion of the piston to the rotary motion of the engine. The obtained 3D part had a hardness of 450–490 HV, yield strength of 754 MPa, a tensile strength of 1189 MPa, and the percentage elongation was found to be 9%. Results were compared with traditional forming operations, such as forging and powder forging, and 3D printing was found to be best suited. The process took about 4.5 to nearly 5 minutes to print each layer of the part model, and the total consumed time was 4 hours 20 minutes.

[31] Roland et al. discuss 3D printing technology by forming a layer of polymerizable liquids at two stages at intermediate levels; after forming with liquids, both the intermediate stages were washed and contacted each other, followed by solidifying and or curing stages, after which the final three-dimensional object was produced. This method can be extended in developing automobile accessories, such as gears, rods, and fasteners; automobile body parts, such as dashboards, bumpers, etc., and electronic circuits and interior design of seats. The effectiveness of this method was tested in imparting a wide range of mechanical properties and associated elastomeric properties, surface hardening properties, or any desired properties by controlling the process parameters.

The ability to impart multiple hardening methods (e.g., dual hardening, where a continuous liquid interface can be used to print the locked shape, and subsequent thermal or another curing can be used to provide elastomeric properties or other desired properties) greatly expands the range of useful 3D products that can be formed. [32] Deshun et al. presented the utility of 3D printing technique using aluminium alloy-based additive manufacturing method. The author has presented a miniature version of an inline engine cylinder consisting of piston, cylinder, connecting rod arrangement made of 3D printed aluminum alloy. The device was made vibration resistant by providing a shock-absorbing property, thereby reducing its wearing and tearing. The device was claimed to have a long life and better dimensional stability.

A simulation-based algorithm base layout tool was created by [33] Willis et al. It includes mechanical and electrical design engines to create mechanical parts and electrical parts, respectively. The 3D printer was also provided with design engine constraints, which are mechanics-based inoperable regions or unbounded configuration in geometry. The following method was able to generate auto parts of the electrical and mechanical configuration. [34] Dacunha et. al. developed several 3D printed mechanical devices, like layshaft covers, polymer aerofoils, piston rods, drive, and transmission shafts. The material used for 3D printing was polyamides, including polyetherimide, polyimide, polyether ketone, polysulfone, etc. Additionally, suitable thermostat materials were provided, like epoxy-cured aromatic amines, polymethacrylates, condensation polyamides, etc.

- LAYSHAFT COVERS are essentially required for covering purposes. They are tubular or cylindrical.
- POLYMER AEROFOILS for gas turbine engines were developed; they were equally functional lightweight hybrid aerofoils having a graphite fiber-reinforced polyetherimide core. The aerofoils exhibited better bond strength relative to their traditional model because of their higher-order structural integrity.
- PISTON RODS are the prime source for power generation. They vary in size, as per power requirement and design considerations. Depending on the amount of force to be produced and their factor of safety, the strength of the material is decided. However, to construct such an assembly, a designer should look for weight and cost of the material. If weight is lowered by selecting material such as titanium, the cost will rise significantly, and if metal, like steel or its types, is chosen, then the cost will be moderate, but the weight of the part will be compromised. The 3D printed part offers the best optimal solution, as suggested by the author.
- DRIVE AND TRANSMISSION SHAFTS transmit torsional power from the engine to the wheels of the transmission system. Due to the high strength and rigidity criteria, drive shafts of transmission lines are often composed of high-strength metals and their alloys. Metal shafts made of alloy steel or high-strength steel, on the other hand, are heavy and expensive. 3D printed shafts have benefits of these components, as well as being lightweight for smoother power transmission, and these rotating components properly balance.

Devesh [35] et al. proposed fused deposition modeling for fabrication of mechanical auto parts, with the base material being acrylonitrile butadiene styrene (ABS), which is a class of thermo-plastics. FDM method is a part of additive manufacturing technology, and it offers advantages such as low-temperature operation, easy material change, and low maintenance cost. The prototype developed was bumper and pillar trim, where the material was tested against bending and compression load. In the analysis, the cylindrical specimen having diameter 12.7 mm and length 25.4 mm and ASTM D790 specimen having size (3.2 × 12.7 × 127) mm, was considered. It was found that the maximum bending strength of the fabricated material was 45.39 MPa and the maximum compressive strength was 32.054 MPa at the prescribed orientation and infill pattern. 3D printing builds parts/components by forming each subsequent layer using materials in fine powder form. This layer-by-layer deposition of materials avoids material wastage and provides design freedom and better control over complexity in design [36]. Juechter et al. [37] investigated titanium aluminide alloy Ti-45Al-4Nb-C to design and develop turbocharger wheels. A turbocharger is a mechanical device that increases the charge density of air entering the cylinder of compression ignition (CI) engines during suction stroke. The entire process chain was investigated from product designing to product development. Optimization of material properties was done by adjusting scanning parameters and the strategy involved during production. Figure 3.6 shown below is the impeller blade of the turbocharger produced with laser-assisted 3D printing.

FIGURE 3.6 Rotor of turbocharger.

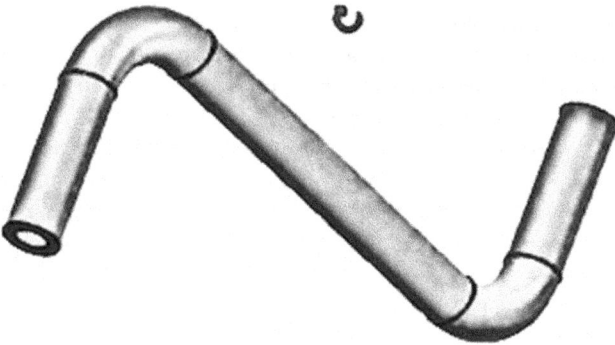

FIGURE 3.7 Flow pipe.

3D printed flow pipes, as shown in Figure 3.7 by selective laser melting, were adopted by the Volkswagen group as an alternative to traditionally manufactured pipes, which were used for PVC coatings. Earlier made pipes were fragile, breaking easily; hence, a more precise flow pipe was developed using additive manufacturing technology, CNC, and electric spark. Computational fluid dynamic simulations were done to determine optimality inflow parameters and design. Additionally, the influencing parameters for fabrication were investigated, such as the orientation of forming surface, the thickness of the pipes, etc., and it was found that optimum conditions for high-quality pipes were forming a surface angle to be 45° and thickness of the pipe nearly equal to 0.15 mm. To further improve the coating performance and surface finish, additional electric spark drilling was performed [38].

Moreover, trying different combinations of nanomaterials in additive manufacturing technology will help to explore all the possibilities in developing fuel cells, solar cells, and batteries and also could help to alter their atomic structure at the microstructural level. Carbon fiber is one such material that is rapidly being used by automobile manufacturers to build automobile roofs and bumpers. The window frame is another significant advancement in material for 3D printing. Carbon fiber among all is lighter and provides substantial deformation resistance. Titanium, however, is by far the most desirable material in the automobile industry for product development in high-end applications, owing to its mechanical properties, such as high strength, corrosion resistance, and low density, as compared to other materials. Additive manufacturing technology using the nanoparticle powder form of these materials has the potential to create wonders and to induce combined properties of different materials, as per design requirements. Additive manufacturing has already set its foot in the automobile sector and has been enlarging its reach with recent scientific developments [39].

3.4   3D PRINTED AUTOMOBILE PARTS

Over decades, automobile manufacturers such as Porsche, BMW, Mercedes, Rolls-Royce, Ford Motors, etc., have been competing to produce better functioning energy-efficient cars. 3D printing has enabled them to compete in the market more strategically and innovatively, providing a flexible ground for manufacturing auto parts that provide their drivers driving comfort and high performance. Recent developments in the product phase are discussed below by various automobile companies. 3D printing technology has revolutionized our industry's ability to design, develop, and produce quality products. In the automobile industry, 3D printing has created a sensation, allowing lighter and more complicated structure development in a shorter period [40]. 3D printing of various car parts and the process involved are given in Figure 3.8.

Various 3D print technologies adopted by different automobile industries are discussed below:

3.4.1   AUTOMAKERS IN THE 3D PRINTING BUSINESS

In 2016, Daihatsu, Japan's oldest car manufacturer company, announced a vehicle customization project for its Copen car type. This type has provided the liberty to its customers to model the front and rear bumper. The most frequent uses of additive manufacturing in the automotive industry are nevertheless prototyping, jigs & fixtures, and tools. However, the last two decades delivered huge progress, and additive manufacturing in automobiles is increasing beyond these applications [41]. Local Motor, for instance, developed the world's first 3D printed electric vehicle in 2014. Local Motor makes not only cars, but also a 3D-printed bus called OLLI, demonstrating the versatility of 3D printing technology [42]. A few examples from automobiles companies using additive manufacturing follow.

3.4.1.1   Ford Motors

Ford Motors is a pioneer when it comes to implementing 3D printing technology, utilizing it to create prototypes and engine parts. Ford Motors group developed a 3D printed cylinder head, which is an essential part of the cylinder block. The company

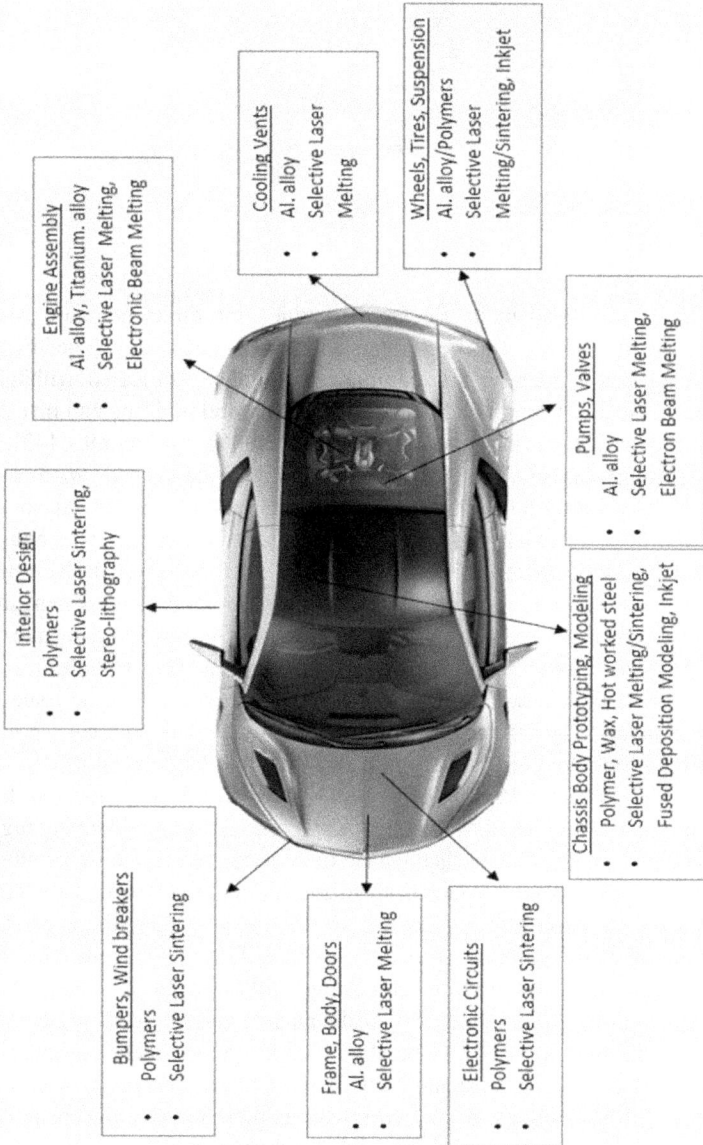

Engine Assembly
- Al. alloy, Titanium. alloy
- Selective Laser Melting, Electronic Beam Melting

Cooling Vents
- Al. alloy
- Selective Laser Melting

Wheels, Tires, Suspension
- Al. alloy/Polymers
- Selective Laser Melting/Sintering, Inkjet

Interior Design
- Polymers
- Selective Laser Sintering, Stereo-lithography

Pumps, Valves
- Al. alloy
- Selective Laser Melting, Electron Beam Melting

Chassis Body Prototyping, Modeling
- Polymer, Wax, Hot worked steel
- Selective Laser Melting/Sintering, Fused Deposition Modeling, Inkjet

Bumpers, Wind breakers
- Polymers
- Selective Laser Sintering

Frame, Body, Doors
- Al. alloy
- Selective Laser Melting

Electronic Circuits
- Polymers
- Selective Laser Sintering

FIGURE 3.8 3D printing of various car parts and the process involved.

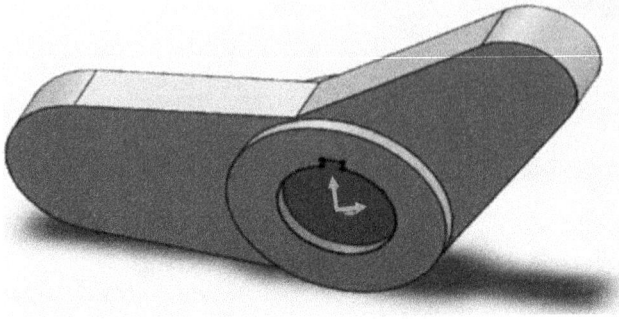

FIGURE 3.9 Lever arm.

has incorporated this head in its EcoBoost engine; this direct-injection gasoline engine is fitted with a series of turbochargers and used in newly developed race cars of the Mustang series. Ideally, this component could be fabricated within three months, whereas its counterpart had been produced by sand molding and investment casting and the manufacturing would take about four to five months [43]. Ford Motors has recently launched its Advanced Manufacturing center, where the development of 3D printing applications is one of their prime focuses. As shown in Figure 3.9, the 3D printed HVAC lever arm created by using carbon digital light synthesis and epoxy 82 material is installed in the 2020 Shelby GT500 model; it was tested successfully, and they are planning to develop brake components as per their standard norms and regulations [44]. Ford Motor group is laying the foundation for a fully functional 3D printed car in the future. In 2010, Ford Motors were struggling to bring their new Explorer model to the market, but some issues with brake noise delayed it by four months of its launch date. The company, however, used 3D printed tools to address the issue, and this time lag was adjusted [45].

Safety has always been a key objective for car makers. It is also essential that the car is theft-proof, to which Ford Motors have developed 3D printed locking lug nuts. These components are installed in each wheel; these parts were made so complex that they could not be easily cloned, which early thieves used to do. Brackets are the parts that hold various components in place, as shown in the figure; these parts have complex contours of the elliptical or parabolic shape. Generally, these types of designs are manufactured by investment castings followed by a series of post-processing operations. 3D printing had made manufacturing much easier for the engineers and with high accuracy. Generally, what happens if these designs are not made perfectly is that the subordinate parts will not fit and the production lot gets rejected [46]. With their unique sound wave feature, these nuts can be custom-made differently (Figure 3.10). It is nearly impossible for thieves to make a wax impression also, making it hard to clone; additional biometric security was also installed [47].

3.4.1.2 BMW

BMW is considered the pioneer of additive manufacturing technology. It has continuously been testing and developing various prototypes over the past several years.

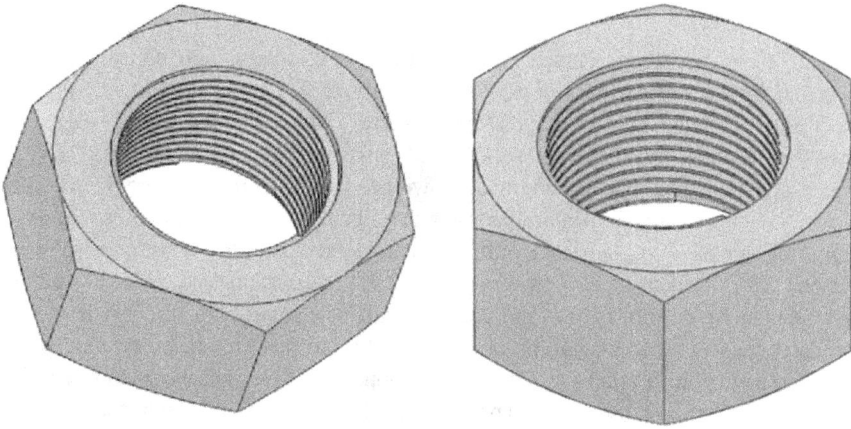

FIGURE 3.10 Locking nuts.

The company has successfully installed many 3D printed parts and now aims to manufacture over 50,000 components and 10,000 spare parts every year, creating an era of digitalized printing. As per previous statistics, since 2010, the company has printed over 1 million components following various 3D printing techniques [48].

3.4.1.3   Volkswagen

The company has invested in the 3D printing business over past years and has installed over 100 3D printing plants to date. Mainly, utilities like jigs and fixtures, prototyping, and tooling are successfully manufactured. It has switched to 3D printed tools and assemblies for manufacturing, and this switch has saved the company thousands of dollars every year. With the success of the pilot program at the Volkswagen Autoeuropa factory in Portugal, Volkswagen is now 3D printing all of its tooling parts. As estimated, this technology saved the company nearly €325,000 in 2017, while improving its net profit and gross productivity with good agreement with the operator [49]. An F1 race car prototype is tested via wind tunnel testing to check its performance against aerodynamic drag and lift. Manufacturers are inclining toward 3D printing to create model-based prototypes. Swiss Alfa Romeo Sauber F1 Team was found to have tested its prototype model, which was 3D printed (60% scaled model) using selective laser sintering technique. Volkswagen Motorsport developed their electric version of I. D. R Pikes Peak race car. A 50% called model was produced for testing purposes; it took a few days instead of weeks [50].

3.4.1.4   McLaren

McLaren has collaborated with Stratasys and has been working on developing fused deposition modeling alongside poly jet-based 3D printing for developing unique 3D printed components made of composite materials. This personalized production scheme aims at creating more functional and visually efficient parts and tooling to enhance performance, productivity, and the brand name in the coming years [51].

3.4.1.5 Porsche

Automobile manufacturer Porsche has tested and developed a fully functional clutch lever mechanism for the Porsche 959 model. A lever is a mechanical device that can produce a mechanical advantage greater than unity. The lever was tested for a three-ton pressure test, and its effectiveness was checked against factory-tooled levers [52]. The 3D printed lever exhibited better performance, owing to which the manufacturer is extending this technology to create 20 more parts for the digital library of Porsche catalog mentioned in their pilot program; this will occur before extending it to other complex parts. Porsche 3D printed seats, as shown in Figure 3.11, are made of polyurethane material, and they can be customized as hard, medium, and soft based on a firmness level. The manufacturer wanted to extend 3D printing of 40 more seat prototypes for their racing cars, and a street-legal model is yet to be developed by 2021. They wanted to perform seat customization as per their customer's choice, comfort level, and body contours [53].

3.4.1.6 General Electric

Developed by General Electric Motors (GE), engine valves, impellers in a gas turbine engine, and steel brackets are shown in the figure below, which shows a 3D printed caliper. These parts are miniature models, and more complex versions of these geometries are yet to be developed. Few pieces of literature are available for the production assembly of these parts and their applicability in real-world problems at the commercial level. Until now, only a few prototypes are 3D printed and tested against the actual model. It is worth noting that these parts, at a larger scale, if produced, can replace conventional tooling and part models [54].

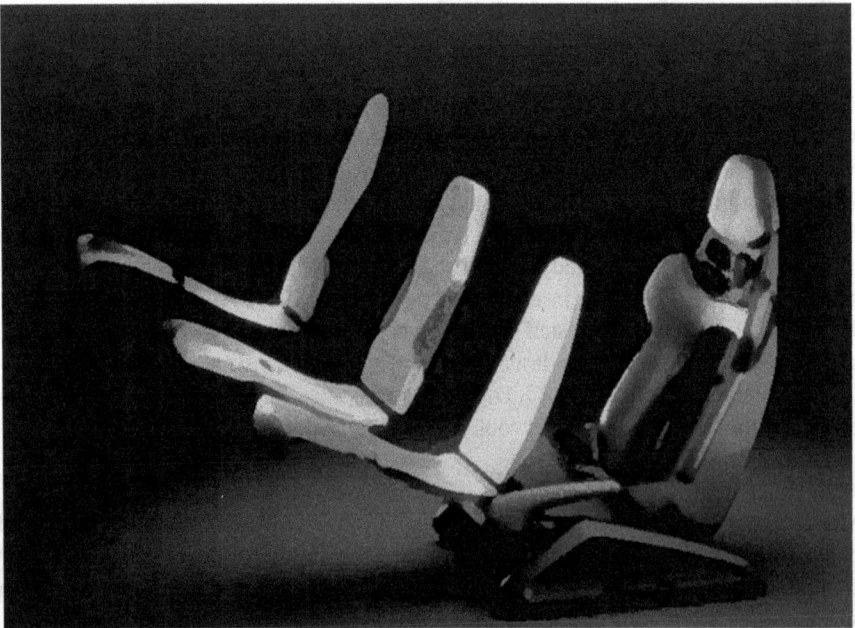

FIGURE 3.11 Porsche 3D printed custom seat.

3.4.1.7   AUDI

Conversely, AUDI teamed up with SLM Solution Group AG to make replacement parts and its prototypes in 2017. AUDI is now manufacturing both spare parts and various prototypes, as per the demand of consumers. Mostly, the rarer spare parts are 3D printed; these are limited in number or their production line has stopped. One such example, developed by 3D printing technology, is the water adapter of the Audi W12 engine, which was manufactured according to the demand of SLM280 [55].

3.4.1.8   Buggati Veyron

Bugatti modeled an eight-piston brake caliper through selective laser-melting technology using a titanium alloy. This 3D-printed caliper is the largest part of its kind fitted on a production vehicle. Bugatti's brake caliper (Figure 3.12) was 3D printed at the required scale using titanium nanoparticles, and it is a massive success, reducing the weight and increasing its stiffness index and strength considerably compared to conventional aluminum metal [56].

3.4.1.9   Rolls-Royce

Rolls-Royce has recently developed and demonstrated the usefulness of 3D printed brackets (Figure 3.13) with the key idea of cost optimization, rapid prototyping, and design optimization. Additionally, they use this technology to print their brand logo and QR code, which is nearly impossible for any subtractive manufacturing process to compete with in terms of time and money [57].

3.4.2   ADDITIONAL APPLICATIONS

For a few additional parts manufactured during race events, recent developments are mentioned below:

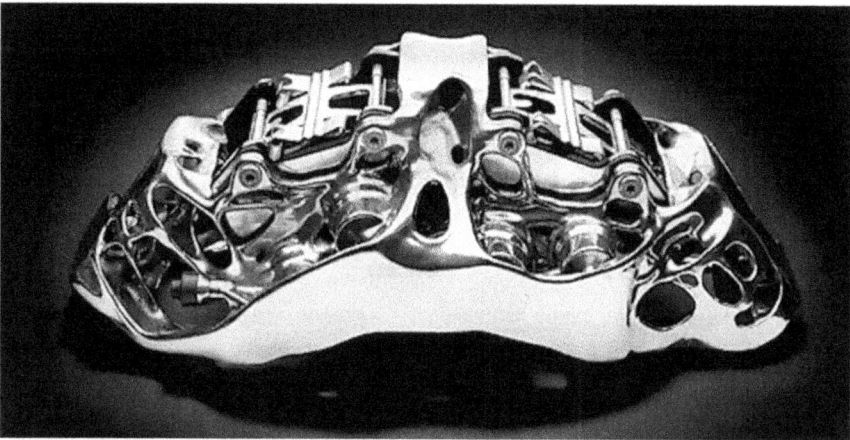

FIGURE 3.12   3D printed brake caliper.

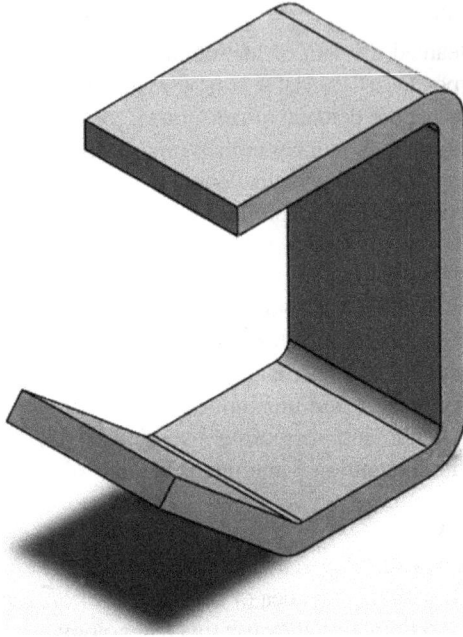

FIGURE 3.13 3D printed Rolls Royce brackets.

3.4.2.1 Cadillac Blackwing V-Series

Developed by General Motors, this bracket, as shown in Figure 3.14, was incorporated in the manual transmission models. 3D printed ducts made of nylon and manufactured by Multijet fusion and brackets made of aluminium powder through binder jetting were introduced [58].

3.4.2.2 Brake Ducts in Aston Martin

As shown in Figure 3.15, auto designer Ian Callum incorporated 3D printing technology to make 25 of these brackets, which were featured in Aston Martin cars. Parts such as front bumpers and brake ducts were also manufactured and tested against deformation, and the results were satisfactory [59].

3.4.2.3 Parking Brake Brackets

Manufactured by the DLP technique, these plastic-based brake brackets used for parking are lighter and more durable. These prototypes were adopted in the Mustang Shelby GT500 models, replacing the older substitutes [60].

3.4.2.4 Gear Lever and Pedals

In 2021, Dakar rally invented a unique gear lever mechanism with three titanium brake medals, as shown in Figure 3.16, all 3D printed using a powder bed technology during an off-road race tournament. The performance of the car's exhaust

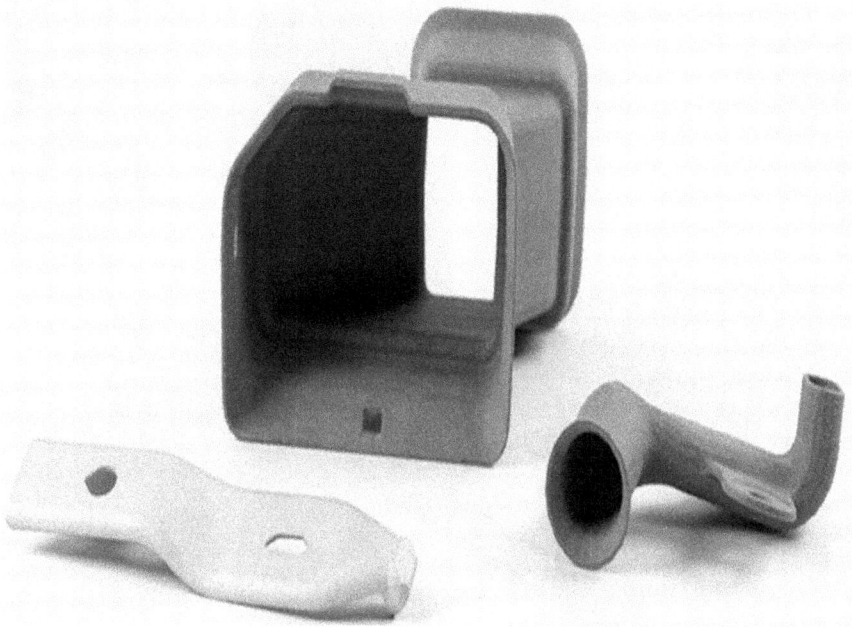

FIGURE 3.14 Brackets for transmission system.

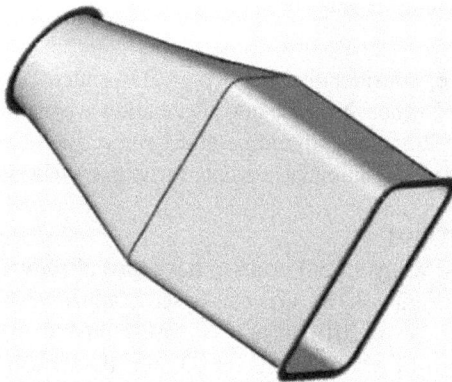

FIGURE 3.15 Brake ducts.

ball joint improved significantly; earlier, it had been susceptible to fracture or breakage [61].

3.4.2.5 3D Printed Ceramic Disc Rotor for Radar Antennae

Not only standard auto parts, but also innovative components could be manufactured by 3D printing technology whose manufacturing processes could not be

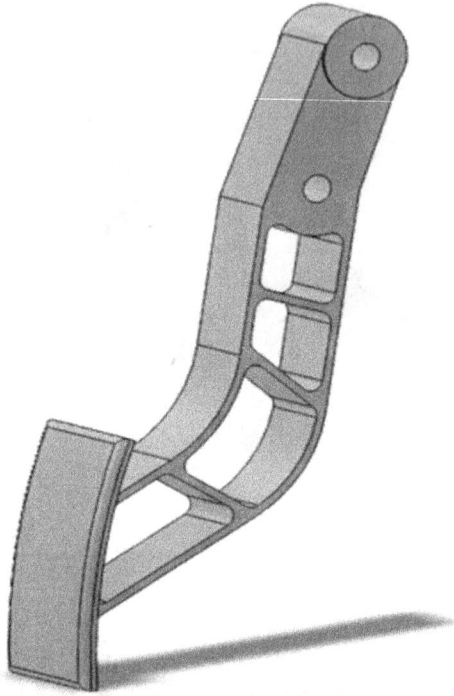

FIGURE 3.16   Titanium made brake pedals.

justified. For instance, consider the following: 3D printed ceramic disc rotors, as shown in Figure 3.17, when brought into the market were proved to be more effective in terms of strength and temperature resistance. Lunewave, a start-up company, is manufacturing antennae models to be used in radars [62].

3.4.2.6   Bike Modelling

Not only cars, but the wings of 3D printing have also already spread to design and develop motorbikes (Figure 3.18). German-based company APWORKS developed a 3D-printed motorcycle in 2016, which is 30% lighter than its earlier designed model using PLA, pro-FLEX materials, etc. Later, the moto group developed the NERA motorcycle, which is a fully functional digitally designed prototype entirely created by 3D printing [63].

3.4.2.7   Opportunities for Small Shop Owners

For small-scale manufacturers such as a custom car shop named Ring Brothers, printing car parts is a medium for them to show their creativity more effectively without being concerned about tedious manufacturing processing, the time investment in getting results. As shown in Figure 3.19, the business owner has developed a smart and innovative custom part like the air vent [64]. These start-up companies are getting fame and can share their design ideas on a larger platform. It has motivated many skilled engineers to push forward their talent and shape imagination to reality.

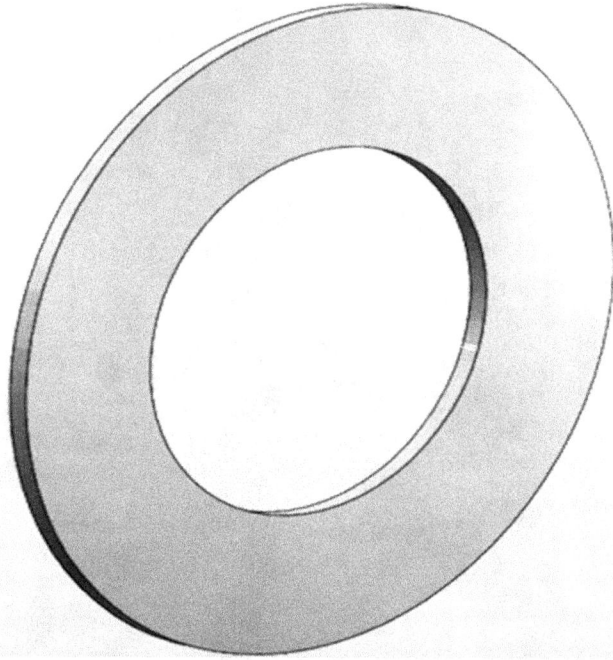

FIGURE 3.17   3D printed ceramic disc rotor.

3.5   ADVANTAGES

Various advantages of 3D printed parts adopted in automobile sectors are as follows:

1. *Rapid prototyping*

   Before making the full-scale model of the actual component, engineers try to develop a scaled miniature version of the same part performing a similar function at desired conditions. During the testing phase, these so-called prototypes are tested against artificial conditions similar to that of the exact environmental condition on which the actual component will perform, just like wind tunnel testing for a car model to test aerodynamic drag and lift. Several prototypes are designed before making the actual model.

2. *Hybrid material formation*

   Among different additive manufacturing technology, selective laser melting can be applied for different materials by trying different combinations of alloys to produce lightweight components for engine and chassis components. Mixing powder of different materials, such as titanium, nickel, and steel, etc., can produce hybrid materials with different strengths and resistances, which the conventional manufacturing method fails to produce, except for powder metallurgy, which is again a complex methodology involving many intermediate steps.

FIGURE 3.18   NEXA electric bike.

3. *Cost efficiency*

   No additional post-processing operations or tooling is required for 3D printed components. Expensive casting and forming operations could be replaced and there is no inventory storage cost is associated because this technology could print components as per requirement or demand. The time consumed to print small complex parts is less as compared to traditionally manufactured counterparts. On large scale production of specific parts, AM reduces cost significantly and manages time also. In other words, it has shortened the design cycle or product development cycle.

4. *Reduced material wastage*

   Compared to subtractive manufacturing where the billet is processed through a series of material removal operations such as milling, turning, drilling, broaching, planing, etc, additive manufacturing uses a unique layer-by-layer formation strategy of the final digitally designed product where materials are added in each subsequent layers, hence very minimal or no loss of material is witnessed.

5. *Lightweight construction*

   Materials such as titanium and aluminum have low density; their powder form can be used to make parts and customized tools, which can improve the ergonomics of the manufacturing operations. These properties have

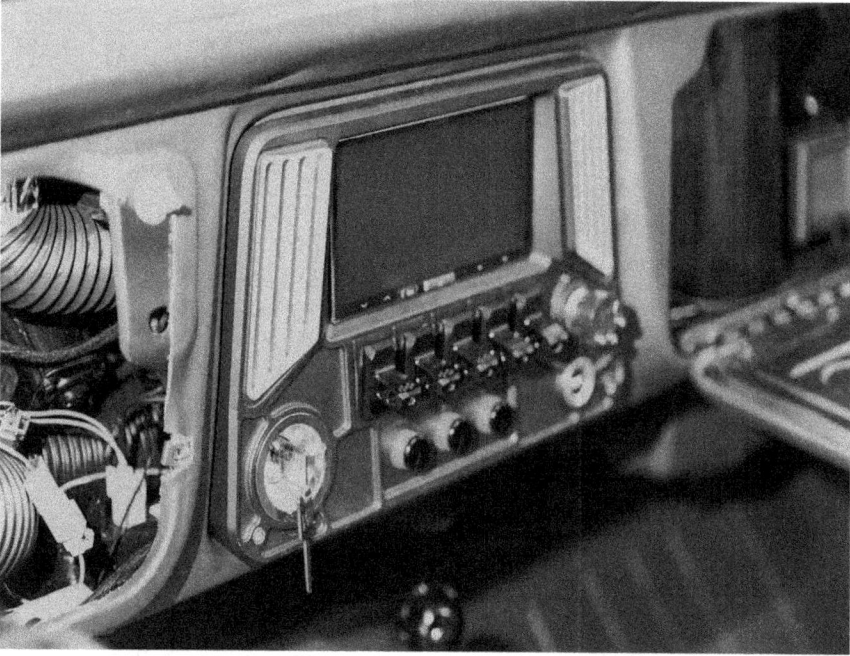

FIGURE 3.19   3D printed air vents.

been extensively used by design engineers to develop aerospace and racing cars parts requiring lightweight but stronger components.

6. *Closer dimensional tolerance*

It is really difficult for a manufacturer to produce a replica of a designed product to its exact dimensioning, even though skilled. There is always some man-made error involved; hence, companies use tolerance level of parts for an acceptable lot in industries. 3D printed parts are made near to exact, and they do not need any human interference during manufacturing. Laser-assisted sintering or melting does the work, and the lasers are designed in such a way that they can accurately print the desired component in stipulated time.

7. *Complex geometries*

It was critical to manufacturing micro-components or small parts with complex contours to be developed through conventional manufacturing processes. It would require highly skilled operators to develop such complicated geometries. However, 3D printing has made it possible for every miniature model or part that can be designed in CAD-based software to be developed in real life. This innovative technology allows the designer to have lesser logistic effort and greater flexibility in designing.

8. *Miscellaneous*

A 3D printer is a typical device that can be installed in our homes. Any model can be programmed using CAD software, and printing could be done. It has become more convenient to design and manufacture any spare part with proper knowledge of design-based ontology. A 3D printer eliminates the fear of theft of confidential archived (.STL) files, which represent CAD geometry. Additionally, it can be sent to the vendor or small-scale manufacturers to 3D print the desired part.

3.6 FUTURE SCOPE

Even though automobile sectors were quick to implement additive manufacturing technology in auto parts and tooling, an ocean of possibilities lies ahead of them, and it needs to be explored in search of excellence. Just like we watch in sci-fi films, 3D printing can generate an entire car model from scratch. However, it lacks technological development to improve cost-effectiveness and time consumed. Considering the developments to date, some prospects could be drawn on which young researchers can work and contribute their role toward global change. Stereolithography and selective laser sintering process are well known to manufacture dashboards and set covers using polymers. They are more controlled and accurate, their application could be extended to aluminum and titanium parts to model springs, dampers, tires, hub caps, etc., and composites and polymers derivative of these complex components should be prototyped and tested. However, SLS technology should be able to manufacture body panels and doors from aluminum alloys [65]. Newly developed advanced designs such as bulletproof windshields and doors could also be manufactured. Titanium and aluminum metals being lightweight can be used to manufacture functional parts of the engine, as well as transmission system following EBM and SLM. Furthermore, external parts such as windbreakers or bumpers can be modeled using polymer materials following SLS. The FDM method, as discussed earlier, can be implemented to produce chassis components and doors using a special class of thermoplastics as it is proven to produce good resistance against compression and bending. Additionally, additive manufacturing through nano printing is best suited for developing MEMS (micro electromechanical sensors); these are chip-based electronic sensors used in almost all digitalized automobile vehicles [66]. Likewise, there is a lot of scope in improving 3D nano printing to make life sustainable and easier, such as multi-material printing, development of new printing strategies, etc.

3.7 CONCLUSIONS

It is well evident that all the automobile companies are inclining toward 3D nanoprinting technology since it sets a range of advantages from product development to product processing. By utilizing nanoparticles in 3D printing applications, materials with higher strength and properties could be manufactured without even worrying about the increase in weight of the product. This was the major motivation for automobile manufacturers in implementing 3D printing technology. Many automakers have successfully tested and developed 3D printed tools and auto parts, saving them a huge sum of money every year. It has bridged the gap between innovation and design, creating multifunctional complex products, ready to be installed in vehicles. A designer now does not have to worry about manufacturing anymore; one has full liberty of developing the product of their choice. The customers could now avail themselves more than they demand from their customized cars. With decreased production costs, both the company and consumers can gain profitable experience. As a result, all automakers are adopting and integrating 3D printing technology into their production processes. Furthermore, all automakers should plan for long-term and short-term goals for incorporating this technology and developing research wings for generating new concepts and ideas. Recently, NASA has developed a multi-layer 3D printing technique that substantially reduces the printing time. The cost-effectiveness of it, however, should be analyzed so that average vehicles could avail the facilities.

REFERENCES

1. Sreehitha, V. (2017). Impact of 3D Printing in Automobile Industries. *International Journal of Mechanical And Production Engineering*, 5(2), 91–94. Retrieved from http://www.iraj.in/journal/journal_file/journal_pdf/2-347-149260399391-94.pd
2. Sarvankar, S.G., & Yewale, S.N. (2019). Additive Manufacturing in Automotive Industry. *International Journal of Research in Aeronautical and Mechanical Engineering*, 7(4), 1–10.
3. Ichidai, Y. (2019). Current Status of 3D Printer Use among Automotive Suppliers: Can 3D Printed-parts Replace Cast Parts? *IFEAMA SPSCP*, 69–82. Retrieved from http://ifeama.org/ifeamaspscp/selected%20papers/13th%20in%20Ulaanbaatar/13th%2006%20%20Y_Ichida_final.pdf
4. Beiderbeck, D., Deradjat, D., & Minshall, T. (2018). The Impact of Additive Manufacturing Technologies on Industrial Spare Parts Strategies. *Centre for Technology Management working paper series*, 1–57. Retrieved from https://pdfs.semanticscholar.org/fde3/38bf35690594dc9a4f2822853bab87fb45dc.pdf
5. Nichols, Megan R. (September/October 2019). How Does the Automotive Industry Benefit from 3D Metal Printing?, *Metal Powder Report*, 74(5). 1–2.
6. Critchley, L. (April 12, 2018). Materials Being Used for the 3D Printing of Cars, https://www.azom.com/article.aspx?ArticleID=15664
7. Alexandrea P. (June 8, 2020). 3D Printing Materials Guide: Plastics, https://www.3dnatives.com/en/plastics-used-3d-printing110420174/
8. Carlota V. (June 6, 2019). All You Need to Know about ABS for 3D Printing, https://www.3dnatives.com/en/abs-3d-printing-060620194/
9. PLA Plastic Material for 3D Printing, https://www.sculpteo.com/en/materials/fdm-material/pla-material/
10. Carlota V. (February 13, 2020). All You Need to Know about Polycarbonate (PC) for 3D Printing, https://www.3dnatives.com/en/polycarbonate-pc-for-3d-printing-110220204/
11. Polyamide 3D Printing: Nylon Plastic Powder Material for SLS, https://www.sculpteo.com/en/glossary/polyamide-definition/
12. Carlota V. (July 16, 2020). All You Need to Know about Polypropylene (PP) for 3D Printing, https://www.3dnatives.com/en/polypropylene-pp-for-3d-printing-160720204/
13. https://www.3dsystems.com/materials/metal
14. Madeleine P. (July 26, 2021). A Closer Look at 3D Printing Materials: Composites, https://www.3dnatives.com/en/composite-materials-3d-printing-260720214/
15. https://www.makerbot.com/stories/design/nylon-3d-printing/
16. https://www.mordorintelligence.com/industry-reports/automotive-3d-printing-market
17. https://www.additive-x.com/blog/selective-deposition-lamination-sdl/

18. https://markforged.com/resources/learn/3d-printing-basics/3d-printing-processes/what-is-electron-beam-melting-ebm
19. https://www.think3d.in/digital-light-processing-dlp-3d-printing-service-india/
20. https://www.protolabs.co.uk/services/3d-printing/multi-jet-fusion/
21. https://www.stratasys.co.in/polyjet-technology
22. https://all3dp.com/2/selective-laser-melting-slm-3d-printing-simply-explained/
23. https://www.protolabs.co.uk/services/3d-printing/stereolithography/
24. https://formlabs.com/asia/blog/what-is-selective-laser-sintering/
25. https://tractus3d.com/knowledge/learn-3d-printing/fdm-3d-printing/
26. Maximize Market Research. (2019). Global 3D Printing Automotive Market and Forecast (2016-2024) by Technology, Input Materials, and Geography. Retrieved from Maximize Market Research PVT. LTD.: https://www.maximizemarketresearch.com/market-report/global-3dprinting-automotive-market/9760/
27. Business Cases: 3D Printing in the Automotive Industry (August 23, 2018). https://www.beamler.com/3d-printing-in-the-automotive-industry/
28. Mohanavel, V., Ashraff Ali, K.S., Ranganathan, K., Allen Jeffrey, J., Ravikumar, M.M., & Rajkumar, S. (2021). The Roles And Applications of Additive Manufacturing in the Aerospace and Automobile Sector, *MaterialsToday: Proceedings*, 47(Part 1): 405–409, doi: 10.1016/j.matpr.2021.04.596.
29. Chinese patent, Finished product function part combined 3D printing method, CN103465636A, 2013-12-25.
30. Chinese patent, Connection rod of automobile engine laser 3D printing technique, CN108057888A, 2018-05-22.
31. Chinese patent, Method for forming a three-dimensional object, CN108475008B, 2020-11-06.
32. Chinese patent, 3D who adopts aluminum alloy to make prints auto parts spare part, CN213575297U, 2021-06-29.
33. United States Patent Willis, Patent No.: US 10,611,090 B2 (45) Date of Patent: April 7, 2020, ELECTROMECHANICAL 3D PRINTING DESIGN SYSTEM.
34. European Patent SpecificatioN, EP 3 019 722 B1, PLATED POLYMER COMPONENTS FOR A GAS TURBINE ENGINE.
35. Yadav, Devesh K., Srivastava, R., & Dev, S. Design & Fabrication of ABS Part by FDM for Automobile Application, 26(Part 2): 2089–2093, doi: 10.1016/j.matpr.2020.02.451.
36. Maghnani, R. (2015). An Exploratory Study: The Impact of Additive Manufacturing on the Automobile Industry, *International Journal of Current Engineering and Technology*, 5(5): 1–4.
37. Juechter, V., Franke, M.M., Merenda, T., Stich, A., Körner, C., & Singer, R.F. (2018). Additive Manufacturing of Ti-45Al-4Nb-C by Selective Electron Beam Melting for Automotive Applications, *Additive Manufacturing*, 22, 118–126, doi: 10.1016/j.addma.2018.05.008
38. Wanga, Di, Wang, Y., Yanga, Y., Lu, J., Xu, Z., Li, S., Lin, K., & Zhang, D. (2019). Research on Design Optimization and Manufacturing of Coating Pipes for Automobile Seal Based on Selective Laser Melting, 273, 116227. https://doi.org/10.1016/j.jmatprotec.2019.05.008
39. https://3d.markforged.com/GA_CarbonFiber.html
40. Reeves, P., & Mendis, D. (2015). The Current Status and Impact of 3D Printing Within the Industrial Sector: An Analysis of Six Case Studies. *Centre for Intellectual Property Policy & Management*, 2015(41), 1–15. Retrieved from https://assets.publishing.service.gov.uk/government/uploads/system/uploads/attachment_data/file/549046/Study-2.pdf

41. 10 Exciting Examples of 3D Printing in the Automotive Industry in 2021. (May 28, 2019). https://amfg.ai/2019/05/28/7-exciting-examples-of-3d-printing-in-the-automotive-industry/
42. Colorado H.A., Mendoza D.E., Lin H.-T.,& Gutierrez-Velasquez E. Additive manufacturing against the Covid-19 pandemic: a technological model for the adaptability and networking, *Journal of Materials Research and Technology*, 1150–1164, https://doi.org/10.1016/j.jmrt.2021.12.044
43. https://corporate.ford.com/articles/products/building-in-the-automotive-sandbox.html
44. https://3dprintingindustry.com/news/3d-printing-enables-high-performance-for-ford-mustang-shelby-gt500-151633/
45. https://media.ford.com/content/fordmedia/fna/us/en/news/2017/03/06/ford-tests-large-scale-3d-printing.html
46. https://www.additivemanufacturing.media/articles/3d-printing-for-production-at-ford-the-cool-parts-show-s2e1
47. Humphries, M. (January 29, 2020). Ford Creates 3D-Printed Locking Wheel Nuts Using Driver's Voice, https://in.pcmag.com/news-analysis/134893/ford-creates-3d-printed-locking-wheel-nuts-using-drivers-voice
48. Industrial-Scale 3D Printing Continues to Advance at BMW Group, 10.12.2020 Press Release, https://www.press.bmwgroup.com/global/article/detail/T0322259EN/industrial-scale-3d-printing-continues-to-advance-at-bmw-group?language=en
49. https://www.volkswagen-newsroom.com/en/press-releases/volkswagen-plans-to-use-new-3d-printing-process-in-vehicle-production-in-the-years-ahead-7269
50. https://3dprintingindustry.com/news/alfa-romeo-doubles-down-on-3d-printing-in-bid-to-move-up-the-f1-grid-186336/
51. https://www.stratasys.co.in/explore/blog/2017/mclaren-racing-additive-manufacturing
52. Chris Bruce, Feb 12, 2018, Porsche Can Now 3D-Print Parts For The 959 Supercar, https://www.motor1.com/news/230257/porsche-3d-print-old-parts/
53. https://www.sae.org/news/2020/03/porsche-3d-printed-seats
54. Duda, T., & Venkat Raghavan, L. (2016). 3D Metal Printing Technology, *IFAC-Papers OnLine*, 49(29), 103–110.
55. Petch, M. (2018). Audi gives update on the use of SLM metal 3D printing for the automotive industry, *3D Printing Industry*. [Online]. Available: https://3dprintingindustry.com/news/audi-gives-update-use-slm-metal-3d-printing-automotive-industry-129376/. [Accessed 2019].
56. Molshiem, W. (January 22, 2018).World Premiere: Brake Caliper from 3D Printer, https://www.bugatti.com/media/news/2018/world-premiere-brake-caliper-from-3-d-printer/
57. Carlota, V. (December 14, 2020). Rolls-Royce Is aAccelerating the Integration of 3D Printing into Its Production Process, https://www.3dnatives.com/en/rolls-royce-3d-printing-production
58. Szymkowski, S. (December 10, 2020). 3D Printing Helps Bring the Cadillac Ct4-v and Ct5-v Blackwing Models to Life, https://www.cnet.com/roadshow/news/cadillac-ct4-ct5-v-blackwing-3d-printing-manual-transmission/
59. Anstee, T. (October 4, 2021). METHOD X 3D Printer for Production of Aston Martin Parts, https://www.electronicspecifier.com/products/3d-printing/method-x-3d-printer-for-production-of-aston-martin-parts
60. Tycho de Feijter for 6th Gear Automotive Solutions. Haarlem. (December 2018). https://www.6thgearautomotive.com/2018/12/08/ford-mustang-shelby-gt500-will-get-3d-printed-brake-line-brackets/
61. Hendrixson, S. (May 18, 2021). Real Examples of 3D Printing in the Automotive Industry, https://www.mmsonline.com/articles/real-examples-of-3d-printing-in-the-automotive-industry

62. Hendrixson, S. (May 1, 2021). The Cool Parts Show Live January 12 to Feature Ceramic Brake Rotor, https://www.additivemanufacturing.media/articles/the-cool-parts-show-live-jan-12-to-feature-ceramic-brake-rotor

63. NERA: World's First Fully 3d Printed E-BIKE, https://bigrep.com/nera-e-motorbike/

64. Accelerating Custom Car Part Development: 3D Printing at Ringbrothers. (April 29, 2019). https://formlabs.com/asia/blog/accelerating-custom-car-part-development-3d-printing-at-ringbrothers/

65. Jiménez, M., Romero, L., Domínguez, I.A., Espinosa, M.D., & Domínguez, M. (2019). Additive Manufacturing Technologies: An Overview about 3D Printing Methods and Future Prospects, Hindawi, 2019, 1–30. doi: 10.1155/2019/9656938

66. Blachowicz, T., & Ehrmann, A., 2020). 3D Printed MEMS Technology-Recent Developments and Applications, *Micromachines*, 11, 434. doi: 10.3390/mi11040434

4 3D Nanoprinting in the Biomedical/Health Care Applications

Farhan Mazahir, Swapnali Birajdar, Deepali Bhogale, Anjali Bhosale, and Awesh K. Yadav
Department of Pharmaceutics, National Institute of Pharmaceutical Education and Research-Raebareli, Uttar Pradesh

CONTENTS

4.1 INTRODUCTION

Massachusetts Institute of Technology (Cambridge, United States) developed three-dimensional (3D) printing in 1992. The CAD was used to design for the production of materials by this new and original technology. However, a terminal computer has a crucial role to change the design of desired material. Also, the print head is used to spray the binding material, and the spraying of raw material is directed by the X-Y movement of the print head. To construct the desired goods, usually a simple and fast layer-by-layer process is adopted. The material is spread on the powder bed. The piston rod assists the Z-vertical motion of the powder bed. To get a construct model of desired thickness, the powder bed is allowed to move to the lower position. Also, the process is repeated several times to obtain the proper and preferred model of any construct. After completion of the process, the powder is detached and

DOI: 10.1201/9781003189404-4

73

product has been leftover (Sachs, Cima, and Cornie, 1990). (https://patents.google.com/patent/US5204055A/en). Recently, additive or prototype manufacturing has been known as three-dimensional (3D) printing technology and can be used to produce biomaterial scaffolds, tissue prostheses according to the customized anatomy of patients, and to print models of tissues. The process of 3D printing is regulated by computer-aided design to deposit materials onto a moving dais. Nowadays, the 3D printing system has been modified with several techniques, such as printing of material with the help of chemical crosslinker, chemically or thermally processed materials, sintered powder materials, and photo-polymerization of liquid monomer. When integrating the imaging technique to the 3D printer, the production of the customized anatomical structure in shape and size is possible, according to the anatomy of the patient at the site of implantation.

4.1.1 FUSED-DEPOSITION MODELLING (FDM)

With the help of this technique, layer-by-layer deposition of thermoplastic biomaterial is possible to construct a 3D model. The cells can be seeded before the printing or after printing. Since the thermoplastic materials are harder, cells are seeded after 3D printing. However, thermoplastic is found to have superior compatibility for musculoskeletal purposing. However, thermoplastic biomaterial seeded with cells has been used for neuron-related tissue engineering. Moreover, FMD is an inexpensive technique (C. M. O'Brien et al., 2015). Refer to Figure 4.1.

4.1.2 SELECTIVE LASER SINTERING (SLS)

This technique is similar to FDM except for a long wavelength of the laser is used. The materials like PVA, PCL, PLA, and HAP have been used to construct the 3D model for biomedical applications. But few powders are found to be compatible. In brief, a layer of beads or powder is deposited on the platform, heated, and allowed to fuse before the deposition of the next layer (Thomas and Willerth, 2017).

4.1.3 STEREOLITHOGRAPHY

This method of 3D printing is suitable for the printing of light-sensitive polymeric biomaterials with the highest resolution; thus, 3D models similar to almost innate tissue-like microenvironment can be obtained. This technique has been used to construct scaffolds for neuronal tissue engineering; they displayed their potential in the differentiation of cells. This technology is also useful to add growth factors and make composite biomaterials with the help of nanomaterials (Edgar, Robinson, and Willerth, 2017). The illustration of different parameters and steps adopted during the 3D printing process is presented in Figure 4.2.

The nanotechnology scales down the dimensions and has different routes for materials and devices. The objective of 3D printing is to allow the production of a solid three-dimensional product with a definite shape, which is difficult to achieve by classical or conventional techniques. Nanomaterials can act as a feedstuff for the 3D printer during the bio-print process. The most widely used nanomaterials during

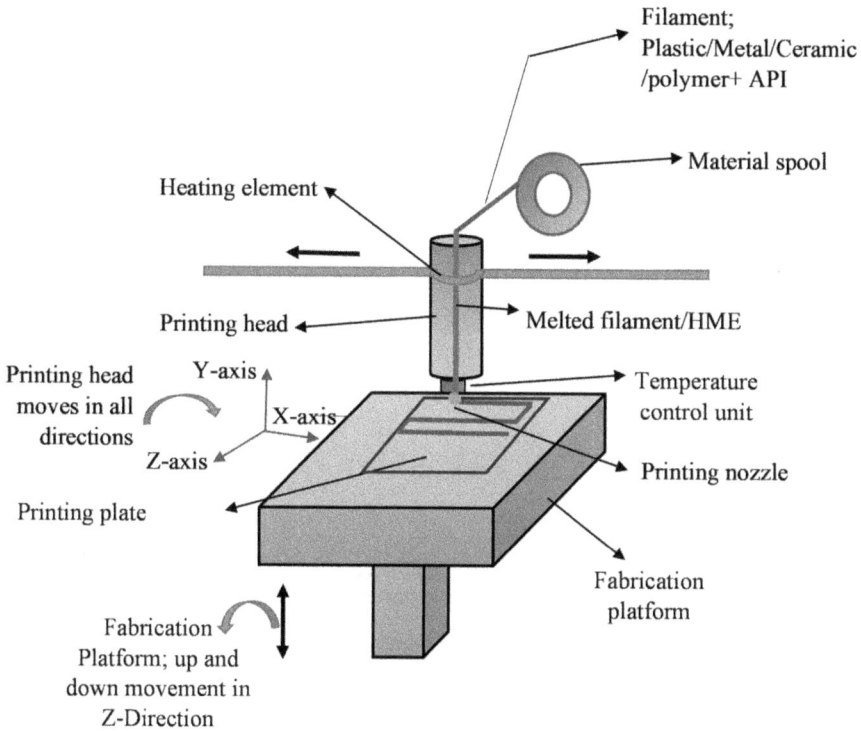

FIGURE 4.1 A fused deposition modeling FDM, 3D printing technology with hot-melt filaments/extrusions (HME) and polymer coating filament.

printing are zirconium oxide and aluminium oxides for generating 3D print models; these nanomaterials acts as an additive during manufacturing, which built the product by layer-by-layer configuration with the help of data templates or blueprints uploaded to the design software with a laser and feed stalk (Nau and Scholz, 2019). Also, the metals can be used as feed stalk as a nanopowder; for example, titanium and aluminum are commonly used materials for feed stalk. Sometimes, these metals are also available in nanosuspension form, and for the production of high electronic conductivity objects, the metal feed stalk is mostly preferred on plastic feed stalk. The reason for choosing metal nanomaterials over bulk metal materials is their melting temperature. The metal nanoparticles melt at a lower temperature than bulk metal materials do. The combination of 3D printing and nanotechnology offers great promise to industry and in the field of biotechnology and medical applications. 3D printed scaffolds are manufactured by using biodegradable, biocompatible, and bioinert materials scaffolds, and antimicrobial nanomaterials are of great interest for medical applications nowadays. Nowadays a 3D printing-based research and developments are based on nano biomaterials for the fabrication of 3D printed tissue models and also for bone regenerations using 3D printed scaffolds. Nanotechnology provides better opportunities for the development of efficient devices for better targeting and treating diseases. The use of nanomaterials, such as gold, silver,

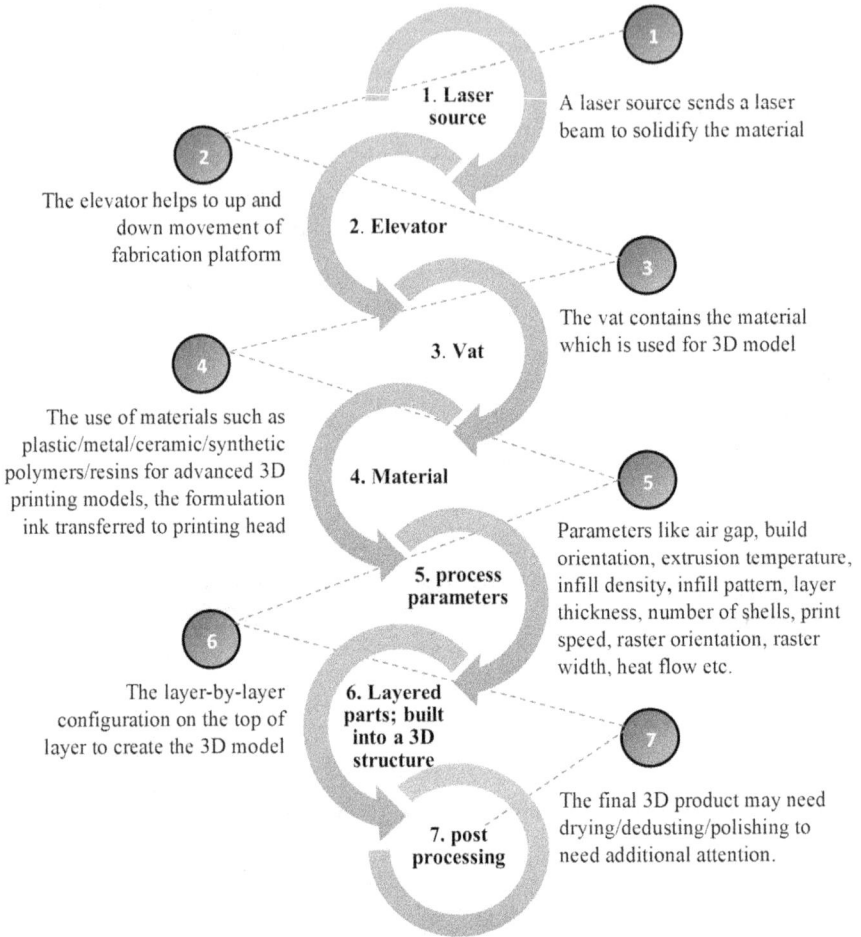

FIGURE 4.2   A 3D printing design and step-by-step illustration of the 3D printing process.

hydrogels, SiO_2-gold nanoshells, water-soluble polymers, fibrin, and starch-based powders, promotes wide application in biomedicines as well as biology. These biomaterials are used for the fabrication of 3D printed scaffolds, such as synthetic biodegradable scaffolds, collagen scaffolds, and also for the fabrication of medical devices including neuron-adhesive patterns and fibrin channels (Boland et al., 2007; X. Li, Liu, et al., 2009; Yeong et al., 2007).

The 3D printing process involves two types of nano biomaterials fabrication techniques.

1. Inkjet printing technology by using NP 2.1 and Z402 printer (Marizza, Keller, and Boisen, 2013).
2. Nanoimprint lithography technique, which typically involves EVG620 nano imprinter and EV520 embosser (Waid et al., 2012).

4.2 NANOFIBERS

Nanofibers are similar to fibers in appearance but have their diameter on the nanometre scale. Nanofibers demonstrate a higher surface-to-volume ratio and remarkable porosity. Many polymers include cellulose, poly-caprolactone (PCL), PLA, and the copolymer of polylactic/glycolic acid (PLGA). Formhals was the first to invent and file a patent for nanofibers in 1934 [https://patents.google.com/patent/US1975504A/en]. Nanofibers have gained their applicability in diverse applications. A few of them are wound healing, drug delivery, tissue engineering, separator membrane, and nano-composites, and many more. 3D printing technology could be used to fabricate nanofibers. For the production of nanofibers by using 3D printing technology, the printer is comprised of multisets of nozzles and an electric field. The nozzles are used to push the polymeric fluid, and an electric field is used to assist the formation of tiny fibers. Synthesis of 3D porous protein scaffolds and their application in the field of tissue engineering as an aid for tissue repair is one of the many applications. To elicit a cellular response, Yu and co-workers prepared gelatin and PCL-infused nanofibers in the tangles of protein scaffolds. The porous design was preferred over the non-porous design to elicit a cellular response (Y. Yu et al., 2016). PCL nanofibers-based dual-scale 3D printed scaffolds have been used by Huang and co-workers to repair bone tissue (Huang et al., 2020). Ambrusad et al. were able to demonstrate the effect on the solubility of drug-loaded nanofibres produced by 3D printers. The dispersion of loratadine, a low water-soluble drug, into amorphous electrospun nanofibers increased the solubility of loratadine up to 26 times in comparison with the plain drug. Also, the dissolution efficacy was found to be increased up to 60% (Ambrus et al., 2019). To address the problem that arises due to the unavailability of organs and injury to the tissues for the transplantation in the field of tissue engineering, improvements in the biomaterials were made. The concept of conductive hybrids comprised of nanofibers and hydrogel promises its usefulness in the field of biomedical engineering. A conductive hybrid consists of alginate and gelatin carrying homogeneously dispersed carbon nanofibers (CNF) in the form of printable scaffolds. Homogeneous dispersion of CNF in hydrogels imparted good conductivity and mechanical strength. Interestingly, the incorporation of CNF increased the shear-thinning ability of hybrid hydrogel and easy 3D printing. Moreover, *in-vitro* evaluation demonstrated the enhanced proliferation of the cells in comparison with control. Due to conducive behavior, CNF incorporated hybrid hydrogel may be used for the tissue engineering-related application to neuronal and cardiac strategies (Serafin, 2021). A biodegradable polymeric composite patch based on polylactide (PLA) and collagen incorporating PLGA nanofibrous membrane were developed to promote tendon graft healing. PLA and PLGA are used due to their biodegradable nature. The *in-vivo* efficacy studies conducted on rabbit bone tunnel showed that PLGA nanofibrous membrane reduced the tunnel enlargement and increased tendon-bone integration phenomena. In addition, the PLA bolt was found to be resilient for tendon-bone integration, biocompatible during progressive bone growth, and having no observable excessive tissue reactions. Based on the outcome, this PLGA nanofibrous-based composite may be effective in humans for the reconstruction of the tendon (Chou et al., 2016).

4.3 CELL PRINTING AND IMPLANTATION

Saenz del Burgo and co-workers developed implantable, porous, alginate in-corporating polyamide-based microencapsulation of cells inside macro-capsule for the encapsulation engineered cells for the de novo biosynthesis of therapeutic agents. The design of implanted microcapsule device was prepared by computer-aided drafting (CAD) and produced by selective laser sintering. The 3D computer image was partitioned into 2D thin layers, and 3D production of the implant of expected internal architecture, external shape, and size was completed with the help of laser-derived fusion of tiny particles of thermoplastic powder, metal, or ceramic. For the assessment of therapeutic suitability of capsule-based implants, two types of cells were used, i.e., C2C12 myoblast cells and BHK fibroblast cells for the de novo biosynthesis of erythropoietin and vascular endothelial growth factor, respectively. The results confirmed that the release of bioactive protein was found to be constant despite the reduction in the metabolic process of encapsulating cells (Saenz Del Burgo et al., 2018). Also, the viability of encapsulated hepatocellular carcinoma cells in alginate capsule later subjected to 3D based extrusion in the form of the methacrylate-based hollow microneedle. The assessment of post-extrusion cell viability in comparison with the non-extruded control sample did not exhibit sig-nificantly. Furthermore, this type of 3D printed cell encapsulated microneedle can be used for wound healing. Additionally, the use of a microneedle may reduce the pain and anxiety encountered by the person during conventional administration of the drug by hypodermic needle (Farias et al., 2018). A 3D printed biocompatible and ingestible capsule for the targeted analysis of gastro-intestinal-fluid to diagnose the specific markers at different pH and pathologies has been proposed. The pH-sensitive coatings of polymers on the capsule allow the permeation of fluid at specific pH to the sensing compartments of capsules and eliminate the clinical intercession for diagnosis of disease. The change in the capacitance relay the data of analysis of GI fluids can be transferred to the external android phone via low energy bluetooth. This user-friendly, real-time and cost-effective analysis of GI fluids of-fers a useful strategy for the diagnosis of gut (Banis et al., 2019).

4.4 DESIGN OF SCAFFOLDS AND PRINTING OF MAMMALIAN CELLS OR TISSUE

Shape, size, porosity, roughness, interconnection, and orientation of fibers in scaffolds should be per the targeted tissue. These characteristics affect the cell viability, adhesion, migration, and differentiation, and the architecture of scaffolds should allow the flow of medium during the *in-vitro* procedure or sufficient blood should be entered into the scaffolds (*in-vivo*) to make available oxygen and nutrient (Yeong et al., 2004). The scaffolds should not elicit an immune, inflammatory response, and if the scaffolds are biodegradable, then their by-products should not be toxic. For the best proliferation, adhesion, and differentiation of cells, the scaffolds must exhibit the same mechanical properties an innate tissue possesses. Furthermore, the rate of biodegradation of scaf-folds to be optimum to ensure the new tissue regeneration (F. J. O'Brien, 2011). These are the few important properties of great importance that a biomaterial should have to support the proliferation of cells in the recipient tissues (Abdulghani and Mitchell, 2019). The shortage of organs is being witnessed despite the high number of willing organ donors. Production or manufacturing of live organs from the own cells of a person could be a long-term solution. 3D living organs can be printed from the bottom up in a medium that could support live cells (Ozbolat and Yu, 2013). The integration of imaging techniques to 3D printing allows the production of organs with customized anatomy, shape, and size for the individual patient (V. K. Lee and Dai 2017). Lee et al., with the help of 3D printer, develop a prototype human skin. Fibroblast and kerati- nocytes cells were chosen as constituent cells for the dermis and epidermis layer. Collagen was the matrix material for the bio-print of skin. Different parameters were selected to maintain the optimum densities of cells in the dermis and epidermis layer, their maximum viability to imitate relevant physiological characteristics of human skin. For stratification and maturation, the epidermal layer of the submerged culture was exposed to an air-liquid interface. Immunofluorescence and histological evidence confirmed that the printed skin could be biological representatives of *in-vivo* human skin. This bio-printed human skin analog could be used as a template for studying pathological conditions (V. Lee et al., 2014). Also, the use of laser-aided bio-print (LABP) has been demonstrated to print fully cellularized skin grafts. LBP is helpful to place different types of cells in precise 3D spatial patterns. Keratinocytes and fibroblast cells were dispensed on the top of the Matriderm®. The *in-vivo* evaluation of different skin grafts was carried out by implanting the skin substituted into thicker skin wounds. The results indicated the grafted skin substitute was connected to the surrounding tissue after 11 days. The cells began to proliferate and differentiate, and the formation of stratum corneum was demonstrated. The partital migration of keratinocytes into Matriderm® was detected (Michael et al., 2013). Refer to Figure 4.3.

4.5 MATERIAL FOR CELL PRINTING

For a better understanding of interactions between cells, the use of cell printing tech- nology could be the best choice. The bio-printed cellular hierarchy may be useful to advance experimental models for exploring the interaction of cells to the scaffold and soluble factor. The cell printing technology seems important for the generation of well- organized 3D organoid cell culture by printing cells in high density and allowing their subsequent cell fusion. These printed micro-tissue models resemble the complex cell-to- cell communications under *in-vivo* conditions and fast and high screening of drugs. So, the best ability of cell printing technology is to generate 3D tissue structure of several cells in a defined matrix to resemble the innate tissue. The success of the therapeutic application of bio-printing heavily depends on the compatibility of biomaterials with cells and bio-printing instruments, i.e., scaffolds materials. These biomaterials should support cellular components during and after the bio-printing. The materials used in traditional bioengineering are being selected for the printing of bio-scaffolds (Skardal and Atala, 2015). Various factors like polymerization process, cytotoxicity, rheology, compatibility, and printing ability of material should be considered. Hence, limited options are available in the case of biomaterials. Generally, two primary categories of biomaterials, namely curable and soft hydrogel as biomaterials, are used for cell printing. For the purpose of mechanically stout scaffolds generally, curable polymers are used, while for the better

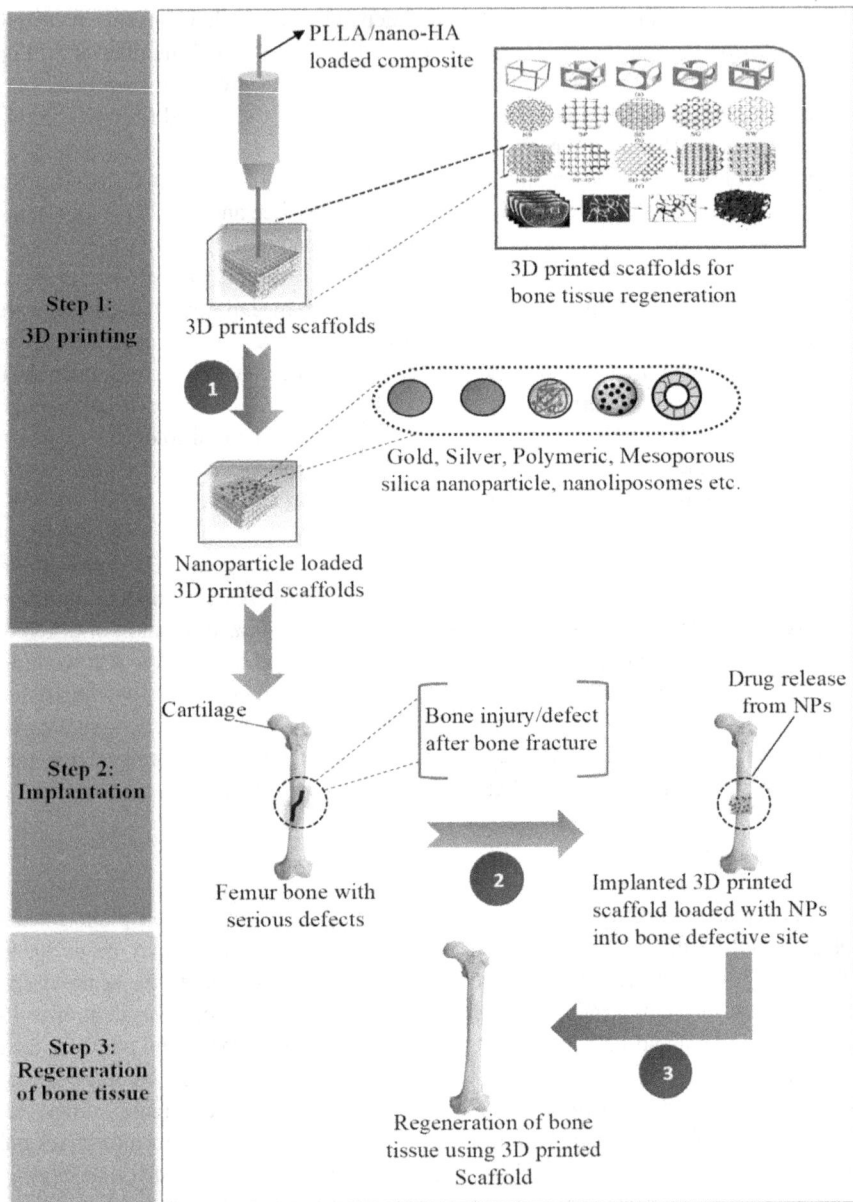

FIGURE 4.3 Scaffold fabrication and regeneration of bone tissues using 3D printed scaffold loaded with nanoparticles, which promote regeneration of defected/injured bones.

microenvironment of incubated cells, the use of soft hydrogel is recommended. But, soft biomaterial lacks sufficient mechanical strength in comparison with curable polymers. However, the soft hydrogel materials were found to be cytocompatible. The synthesis of cure polymer involves harsh polymerization conditions; hence, seeds can be incorporated after proper washing of scaffolds obtained from cure polymers. The printability and quality of scaffolds can be determined on the basis of polymerization technique, mechanical properties, melting point, and possible chemicals. Hydrogel is being used as a printable biomaterial reform (Murphy and Atala, 2014; V. K. Lee and Dai, 2017). Hydrogels are synthesized with the help of peptides or polymers. The original liquid structure is crosslinked to obtain solidified macromolecules of the hydrogel. Based on their origin, the hydrogel can be classified into two broad categories: synthetic hydrogel derived from polymers, and other hydrogels obtained from naturally occurring materials. These naturally occurring materials can be modified for their intended use. So, they are being used in the form of combination or composites (Wüst et al., 2014). Synthetic hydrogels provide precise tuning of properties of natural hydrogels. The fine adjustment of properties of the synthetic hydrogel is made by the change in molecular weights, molecular distribution, and crosslinked densities. Synthetic hydrogels PEG-based cell printing biomaterials include PEG diacrylamide (PEGDA) or diacrylate gels, which are preferred as hydrogel materials. The mechanical characteristic of PEG-based acrylamide can be changed easily with the help of crosslinking via UV or exposure to mimic the stiffness of tissues. However, this material exhibits brittle nature and cannot be suitable for surgical procedures. The PEG-based hydrogels lack motifs to establish an interaction between cell and material. This type of problem can be solved by modifying the PEG-based hydrogel with the help of conformational motifs or nature-derived peptides. The naturally occurring materials like fibrin, matrigel, collagen, and alginate are commonly used for cell printing (Guillemot, Mironov, and Nakamura, 2010; Pati et al., 2015; Seol et al., 2014).

The advancements in photochemistry help in the crosslinking of methacrylated gelatin or naturally occurring HA after printing to form and retain 3D patterns (Billiet et al., 2014; Skardal et al., 2010). But neither synthetics nor naturally occurring biomaterial fulfill the all-basic criteria for cell printing nor resemble ECM (Colosi et al., 2016). For an illustration of the printed vascularized heart, refer to Figure 4.4.

4.6   REGULATORY ASPECTS RELATED TO 3D PRINTING

The medical devices made by 3D printing processes are regulated under Food and Drug Administration, or FDA; the FDA does not directly regulate 3D printing. The regulatory reviews are reliant on the type of devices, potential use, and its latent safety and hazards of biomaterial used for the preparation of 3D printing devices. the FDA's Centre for Devices and Radiological Health regulates the devices and 3D printed products and is classified into three regulatory categories.

- **Class 1:** Includes products/devices having a low risk; for example, surgical instruments, bandages, etc.
- **Class 2:** The class 2 devices are considered as moderate-risk devices; for example, infusion pumps.
- **Class 3:** The class 3 devices are at potentially high risk; for example, these are life-sustaining products that potentially improve human health, such as a pacemaker device (https://www.accessdata.fda.gov/scripts/cdrh/cfdocs/cfcfr/CFRSearch.cfm?fr=860.3).

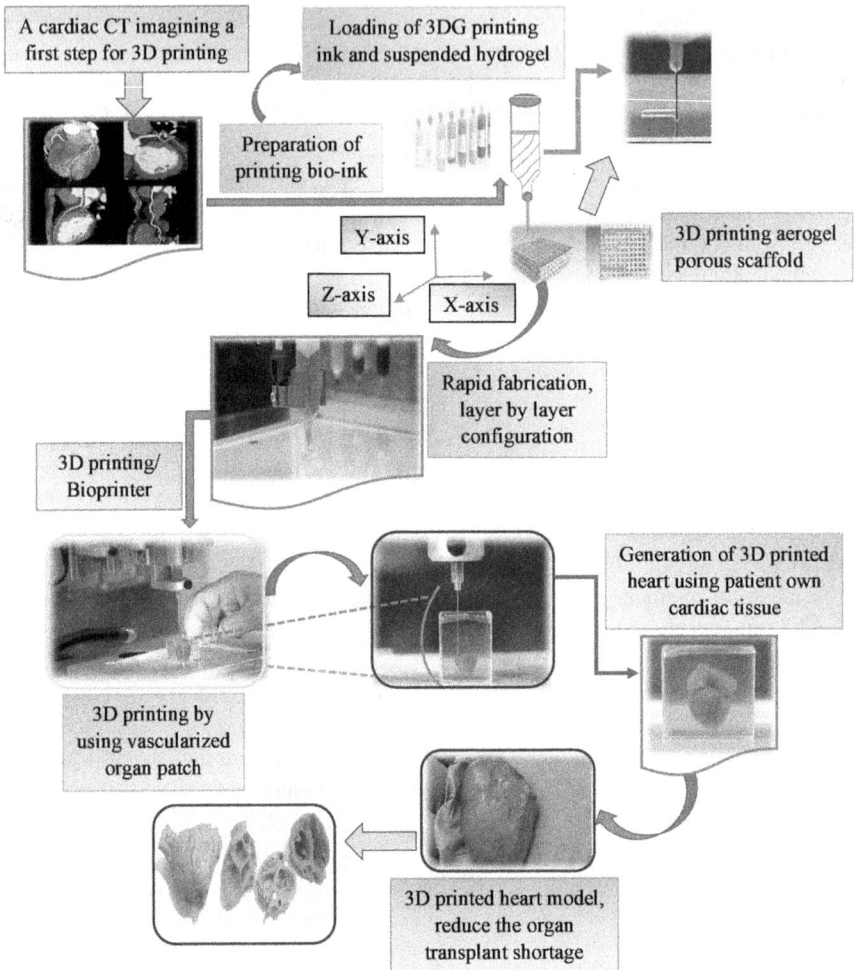

FIGURE 4.4 A schematic representation of the three-dimensional printing (3D) process (the generation of 3D bio-printed vascularised heart).

Mostly the 3D printed devices in class 1 and class 2 come under the category of 510(K). Or it is also called a premarket approval submission review of the federal Food and Drug Cosmetics Act as these devices are exempted from FDA approval review before entering the market. The class 1 and class 2 products must meet the ideal high-quality standards established by FDA before entering the market. This premarket review, or 510(K) review, reduces the need for clinical trials. The class 3 devices extensively need clinical trials, and these class 3 devices must apply with full details from clinical trials to regulatory authority before entering the product into the market. Based on the details of submission from clinical trials, the FDA authority then determines the product is safe with therapeutic potential and its potential use in humans (https://www.fda.gov/medical-devices/premarket-submissions/premarket-approval-pma). The regulations made by the FDA quality system and the medical devices of good manufacturing practices for industrial devices should be instigated throughout the manufacturing process of 3D printed devices; and this should be done from the level of feedstuff, substrate, and biomaterials used in manufacturing processes of 3D printing devices. The regulatory considerations set by the FDA for devices should follow each step from raw materials to post-manufacturing processes, such as the regulations on consideration of raw materials, technical considerations, consideration of post manufacture quality assurance, and consideration of post printing considerations: cleaning and finishing, biocompatibility, and sterilization (Morrison et al., 2015).

The FDA also regulates the software programs that are projected for the generation of 3D printed models, as well as 3D printed scaffolds. The FDA authority released some guidelines in 2017, in which the information should be provided on 3D printed products or devices and the type of information that must be provided for application on submission of 3D printed devices; for example, the devices for cranial implants and devices for replacement of joints and bone regenerations. The overall information for submission should represent the detailed information on devices, the process of manufacturing, and testing considerations during the post-operational 3D printing process. The Radiological Society of North America has released the guidelines for employing 3D printing, its potential safety for human use, and processes of producing 3D printed anatomical models, which are generated from medical imagining and scanning. Also, the Centre for Drug Evaluation and Research (CDER) or the Centre for Biologics Evaluation and Research (CBER) of the FDA regulates 3D printed devices. The role of CDER is to conduct research on 3D printed models/ devices and to understand its potential role in developing drugs. CDER conducts its research on the pharmaceutical quality of 3D printed devices by coordination among pharmaceutical utilizers to understand the potential usage of 3D printing technology. Similarly, the role of CBER is to research the use of 3D printing biomaterials and to coordinate with stalk holders who are involved in this 3D printing process (Chepelev et al., 2018).

In the United States, the chemicals used in industrial processing are regulated by the U.S. EPA (the United States Environmental Protection Agency) under TSCA (Toxic Substances Control Act). The TSCA's regulations are based on evaluating the probable supplementary risk and the action of nano-form chemicals, which are listed in the TSCA inventory. The regulations and the evaluation of the safety of nano biomaterials in North America and Europe are mainly based upon its intended use of nanomaterials. Further, in the United States, the OSHA (Occupational Safety and Health Administration) regulations and the NIOSH (National Institute for Occupational Safety and Health) regulations govern nanomaterials. OSHA has regulations over protection to workers who are involved in nanomaterials and the 3D printing industry under the standard of hazards communications (29 CFR 1910.1200).

4.7 LIMITATIONS

At the same time, there are several disadvantages to the adoption of 3D printing technology in the manufacturing industry. The operational cost may be high. As it has been mentioned, no hydrogel is full of all the qualities required for tissue engineering. So, mechanical strength as well as reproducibility could be major challenges. The possibility of cell agglomeration and low cell bioavailability due to the generation of heat is limited bioprinting of tissue. The layer-by-layer designing of scaffolds requires a long operational time.

4.8 CONCLUSION

Lithography is a sophisticated strategy for the fabrication of a smart drug delivery system when combined with self-assembly. Fabrication of a device with the combination of monodisperse porosity, improved diffusion, and integration of electronic and sensing components can be achieved easily with the 3D printing technique. The design of layer-by-layer scaffolds to mimic the three-dimensional tissue environment can be produced easily with the help of 3D printers. The layer-by-layer design of scaffolds demonstrated their efficacy in the field of regenerative medicine. In addition, the use of 3D for the design and production of prosthetic implants to accommodate the anatomy of the individual patient attracts the tremendous attention of researchers toward the 3D printers. In brief, 3D printing technology has emerged as a flexible and robust technique in the advance manufacturing industry. The 3D printer has been widely used in many countries, mainly in the manufacturing industry.

REFERENCES

Abdulghani, Saba, and Geoffrey R. Mitchell. 2019. "Biomaterials for in Situ Tissue Regeneration: A Review." *Biomolecules*. MDPI AG. 9 (11): 750, 1–24. doi: 10.3390/biom9110750

Ambrus, Rita, Areen Alshweiat, Ildikó Csóka, George Ovari, Ammar Esmail, and Norbert Radacsi. 2019. "3D-Printed Electrospinning Setup for the Preparation of Loratadine Nanofibers with Enhanced Physicochemical Properties." *International Journal of Pharmaceutics* 567 (August). 10.1016/j.ijpharm.2019.118455

Banis, George E., Luke A. Beardslee, Justin M. Stine, Rajendra Mayavan Sathyam, and Reza Ghodssi. 2019. "Gastrointestinal Targeted Sampling and Sensing via Embedded Packaging of Integrated Capsule System." *Journal of Microelectromechanical Systems* 28 (2): 219–225. 10.1109/JMEMS.2019.2897246

Billiet, Thomas, Elien Gevaert, Thomas de Schryver, Maria Cornelissen, and Peter Dubruel. 2014. "The 3D Printing of Gelatin Methacrylamide Cell-Laden Tissue-Engineered Constructs with High Cell Viability.", *Biomaterials* 35 (1): 49–62. 10.1016/j.biomaterials.2013.09.078

Boland, Thomas, Xu Tao, Brook J. Damon, Brian Manley, Priya Kesari, Sahil Jalota, and Sarit Bhaduri. 2007. "Drop-on-Demand Printing of Cells and Materials for Designer Tissue Constructs." *Materials Science and Engineering C* 27 (3): 372–376. 10.1016/j.msec.2006.05.047

Chepelev, Leonid, Nicole Wake, Justin Ryan, Waleed Althobaity, Ashish Gupta, Elsa Arribas, Lumarie Santiago, et al. 2018. "Radiological Society of North America (RSNA) 3D Printing Special Interest Group (SIG): Guidelines for Medical 3D Printing and Appropriateness for Clinical Scenarios." *3D Printing in Medicine* 4 (1): 11. 10.1186/s41205-018-0030-y

Chou, Ying Chao, Wen Lin Yeh, Chien Lin Chao, Yung Heng Hsu, Yi Hsun Yu, Jan Kan Chen, and Shih Jung Liu. 2016. "Enhancement of Tendon–Bone Healing via the Combination of Biodegradable Collagen-Loaded Nanofibrous Membranes and a Three-Dimensional Printed Bone-Anchoring Bolt." *International Journal of Nanomedicine* 11 (August): 4173–4186. 10.2147/IJN.S108939

Colosi, Cristina, Su Ryon Shin, Vijayan Manoharan, Solange Massa, Marco Costantini, Andrea Barbetta, Mehmet Remzi Dokmeci, Mariella Dentini, and Ali Khademhosseini. 2016. "Microfluidic Bioprinting of Heterogeneous 3D Tissue Constructs Using Low-Viscosity Bioink." *Advanced Materials* 28 (4): 667–684. 10.1002/adma.201503310

Edgar, John M., Meghan Robinson, and Stephanie M. Willerth. 2017. "Fibrin Hydrogels Induce Mixed Dorsal/Ventral Spinal Neuron Identities during Differentiation of Human Induced Pluripotent Stem Cells." *Acta Biomaterialia* 51 (March): 237–245. 10.1016/j.actbio.2017.01.040

Farias, Chantell, Roman Lyman, Cecilia Hemingway, Huong Chau, Anne Mahacek, Evangelia Bouzos, and Maryam Mobed-Miremadi. 2018. "Three-Dimensional (3D) Printed Microneedles for Microencapsulated Cell Extrusion." *Bioengineering* 5 (3): 1–26. 10.3390/bioengineering5030059

Guillemot, Fabien, Vladimir Mironov, and Makoto Nakamura. 2010. "Bioprinting Is Coming of Age: Report from the International Conference on Bioprinting and Biofabrication in Bordeaux (3B'09)." In *Biofabrication*. 2. 1–7. 10.1088/1758-5082/2/1/010201

Huang, Boyang, Enes Aslan, Zhengyi Jiang, Evangelos Daskalakis, Mohan Jiao, Ali Aldalbahi, Cian Vyas, and Paulo Bártolo. 2020. "Engineered Dual-Scale Poly (ε-Caprolactone) Scaffolds Using 3D Printing and Rotational Electrospinning for Bone Tissue Regeneration." *Additive Manufacturing* 36 (December): 1–11. 10.1016/j.addma.2020.101452

Lee, Vivian K., and Guohao Dai. 2017. "Printing of Three-Dimensional Tissue Analogs for Regenerative Medicine." *Annals of Biomedical Engineering*. Springer New York LLC. 10.1007/s10439-016-1613-7

Lee, Vivian, Gurtej Singh, John P. Trasatti, Chris Bjornsson, Xiawei Xu, Thanh Nga Tran, Seung Schik Yoo, Guohao Dai, and Pankaj Karande. 2014. "Design and Fabrication of Human Skin by Three-Dimensional Bioprinting." *Tissue Engineering – Part C: Methods* 20 (6): 473–484. 10.1089/ten.tec.2013.0335

Li, Xiaoming, Xinhui Liu, Wei Dong, Qingling Feng, Fuzhai Cui, Motohiro Uo, Tsukasa Akasaka, and Fumio Watari. 2009. "In Vitro Evaluation of Porous Poly(L-Lactic Acid) Scaffold Reinforced by Chitin Fibers." *Journal of Biomedical Materials Research - Part B Applied Biomaterials* 90 B (2): 503–509. 10.1002/jbm.b.31311

Marizza, Paolo, Stephan Sylvest Keller, and Anja Boisen. 2013. "Inkjet Printing as a Technique for Filling of Micro-Wells with Biocompatible Polymers." *Microelectronic Engineering* 111: 391–395. 10.1016/j.mee.2013.03.168

Michael, Stefanie, Heiko Sorg, Claas Tido Peck, Lothar Koch, Andrea Deiwick, Boris Chichkov, Peter M. Vogt, and Kerstin Reimers. 2013. "Tissue Engineered Skin Substitutes Created by Laser-Assisted Bioprinting Form Skin-Like Structures in the Dorsal Skin Fold Chamber in Mice." *PLoS ONE* 8 (3): 1–12. 10.1371/journal.pone.0057741

Morrison, Robert J., Khaled N. Kashlan, Colleen L. Flanangan, Jeanne K. Wright, Glenn E. Green, Scott J. Hollister, and Kevin J. Weatherwax. 2015. "Regulatory Considerations in the Design and Manufacturing of Implantable 3D-Printed Medical Devices." *Clinical and Translational Science* 8 (5): 594–600. 10.1111/cts.12315

Murphy, Sean V., and Anthony Atala. 2014. "3D Bioprinting of Tissues and Organs." *Nature Biotechnology*. Nature Publishing Group. 10.1038/nbt.2958

Nau, Katja, and Steffen G. Scholz. 2019. "Safe by Design in 3D Printing." In *Smart Innovation, Systems and Technologies*, 155: 341–350. Springer Science and Business Media Deutschland GmbH. 10.1007/978-981-13-9271-9_28

O'Brien, Christopher M., Benjamin Holmes, Scott Faucett, and Lijie Grace Zhang. 2015. "Three-Dimensional Printing of Nanomaterial Scaffolds for Complex Tissue Regeneration." *Tissue Engineering – Part B: Reviews* 21 (1): 103–114. 10.1089/ten.teb.2014.0168

O'Brien, Fergal J. 2011. "Biomaterials & Scaffolds for Tissue Engineering." *Materials Today*. Elsevier B.V. 10.1016/S1369-7021(11)70058-X

Ozbolat, Ibrahim, and Yin Yu. 2013. "Bioprinting towards Organ Fabrication: Challenges and Future Trends." TBME-01840-2012, 1–9. http://www.personal.psu.edu/ito1/assets/files/Bioprinting_towards_Organ_Fabrication_Ch.pdf

Pati, Falguni, Dong Heon Ha, Jinah Jang, Hyun Ho Han, Jong Won Rhie, and Dong Woo Cho. 2015. "Biomimetic 3D Tissue Printing for Soft Tissue Regeneration." *Biomaterials* 62. 10.1016/j.biomaterials.2015.05.043

Sachs, E., M. Cima, and J. Cornie. 1990. "Three-Dimensional Printing: Rapid Tooling and Prototypes Directly from a CAD Model." *CIRP Annals* 39 (1): 201–204. 10.1016/S0007-8506(07)61035-X

Saenz Del Burgo, Laura, Jesús Ciriza, Albert Espona-Noguera, Xavi Illa, Enric Cabruja, Gorka Orive, Rosa María Hernández, Rosa Villa, Jose Luis Pedraz, and Mar Alvarez. 2018. "3D Printed Porous Polyamide Macrocapsule Combined with Alginate Microcapsules for Safer Cell-Based Therapies." *Scientific Reports*. 8 (1): 1–14. 10.1038/s41598-018-26869-5

Seol, Young Joon, Hyun Wook Kang, Sang Jin Lee, Anthony Atala, and James J. Yoo. 2014. "Bioprinting Technology and Its Applications." *European Journal of Cardio-Thoracic Surgery*. 46 (3), September 2014, 342–348. https://doi.org/10.1093/ejcts/ezu148

Serafin A., Murphy C., Rubio M.C., and Collins M.N. 2021. "Printable Alginate/Gelatin Hydrogel Reinforced with Carbon Nanofibers as Electrically Conductive Scaffolds for Tissue Engineering." *Materials Science & Engineering. C, Materials for Biological Applications* 122 (March): 111927. 10.1016/j.msec.2021.111927

Skardal, Aleksander, and Anthony Atala. 2015. "Biomaterials for Integration with 3-D Bioprinting." *Annals of Biomedical Engineering* 43 (3): 730–746. 10.1007/s10439-014-1207-1

Skardal, Aleksander, Jianxing Zhang, Lindsi McCoard, Xiaoyu Xu, Siam Oottamasathien, and Glenn D. Prestwich. 2010. "Photocrosslinkable Hyaluronan-Gelatin Hydrogels for Two-Step Bioprinting." *Tissue Engineering – Part A* 16 (8): 2675–2685. 10.1089/ten.tea.2009.0798

Thomas, Michaela, and Stephanie M. Willerth. 2017. "3-D Bioprinting of Neural Tissue for Applications in Cell Therapy and Drug Screening." *Frontiers in Bioengineering and Biotechnology*. Frontiers Media S.A. 10.3389/fbioe.2017.00069

Waid, Simon, Heinz D. Wanzenboeck, Michael Muehlberger, and Emmerich Bertagnolli. 2012. "Optimization of 3D Patterning by Ga Implantation and Reactive Ion Etching (RIE) for Nanoimprint Lithography (NIL) Stamp Fabrication." *Microelectronic Engineering* 97 (September): 105–108. 10.1016/j.mee.2012.02.028

Wüst, Silke, Marie E. Godla, Ralph Müller, and Sandra Hofmann. 2014. "Tunable Hydrogel Composite with Two-Step Processing in Combination with Innovative Hardware Upgrade for Cell-Based Three-Dimensional Bioprinting." *Acta Biomaterialia* 10 (2): 630–640. 10.1016/j.actbio.2013.10.016

Yeong, Wai Yee, Chee Kai Chua, Kah Fai Leong, and Margam Chandrasekaran. 2004. "Rapid Prototyping in Tissue Engineering: Challenges and Potential." *Trends in Biotechnology*. 10.1016/j.tibtech.2004.10.004

Yeong, Wai Yee, Chee Kai Chua, Kah Fai Leong, Margam Chandrasekaran, and Mun Wai Lee. 2007. "Comparison of Drying Methods in the Fabrication of Collagen Scaffold via Indirect Rapid Prototyping." *Journal of Biomedical Materials Research - Part B Applied Biomaterials* 82 (1): 260–266. 10.1002/jbm.b.30729

Yu, Yinxian, Sha Hua, Mengkai Yang, Zeze Fu, Songsong Teng, Kerun Niu, Qinghua Zhao, and Chengqing Yi. 2016. "Fabrication and Characterization of Electrospinning/3D Printing Bone Tissue Engineering Scaffold." *RSC Advances* 6 (112): 110557–110565. 10.1039/C6RA17718B

5 3D Printing of 2D Nanomaterials

Muhammad Suleman and Naila Nasir
Department of Nanotechnology and Advanced Materials
Engineering, Sejong University, Seoul, South Korea

Muftooh Ur Rehman Siddiqi
Department of Mechanical, Biomedical and Design
Engineering, School of Engineering and Technology, Aston
University, Birmingham, England

Muhammad Usman and Sundus Tariq
Department of Mechanical Engineering, CECOS University,
Peshawar, Pakistan

CONTENTS

DOI: 10.1201/9781003189404-5

5.1   INTRODUCTION

3D printing refers to the using of printing methods for the construction of objects and structures. In 3D printing, a material is created from the model that is designed in the computer. A variety of processes are involved in the 3D printing process, and these include the deposition, solidification, and joining of materials to form a structure. Typically, the 3D printing process consists of layer-by-layer printing where the different types of materials are fused together. These materials may include powders, plastics, and liquids, and together, they make up a structure [1].

With the developments in the field of printing and computation, 3D printing has now developed as a method with which the accurate manufacturing of a structure is made possible. Initially, the methods that were available for 3D printing were primitive, and the technology was only used on a limited scale where the object size was also limited.

Automation also helped in the printing where the fabrication of the devices was made easier as the technology improved. 3D printing became a better solution for the production of complex shapes at a larger scale [2].

Some of the structures might be way too complex and their construction by hand might be impossible; therefore, 3D printing can be very helpful in such situations, where the complex structures are manufactured with ease without the typical problems faced when printing of geometrical shapes.

3D printing involves some very basic steps, and these steps, when followed properly, result in the formation of 3D printed models. The first step in 3D printing is modelling, where a suitable model is designed of the material and this model is then used for the printing process.

The modelling phase results in the formation of a 3D model file that can be used to rectify errors. The files used in this case can help in the removal of errors and defects from the designed model, and this will help in the accurate manufacturing of the object using the 3D printing method [3].

Finishing is the final step where the design of the object is completed. The use of materials is done in accordance with the requirement, and, depending on the size and shape, the printing process is finalized. There are also some techniques that use the additive materials for the manufacture of objects. These additive materials are used for ensuring that the desired combination of materials is made properly, which will ensure that the final design has all the essential components in the structure and has the desired properties needed in the final design of the object [4].

In case of 2D materials and the structures based on 2D materials, 3D printing is achieved as various major components of printing are modified together, resulting in the formation of complex macrostructures. The advantage of using 2D materials is that they provide face-to-face stacking of the intersheets and the large contact area where the unique features of 2D materials are added to the macrostructures.

The exciting features 2D materials provide can be incorporated into existing techniques available with 3D printing. They also can be applied to introduce new features into the macrostructures that can be developed using 3D printing.

A number of methods are used for 3D printing. Each of these methods provides the fabrication of microstructures having a 3D nature. One such method is the inkjet printing method in which the material deposition is done using droplets, and using these droplets, 3D microstructures are developed. Using inkjet printing, both organic and inorganic methods are utilized, and, for each of these methods, the fabrication of 3D macrostructures can be achieved [5].

E-Jet printing is the method of printing in which a higher resolution can be achieved by using the inkjet system. Instead of physically pushing the liquids, they are pushed using the electric fields. The liquid ink is ejected into the substrate, and for that, there is application of a high power supply that ensures that the opposing conducting support is provided to the ink as the pressure is applied. The materials that are used for E-jet orienting are needed in flowing form and are made by using insulating polymers [6].

5.2   2D MATERIALS FOR INKS

5.2.1   Graphene

Graphene was the first of the 2D materials that was developed and is actually a single layer of graphite. Graphene has an atomically thin layer of carbon atoms that are arranged in a hexagonal shape [7]. The properties of graphene, such as the high electrical conductivity, excellent thermal strength, and optical transparency make it an ideal material for use in the development of materials needed for 2D material-based 3D printing [8].

The development of graphene oxides is also another popular method in which the oxides of graphene are developed and are then used to stack up poof layers for the ink-based printing applications. The reduced graphene oxides that are chemically derived from graphene are also used as inks for the 3D printing of 2D materials-based nanostructures. This also allows the efficient deposition of material on the substrate. This deposition is facilitated by the use of functional groups that are present in the compound [9].

5.2.2   MXene

MXene is a 2D material in which there is a combination of single layers of metal carbides. The formula for this material is in the form of nitrides of transition metals. The word mXene was used because there is a single layer and the precursor of graphene is in its maximum state [10]. The exfoliation of mXene is done by the use of a wet-etching method in which hydrofluoric acid is used. The hydrophilic surface of mXene, however, allows for the excellent properties in terms of storage devices, sensors, shielding devices, and in transistors [11].

5.2.3 MoS$_2$

Molybdenum disulfide is also another 2D material that is capable of providing excellent properties in terms of the band gap and the ON/OFF ratio. MoS2 has a similar structure to graphene, which is in the form of a hexagonal structure [12]. These properties of MoS2 provides evidence that this material is ideal for use in applications like switching devices, which include transistors, MOSFETs, and FET devices [13].

5.2.4 BLACK PHOSPHOROUS

Black phosphorous is also another 2D material that exhibits excellent properties in terms of the band gap, mechanical properties, and also in the form of thermal and electrical properties [14]. This material has a lot of advantages in the fields of electrical and optoelectronics engineering. However, it also suffers from the problem of oxidation [15].

5.2.5 HEXAGONAL BORON NITRIDE (hBN)

Hexagonal boron nitride, also termed as hBN, is the isomorph of graphene, which has a layered structure similar to graphene. In hBN, there are BN atoms connected together in a layered structure and form a honeycomb structure. This 2D nano-material also has excellent properties in terms of the thermal stability, electrical conductivity, and also in the form of mechanical robustness, which makes it ideal for use in electronic devices like FETs and MOSFETS.

Of all the 2D nanomaterials, hexagonal boron nitride (hBN) is important since the properties it offers are mostly electrical insulation, as well as better thermal conduction. The hBN structure has the hexagonal bonding of boron and nitrogen atoms. This forms 2D layers, which are stacked together, and van der Waals forces are present, which ensure that the 2D layers are connected together properly. The properties of hBN lie in a region that makes it electrically insulating, but the thermal conductive properties are very good [16].

These properties of hBN make it ideal for use in applications such as the magnetic alignment, stretching, chemical treatments, and also in nanocomposites, which are enhanced for thermal conductivity. These devices are excellent for use in flexible electronic devices [17].

The use of 3D printing can be advantageous in this situation since the 3D printing provides the opportunity of creating complex structures, which are created easily and have very little waste created by the technique. The traditional methods that are available do not provide this level of accuracy of design, and therefore, these are the preferred methods [18].

The 3D printing method also has some constraints associated with it, and therefore, this method cannot be applicable everywhere. Most of the designs that are made using 3D printing have the materials used in the form of metals and polymers, and also in the form of metals that have constraints in terms of the mechanical and thermal properties of the materials used [19].

The use of 2D materials-based nanomaterials is considered ideal for such situations since these are materials are considered to be superlative 2D nanomaterials, which have excellent mechanical, optical, and electrical properties. Considering all these advantages, the 2D nanomaterials are considered the ideal materials in these situations since these materials provide excellent conduction properties for the electronic devices. The 2D materials that are used commonly for nanocomposites include graphene, molybdenum disuplhide, and hBN. These advantages are obtained in the form of more control over the surface area of the manufactured surface. The use of 3D printable devices enables their use in printed electronic devices, as well as in the field of batteries and electrical engineering [20].

However, there are many challenges in the actual realization of materials that are capable of 3D printing. For the proper utilization of 2D materials in 3D printing technology, there is need for formulating inks based on these 2D nanomaterials. For this to be achieved successfully, there is need to formulate inks that are stable and also have the properties like stability of band structure and thermal conductivity [21].

There is usage of solvents that are organic as well as inorganic for the proper fabrication of inks based on 2D materials. The addition of solvents for formulating inks can also be challenging as the additives may reduce the conductivity of the 2D materials. There is also the usage of binders, which are used primarily for increasing the rheological properties of the inks. The additives of these binders have a major impact on the properties of the 2D materials and therefore are used after proper optimization is done of their properties [22].

For the processing of inks based on 2D materials, there are inks that are fabricated, and these inks are developed using a three-step mechanism. The first step is to find the suitable concentration of the 2D material that is to be used [23].

There are also casting operations that are used for ensuring that the 3D shaping of the nanocomposites of hBN can be made possible. hBN can also be used as a material that is 3D printable. This transformation is achieved in the form of PVA of polyvinyl alcohol that is used for the improvement of thermal properties.

The material developed can 3D printed, but the printing is demonstrated to be of single fibers. There is no particular proof that the material that is developed is self-supporting or has the ability to be printed in multiple layers. The printing of materials in 3D printing methods can be a huge challenge as the printing in multiple layers can be challenging.

5.3 PRACTICAL APPLICATIONS OF 3D PRINTING FOR 2D MATERIALS BASED NANO STRUCTURES

5.3.1 HEXAGONAL BORON NITRIDE BASED NANOCOMPOSITES

For the development of nanoscale electronic devices, a number of considerations related to the technology of integrated circuits are done. With the increasing speed, the dissipation of power and the increasing density, there is need of proper thermal management of the electronic devices [24]. Therefore, the thermal management of electronic devices has always been a key consideration during the manufacturing phase. The technology is also evolving toward flexible electronics and that also puts a major constraint on the manufacturing techniques that are available nowadays. These constraints include the constraints of weight, flexibility, and stretchability, as well as the most important constraint of manufacturing cost.

Considering all these constraints, there is need for the development of electronic devices that can be developed into complex patterns and can be molded into three-dimensional structures. These 3D architectures are used frequently in a number of applications, such as the bioelectric implants [25]. The polymer materials that are normally used in this regard are the ideal materials for such a scenario, but the problem exists of the low thermal conductivity, which is always present in these materials. One such material that can be used as a replacement to polymers is the hBN. The work done in [26] introduces the use of a 3D printable nanocomposite, which is based on hBN and can be used for the development of nanocomposites that have high thermal conductivity. The application of this nanomaterial extends as it can be used in situations where there is high flexibility and stretchability that is required for the device to be used. The mentioned properties are used extensively in the bioelectronics devices of the future, and the use can be extended to the devices that are based on the bioelectronics [27].

The work done by Guiney et al. [26] shows the use of a hBN polymer ink, which is 3D printable and has the capability of usage with other particle systems that include the use of nanotubes of carbon and graphene, as shown in Figure 5.1. The

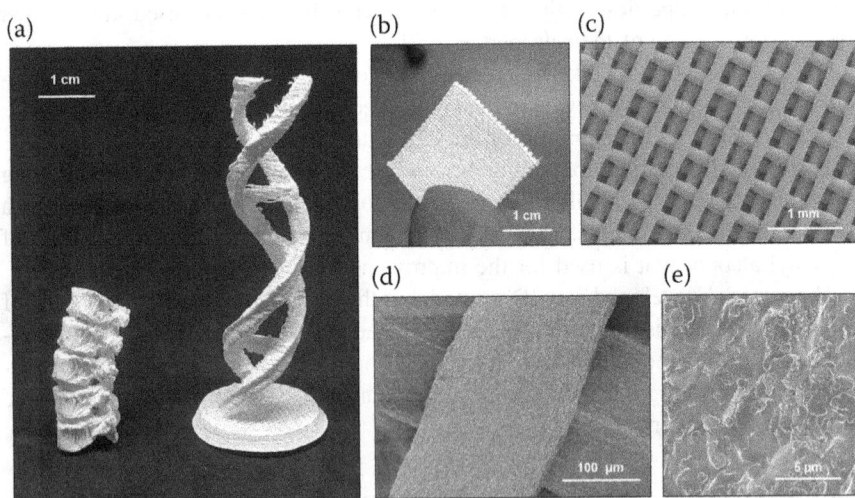

FIGURE 5.1 3D printed nanostructure of hBN (a) Photograph showing the complex structures of helix structure printed by using nanocomposite ink of hBN with 60% volume. (b) Photos (c-e) SEM images of hBN with 3D printing with 40% hBN scaffold. Image shows high uniformity of the images over a large range of height and area. There are also visible span gaps in the inner structure of the grid as can be observed in the photographs of the SEM images as well as the double helix bridge. The surface morphology is observed to be smooth as the coating of PLGA is uniform and the hBN flakes are dispersed within the polymer matrix. Reprinted (adapted) with permission from reference [14]. Copyright [2018] American Chemical Society.

materials that are used in the 3D printable ink include hBN and poly lactic co glycolic acid (PLGA). The use of PLGA can ensure that it can work as an elastomer and be used as a binding polymer for the hBN particles.

The constructs of hBN that are made using the 3D printing process are found to be mechanically robust and can be handled as soon as the printing process finishes. The hBN scaffolds that are obtained this way are also flexible and can be folded and rolled accordingly. The folding can be done without the damage to the mechanical properties, which can help in achieving a wide range of shapes and structures using the 3D printing technology.

The use of 3D printing for the manufacture of hBN-based nanocomposites will open the doors for advanced manufacturing techniques that will enable the further prototyping of these nanocomposites, which can boost the manufacturing industry. Since the materials used are nanocomposites, they can be used for manufacturing other nanocomposites with different 2D materials because the manufacturing process is facilitated by the 3D printing technology.

5.3.2 Use of 3D Printing Based (MXene) Ink for Fabrication of Micro-Supercapacitors

Supercapacitors have gained a lot of popularity in recent years, and the reason behind this gain is the development of portable electronics, which have a wide range of use for Supercapacitors. The wireless sensors that are used also have the requirement of electrodes, which means that for the development of batteries and capacitors that are used in self-powered systems, there is need for direct fabrication of the devices, which is done on various substrates.

The architecture of the device that is manufactured also has an influence on the performance as the electrodes show higher power performance when ion transportation is better. The architecture of super electrodes also carries a lot of importance when the micro super capacitors are fabricated. Considering all these facts, it can be inferred that the three-dimensional architectures of the micro super capacitors carries a lot of importance because this architecture allows more materials to be loaded on the device on a particular area.

This also shows that the fabrication of devices in which scalable fabrication is done with 3D architecture have the challenge of electrode material selection. The ink used should be made more printable and should also exhibit viscous properties to ensure that the layer-by-layer shape is retained and the fluidity remains in the adhesion that will enable the 3D printing of the electrodes. This is represented in Figure 5.2.

For the proper fabrication of micro super capacitors, there is need of inks based on 2D materials that can be used by the 3D printer. The ink that is used in the work done by Orangi et al. [28] is highly concentrated and is based on MXene. The viscoelastic nature of the vaccine allows the extrusion printing to be done at room temperature. The ink can be used on a variety of substrates, including papers and polymers films. The deposited layers that are used determine the height that is achieved of the electrodes, as well as the active materials that are loaded.

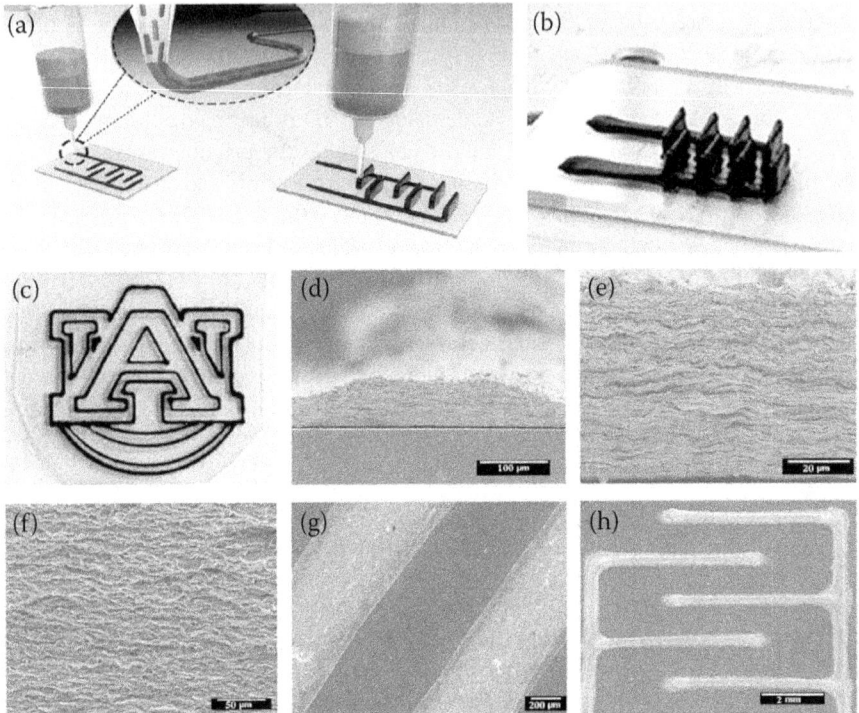

FIGURE 5.2 (a) Schematic representation of 3D printing MSCs with the development of interdigital architectures. The sheer stress that is present in the nozzle results in the alignment of the flakes which results in horizontal alignment. The additional layers might be used for increasing the height of the electrodes. (b) The optical image taken of the device made with MSC-10 and printing done on a glass substrate prior to drying. Reprinted (adapted) with permission from reference [15]. Copyright [2020] American Chemical Society.

The micro super capacitors that are fabricated using the 3D printed method showed excellent properties in terms of the electromechanical performance. The energy densities, as well as the capacitances values, were also obtained to be good; this shows that the MXene inks have high potential in fabricating the devices of the future, including the electronic and the electrochemical devices.

The applications of the MXene based inks that are additive free include a lot of applications apart from the energy storage applications, and these applications are used in sensors, electromagnetic devices, and electronic and biomedical applications.

5.3.3   3D Printing of MoS2-Graphene Based Aerogels for Anodes of Sodium Ion Batteries

3D printing is a technique that is applied for making the prototypes of materials, architectures and are also used for the fabrication of electrodes for Li Ion batteries. 3D printing is also used for the technology in which layer-by-layer material processing is done to achieve the control in the internal as well as the external microstructure. With 3D printing, the materials are able to provide the macro architect and are also able to maintain an internal structure which is porous.

The work done by Brown et al. [29] shows the development of an aerogel, which uses graphene and MoS2. The method of printing used in this case is at freezing temperature as the substrate is cooled down to a temperature of −30 °C.

This temperature results in the freezing of the aqueous droplets of ink used for the development of MoS2 Nanosheets that are plced with the graphene. The work shows the use of a 3D freeze-printing method, which does the printing of aerogel based on MoS2 and rGO. The work shows the integration of inkjet printing and also the casting method based on freezing through which the control is achieved of the hybrid structure that is to be created using 3D printing. The graphene network that exists in these 3D aerogels has shown the behavior to improve the conductivity properties as well as the mechanical strength. This work also shows that how the 3D printing method can be used in the future for the fabrication of electrode materials that are microporous and can be used for the development of energy storage devices. Figure 5.3 shows the details of this process.

The work done by Liu et al. [30] uses the fused deposition modelling for 3D printing. This method is used widely as it promises high accuracy. This method of modelling also has simple technique of fabrication and the cost of material is also low. The safety is also an advantage of this method as the process is easier to implement as compared to other methods used for the 3D printing process using thermoplastic materials. The composites that are obtained by the use of FDM method have remarkable properties in terms of mechanical strength and also in terms of the anisotropic behavior of the composites that are used. It also shows the use of fused deposition modelling for the preparation of thermoplastic polyurethane samples that are filled with hBN platelets. The work also shows that the FDM printing process can be used in the future as an accurate method to ensure that the composites are designed properly in which oriented fillers are used. This 3D printing mechanism is shown in Figure 5.4.

5.3.4 3D PRINTING OF 2D MATERIALS BASED ON ALIGNMENT

2D materials like graphene, boron nitride and molybdenum disulfide have excellent optical, electrical, and thermal properties. Usually, these structures are assembled horizontally. In such orientation, the 2D layer is aligned along the surface of the substrate. Using these structures, the aligning can be done properly, which ensures that the interaction of 2D materials is increased; this ensures that there are outstanding properties achieved in terms of the thermal properties as well as the mechanical properties. In a lot of applications, such as the batteries and charging applications, the use of vertically aligned structures is highly desirable since it ensures that the desired properties of ion charging and discharging of batteries are achieved. The work done by Liang et al. [31] shows the use of a 3D printing method in which the fabrication of vertically aligned sheets is done. The printing is done in multiscale, and the design is shown of a boron nitride (BN) nanosheet, which is fabricated successfully. The ink that is used for the fabrication of the vertical structure has shear thinning properties, and the modulus of storage is also very high.

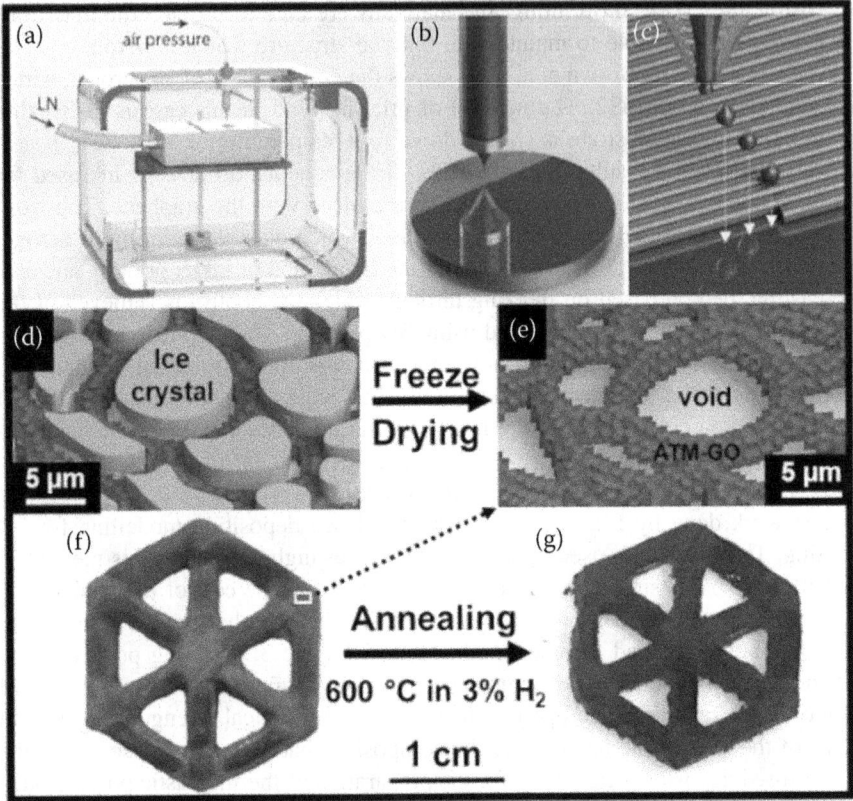

FIGURE 5.3 Schematic representing the inkjet printing setup with the drop-on-demand methodology. (a) The printing process representation of ATM-GO Droplets. (b-c) Formation of the ice template. (d) Image representing the ice crystal during printing and the resulting gel of ATM-GO after the implementation of freeze drying. (e) Macrostructure showing the 3D printed ATM-GO aerogel observed after the process of free drying. (f) The observation of aerogel made with MoS2-Rgo after doing the process of reductive thermal annealing. Reproduced with permission from Reference [29].

The structure that is obtained by the vertically aligned structure has outstanding properties in terms of the thermal conductivity as the value of thermal conductivity is very high in this case. The advantage of vertical printing of nanostructures include the direct printing of nanostructures where there is no need of any solidification treatments. This ensures that the scalable fabrication is achieved for the vertical nanostructures. This also means that vertical nanostructures of nanosheets of boron nitride are achieved at a macroscale as well as at a microscale. The adjusting of the printing parameters can also be helpful in this regard as the desired properties in the structures can be achieved as the printing parameters are changed. The thermal properties of these materials also indicate that the thermal management and the cooling capability is also high, which shows that these materials can be used in various thermal management applications. This shows that how 3D printing

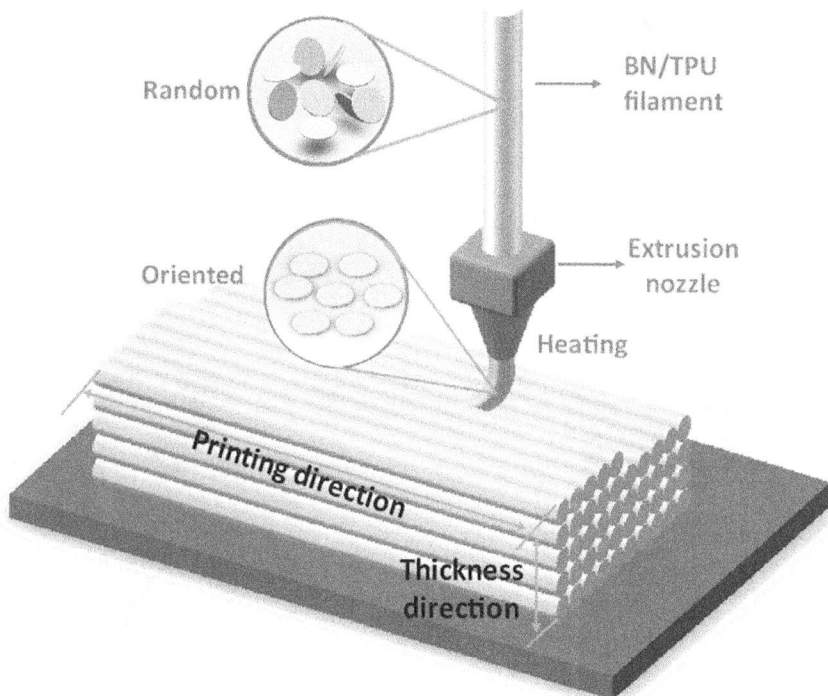

FIGURE 5.4 Illustration of the 3D printing schematic of the fillers which are progressively aligned within the nozzle as the deposition happens of the composite ink. Reproduced with permission from Reference [30].

based on extrusion can be used for the direct assembly of nanomaterials that have 2D structures aligned into them. This methodology is shown in Figure 5.5.

As the technology based on 3D printing is developed, there is increased usage of 2D materials and 2D material-based inks for the development of macrostructures using 3D printing technology. The superior properties that are provided by the 2D materials make them ideal candidates for use in 3D printing-based applications.

From the design point of view, the excellent mechanical properties that are provided by the 2D materials are the major advantage that they bring to the world of 3D printing. The world of design engineering relies heavily on materials that have excellent strength and are also light weight. This can be demonstrated from the fact that the value of young modulus of graphene is higher than the aerogel, which is made of lattice. This means that for the same value of density, there is a major difference in young modulus of these two materials, which shows how much these materials can be used in the 3D printing world due to their excellent mechanical properties [32].

There are also other parameters that are controlled through the use of 3D printing. The formation of interconnected networks of 2D materials can be achieved by the use of alignment in which the alignment is done of the 2D nano sheets to get the required outcome from the designed structure. This can be done to get higher values of thermal conductivity as well as of electrical conductivity [33].

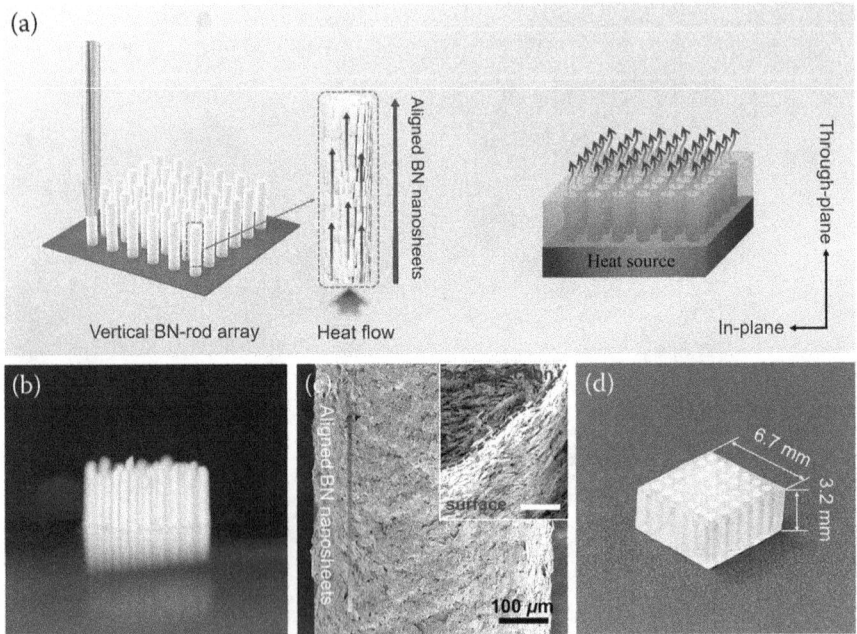

FIGURE 5.5 3D vertical printing based on Extrusion of a BN array. (a) Schematic representing the vertical 3D printing of a BN array and the diagram showing the magnified image of the BN rod (middle inset), the illustration of the BN Nano sheets aligned vertically making the structure and resulting in the process of heat conduction, in the vertical direction (right inset). (b) Side view of the optical image taken of a BN array in which there is usage of a 6-by-6 rod array of BN. The distance, height and the diameter were observed to be 0.4 mm, 4 mm and 0.6 mm, respectively. (c) Representation of the SEM morphology of BN rod. Showing of the BN Nano sheets packing in the vertical direction and orientation with the shear force exerted by BN ink filaments as they are emitted from the nozzle during the process of 3D printing. The inset represents the SEM image with cross sectional view as the BN Nano sheets are represented with a vertical alignment. (The scale bar is 50 um). (d) The sample representing the PDMS/BN array which is fabricated by the encapsulation of PDMS matrix and a 6-by-6 BN rod Array. Reprinted (adapted) with permission from reference [18]. Copyright [2019] American Chemical Society.

This can be achieved in materials where there is requirement of high values of thermal conductivity and also of the electron energy density. This is therefore important that all the 2D materials are printed accordingly. This means parameters like the shape, thickness, and dimensions of the electrode that is being printed must be adjusted for the applications in the 3D printing technology.

In the biological applications, the need of multifunctionality becomes even higher as the required features are printed easily using the 3D printing technology. The complex structures inside a tissue are easily printed as the 3D printing technology is utilized. This ensures that all the biological functions are integrated into the structure that is printed using the 3D printer [34].

5.3.5 3D-Printed Graphene for Biological Biosensors

3D printing techniques are also used in scenarios where electrodes are used in various application in the field of electrochemistry. The common usage of such applications include the use in storage devices as well as sensors and electricity generation applications. The work done by Marzo et al. [35] shows the use of 3D printed graphene-based electrode. Figure 5.6 shows the schematic of the proposed technology.

The material used for the manufacture of electrodes includes the graphene and polylactic, and the technique that is used for manufacturing is the additive manufacturing technique. The use of gold nanoparticles is also demonstrated, and the

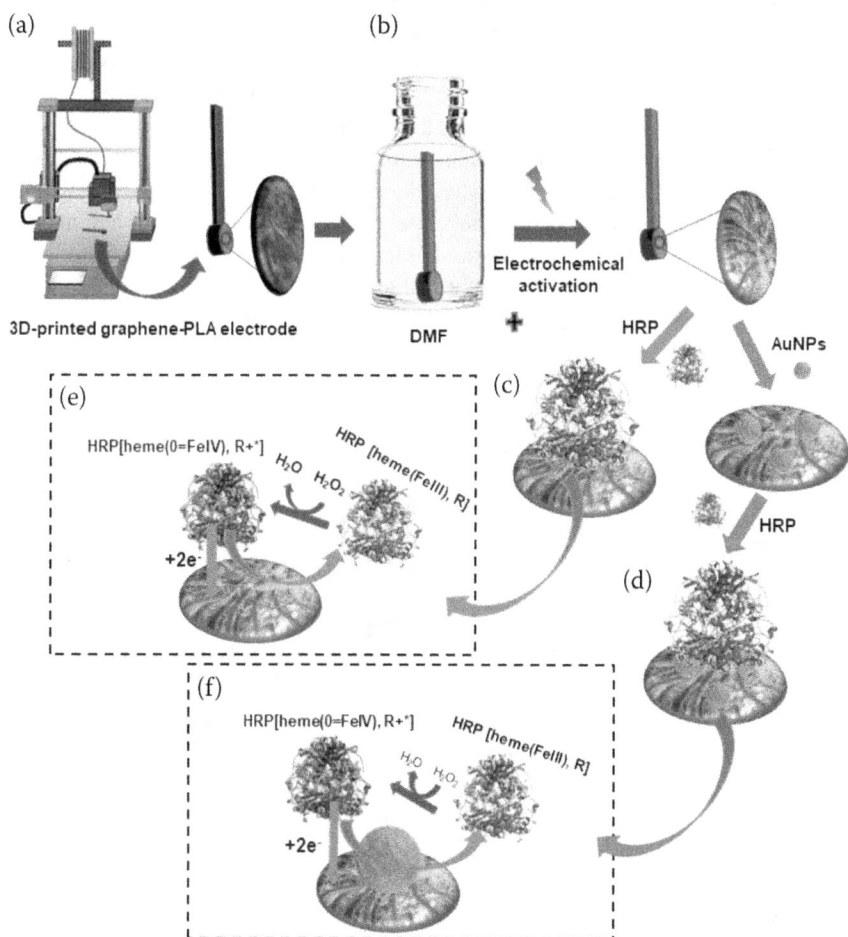

FIGURE 5.6   Fabrication of biosensor based on graphene-PLA. (a) Electrode printing with 3D printing method. (b) Activation by DMF and by electrochemistry. (c) Use of HRP Enzyme for modifying 3D-printed electrode. (d) Modifying of the 3D-printed electrode by the use of gold NPs and also with the use of HRP enzyme. (e) and (f) show the mechanism with which the detection of hydrogen peroxide is done. Reproduced with permission from Reference [35].

homogeneity in the transfer of electrons is also demonstrated. The work demonstrates the use of biosensors, which are printed using the 3D printing technology. The implications of the 3D printing technology are shown for fields like bioelectronics and biomedical engineering.

5.3.6 3D PRINTING OF PVA/HEXAGONAL BORON NITRIDE COMPOSITES FOR BONE TISSUE ENGINEERING

The work done by Aki et al. [36] shows the fabrication of a composite based on hexagonal boron nitride, polyvinyl alcohol, and bacterial cellulose. The work done indicates the use of 3D printing for the fabrication of structures that are necessary for the field of bone tissue engineering. The proposed methodology is shown in Figure 5.7.

The behavior of pseudo-capacitance is observed when the electrode is designed using 3D printing. The electrode is loaded with MoS2 to get better conductive behavior. The pseudo-capacitance also adds an advantage of high-energy density, which is provided by the pseudo-capacitive material. The work shows the extrusion based printing of a graphene electrode, which is loaded with MoS2. The printed structure shows remarkable performance and behavior, which is observed, and the device also has better performance as compared to the devices that were made previously. The network that is present inside the material fabricated has efficient properties in terms of ion diffusion and also in terms of electron transport.

5.4 EXTRUSION BASED 3D PRINTING OF 2D MATERIALS

The use of 2D materials is done extensively in many applications of electrical engineering, which include the use in batteries, optoelectronics devices, digital electronics,

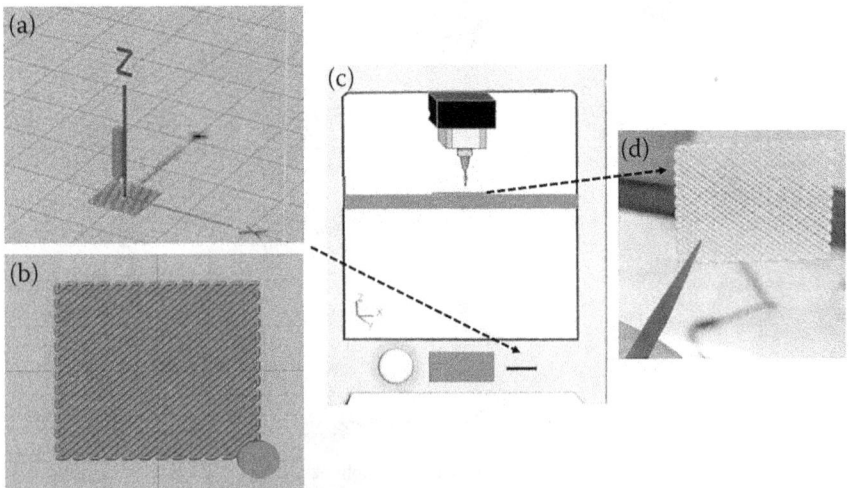

FIGURE 5.7 Model in solid form representing the simulation of the process of printing. (a-c) Shows the image of the 3D printing device. (d) Represents the image of hexagonal boron nitride which is doped by the use of PVA matrix having a wt% of 12. [36].

and sensors. The use of 2D materials has revolutionized the world in these field and some of the applications of 2D materials based printing are discussed next.

5.4.1 3D Printing of Nanomaterials-Based Electrodes for Batteries

Batteries have a huge range of applications in the field of electricity and power generation. The batteries that are used in grid stations and electric vehicles have huge power potential and are able to provide the required powered depending on the voltage and current requirement. The basic materials that are used in the batteries include electrodes, electrolytes, and separating materials, as shown in Figure 5.8. These materials are then covered with plastic coatings. Generally, the materials that are used for the fabrication of these components in batteries include polymers, filters, and binders, which ensure that the anode and the cathode are designed accordingly. The fabrication of these materials is done to ensure that the planer form of structure is achieved for better performance [37].

These electrodes, when based on the extrusion technology, can be printed, and for that, printable inks are used. Some modifications are done in the structure of the inks that are used, and these include the changes in the form of viscosity and the composition of the inks used. There are also some printing parameters that are changed, and these include the changes made in the pressure, the shape, speed,

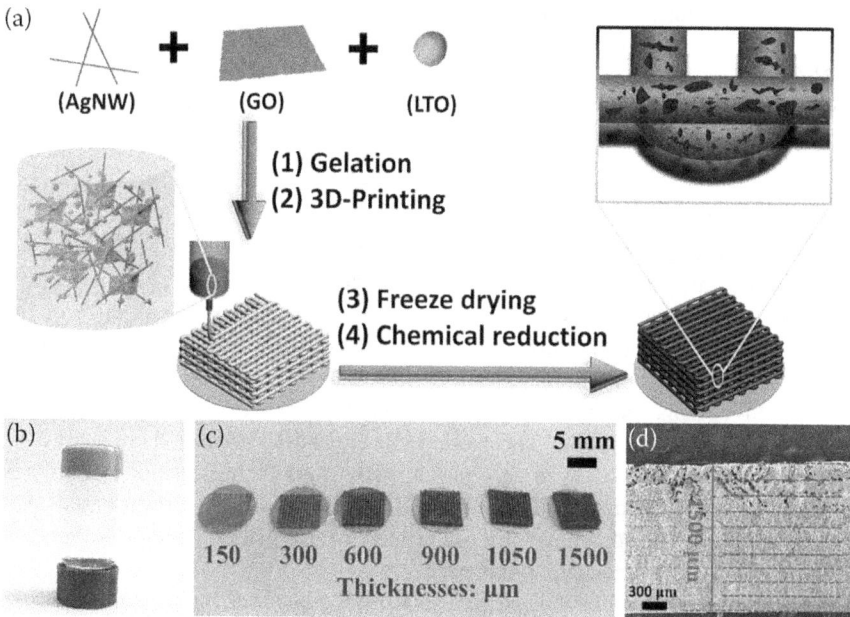

FIGURE 5.8 (a) Electrode structures printed with 3D printing. (b) Optical Image of rGO-AgNWs-LTO Ink. (c) rGO-AgNWs-LTO electrodes of various thicknesses achieved by 3D orienting through freeze drying. (d) Cross-sectional view of the printed structure by use of SEM. Reproduced with permission from Reference [37].

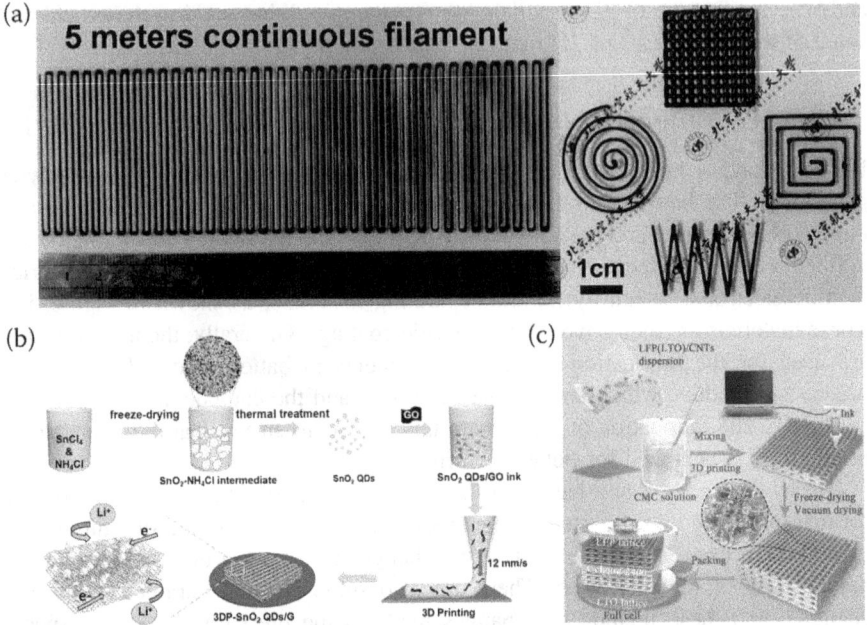

FIGURE 5.9 (a) Images representing the filament printed by 3D printing and also represent the patterns that are fabricated by the use of SnO2 QDs/GO ink. (b) Representation of the SnO2 QDs/G electrodes with micro-lattice architectures. (c) 3D printing process of a full LIMB with LFP and LTO electrodes having the architecture of microlattice. Reproduced with permission from Reference [38].

thickness, and the dimensions of the electrode that is being printed [38]. The printed filament and methodology are shown in Figure 5.9.

Apart from single printing of electrodes, there are also 3D printed electrodes that are used, and these provide better properties as compared to their 2D counterparts. There is also an advantage of high loading density and the short distance of diffusion that is important in this case and is provided by the 3D versions of the electrodes that are available [39].

The conventional batteries that are used have electrolytes in which there is a problem of safety and leakage. These problems are caused because there is always a chance of leakage and safety in case of batteries where there is use of electrolytes.

The problems just discussed can be easily solved when extrusion-based electrolytes are used. The advantage of these printers is also that they provide safety and have excellent electromechanical properties.

Based on these fundamentals, several 3D printed batteries have been developed. The work done by Li et al. [40] shows the use of 3D printing for the development of electrodes for a Li ion battery. The electrodes that were fabricated were the interdigitated electrodes, and they were fabricated by using a layer-to-layer technology. The direct printing involved the use of 3D printer for printing electrode and cathode.

For a battery electrode to be printed, the importance of electrical conductivity is paramount, and therefore, there is need to use inks in which there is better electrical

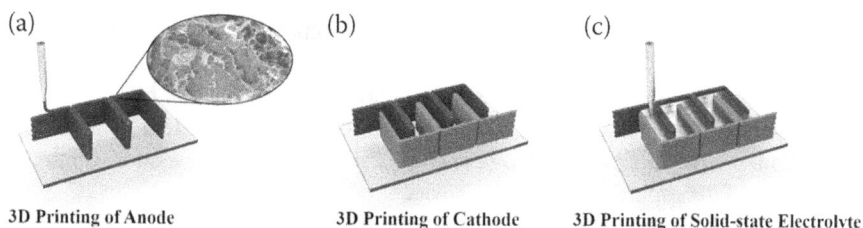

FIGURE 5.10 Interdigitated electrodes obtained by 3D printing. (a) Use of LTO/GO ink for anode fabrication by the use of layer-by-layer printing method. Porous electrode shown in the inset image by SEM. The porous nature of graphene oxide sheets is clearly visible. (b) The printing of cathode structure represented by LFP/GO ink. The development of interdigitated architecture is achieved by printing cathode and electrodes. (c) Insertion of the composite ink between the channel of the annealed electrodes. Reproduced with permission from Reference [41].

conductivity and better performance. The work done by Hu et al. shows the use of reduced graphene oxide for the improvement of electrical conductivity of the electrodes of the battery.

The work done in this case shows the use of aqueous graphene oxide where these inks were used to ensure that the Li-ion battery electrodes are designed. The work done shows the development of reduced graphene oxide inks, which were composed using the sheets of graphene that were concentrated and also were made using active electrode materials. In this fabrication process, water is used as a solvent where it plays a role of a nonflammable solvent, which is also nontoxic and provides the service of low manufacturing cost as well as simplicity into the manufacturing process. The use of GO is also important in this case as it promises higher surface area and also helps in the loading of other materials on the electrodes. They also have the advantage of providing better stress management by the electrode as the sheet of GO is used in the fabrication process. The accommodation of electrodes into compact space is also promised by the use of this technology, and therefore, these methods are being used extensively nowadays [41]. The 3D printed anode and cathode are shown in Figure 5.10.

5.4.2 EXTRUSION BASED PRINTING FOR ELECTROCHEMICAL CAPACITORS

The power density and higher cyclic stability that is provided by the electrochemical capacitors make them a better option to be used as compared to the solid state capacitors. The popularity of the electrochemical capacitors is also increasing these days as they can be used as a replacement of batteries and also the electrolytic capacitors that are used nowadays. The electrodes used in the electrochemical capacitors have the advantage that they can be fabricated using 3D printing approaches that are based on extrusion [42]. The 3D printing techniques can be used for the programming of microelectrodes that are used in these capacitors. The materials that can be used in the electrodes fabricated using the 3D printing technology can be different, and therefore, the electrochemical capacitors that are achieved in this way are patterned and are more symmetrical in their design [43].

The work done by Sun et al. shows the use of symmetrical behavior in which the extrusion-based printing is used and printed by using 3D printing. The work shows that how microelectrodes can be used for this process and what can be achieved by using graphene-based microelectrodes. The method proposed by them showed the fabrication of interdigitated inks based on the microelectrodes and which the symmetrical fabrication of micro-super capacitors is achieved.

The electromechanical results that were conducted show how important the use of 2D materials in the electrodes was since they provide ideal capacitance as the thickness is at atomic level.

The flexibility properties of the electrodes that are fabricated this way are also good, and there is also an advantage of the control of electrode layers as the layer's thickness can be controlled to get the better control of the capacitance that is needed. As the properties of power and density are high in this case, there is high usage of this electrochemical behavior as there are graphene oxide films that are used [44]. The work by Jiang et al. [45] showed how the electromechanical capacitors are used in the energy storage devices. Here, these devices are fabricated by higher power consuption.

5.4.3 EXTRUSION PRINTED SENSING DEVICES

The 2D materials are considered to be ideal materials for use in sensors and sensors-based applications as they offer high aspect ratio and the functional groups present at their surface are also high. The sensitivity of these devices is also high, which makes them ideal for use in certain applications. There are also various additive manufacturing techniques that are used, and each of the techniques has an advantage that they provide unique fabrication of the 2D nanomaterials into desired architectures, which allows for the manufacturing of different types of sensors using 2D materials. These sensors include the temperature sensors, vapor sensors, and also gas sensors [46].

5.4.3.1 Temperature Sensors

2D materials are also very popular in applications where high sensitivity, quick response, and stability of temperature is required for measurement. For the fabrication of high-sensitivity temperature sensors, the use of graphene-based inks is very popular; the ink based on graphene nano platelets is composed and is used as a stretchable matrix in situations where the ink writing is done to make different structures like hexagonal structures and composites that require high sensitivity [47]. The proposed methodology is shown in Figure 5.11.

5.4.3.2 Chemical Sensors

The work done by Kim et al. [48] shows the use of graphene oxide ink for the fabrication of nanowires, which are grown by a meniscus guided approach. The printed GO wires are adjusted accordingly and are then converted into reduced graphene oxide. This conversion is achieved by the use of chemical as well as thermal treatments. The structures can be then molded accordingly, depending on the direction and the location. The use of these reduced graphene oxide nanowires

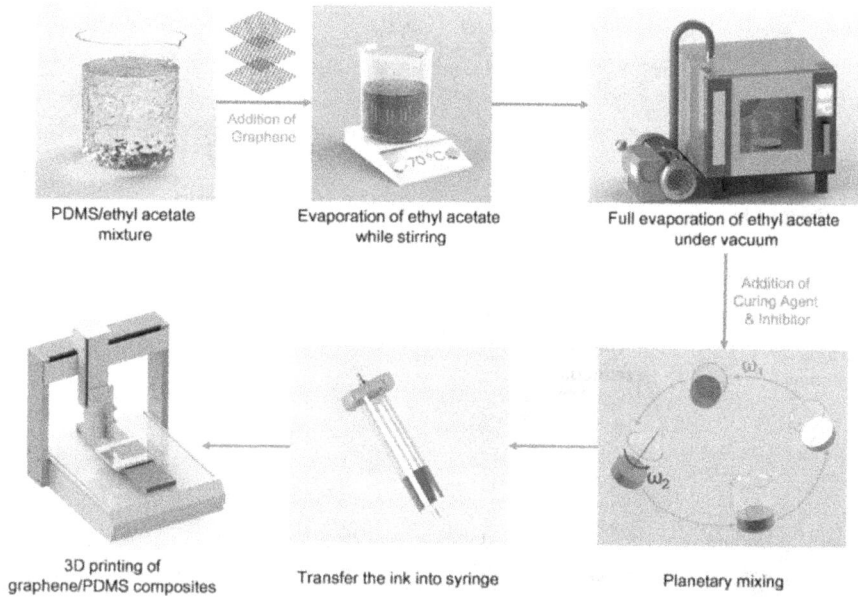

FIGURE 5.11 Process of fabrication of composites of PDMS and graphene. Reprinted (adapted) with permission from reference [11]. Copyright [2019] American Chemical Society.

can then be done in transducer applications where they are used as 3D transducers and are able to operate after the detection of CO2. The electrical conductivity of the reduced graphene oxide is dependent on the concentration of CO2, and therefore, the current flow is detected as the charge carriers change their locations. The electron transfer happens between the rGO and the CO2, which is the reason of increased electrical conductivity.

5.4.3.3 Strain Sensors

The strain sensors, which are wearable and bendable, have various applications in electronics and photonic devices. The work done by An et al. [49] showed the fabrication of an aerogel, which is based on graphene and has high sensitivity and conductivity. The fabrication was done as the solution of graphene oxide was prepared, and for that, extrusion printing was used. After the printing was done, the structures of graphene oxide were subjected to freezing and then were freeze dried. These GO structures were then sandwiched between the layers of PDMS and were also chemically reduced. The microstructure that is achieved of the aerogel of graphene can then be adjusted, and for that, there is tuning done of the concentration of GO that was available initially.

In another study that was performed by Zhang et al. [50], there was a combination that was made of silica and the GO solution. This 3D printing methodology is shown in Figure 5.12. This was done to ensure that the thinning and the stress behavior is improved for the ink that is used. The ink that was prepared was then

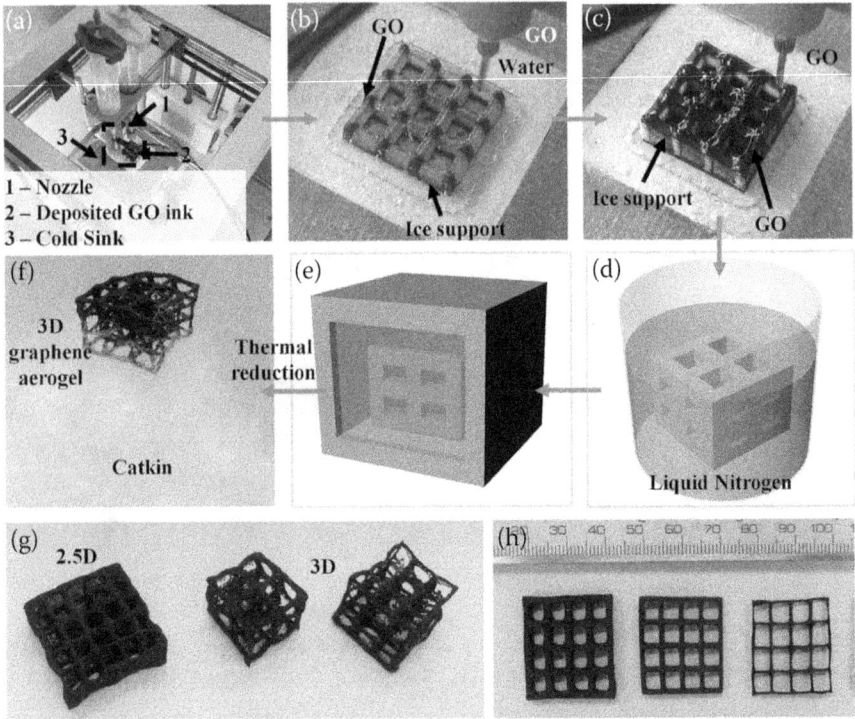

FIGURE 5.12 Graphene aerogel (GA) made by 3D Printing. The process of 3D printing of graphene aerogel (GA). (a) Setup used for 3D printing. (b) Printing of ice support by 3D printing method. (c) GO suspension made by 3D printing. (d) Immersion of iced structure after printing into liquid nitrogen. (e) Process of freeze drying. (f) Reduction of GA on Catkin. (g) 3D structure of GA. Right: Overhang structures based on 3D architecture. Left: 2.5 structure. (h) Gas represented with different thicknesses. Reproduced with permission from Reference [50].

printed, and 3D macroscopic structures were then achieved. The printing was done within an organic bath to ensure that the water solution does not evaporate.

The drying of the printed ink of GO was then done and, after that, the reduction process was done to ensure that the viscosity is increased. The resistance of the printed gel of rGO in 3D form was done and, upon observation, it was observed that the aerogel that is printed is highly sensitive to any sort of deformation. This demonstrates the strain measurement properties of the rGO aerogel and also indicates that the recoverability properties are better as the strain is loaded and unloaded.

This it is clear that the 3D printing of electronic devices based on extrusion can be very beneficial as the interaction is increased between the humans and the designed devices. There is however a lot of room for improvement as the technology is still in very early stages, and it has still a long way to go before it is implemented fully for the achievement of 3D printed nanostructures based on 2D materials. If these electronic devices are fabricated properly, there is a wide range in which these devices can be used, including fields like bioelectronics, optoelectronics, and nanotechnology.

5.5 CONCLUSION

3D printing is definitely the future of material fabrication as there has been significant progress made in this field in the past few years. The customization that is offered by 3D printing technology is also very high, which gives the indication that the range of the material and structures to be printed can be changed from nanometers range to centimeters range.

This can also help to introduce new methods of printing where the inkjet methods are used, and based on them, the fabrication is done in which the methods like curving, rolling, and planar fabrication are used. There can also be other fabrication methods that can be introduced, which include the algorithms derived from real-world techniques. These include the folding of printed 3D structures in the form of structures in which there is use of chip-based structures for the 3D assembly of materials.

From the literature study of the work done on 2D materials, it can be concluded that the technology of 2D materials based 3D printing is still in its very early stages, and there is a long way to go for this technology as more and more innovative techniques are introduced into the world of design engineering. This technology can be further extended to design other 2D materials and their heterostructures for the applications of the futuristic technologies that will be developed based on 3D printing. This will also introduce new and exciting features into the newly developed 2D materials and will open new gates of research and innovation. The further analysis of materials based on 2D materials will open new doors for structural analysis and will help in developing a cost-effective method for the design of materials in which there are ultra-strong properties and the performance is also better like the ultra-strong materials such as graphene.

REFERENCES

1. Lipson, H., & Kurman, M. (2013). *Fabricated: The new world of 3D printing.* John Wiley & Sons.
2. Redwood, B., Schöffer, F., & Garret, B. (2017). *The 3D Printing Handbook: Technologies, Design and Applications.* 3D Hubs.
3. Shahrubudin, N., Lee, T. C., & Ramlan, R. (2019). An overview on 3D printing technology: Technological, materials, and applications. *Procedia Manufacturing, 35,* 1286–1296.
4. MacDonald, E., & Wicker, R. (2016). Multiprocess 3D printing for increasing component functionality. *Science, 353*(6307): 1–10.
5. Chen, J., Zhang, K., Zhang, K., Yang, L., & Jiang, B. (2021). The research progress in recording layer of the inkjet printing materials. *Journal of Applied Polymer Science, 138*(35), 50894.
6. Lai, P. L. (2021). The E-jet printing of organic inks on organic substrates. thesis. DOI: https://doi.org/10.31274/etd-20210609-96 URI: https://dr.lib.iastate.edu/handle/20.500. 12876/Nr1V5Qoz
7. Wang, M., Huang, M., Luo, D., Li, Y., Choe, M., Seong, W. K., … & Ruoff, R. S. (2021). Single-crystal, large-area, fold-free monolayer graphene. *Nature, 596*(7873), 519–524.
8. Tran, T. S., Dutta, N. K., & Choudhury, N. R. (2018). Graphene inks for printed flexible electronics: graphene dispersions, ink formulations, printing techniques and applications. *Advances in Colloid and Interface Science, 261,* 41–61.

9. Parvez, K., Worsley, R., Alieva, A., Felten, A., & Casiraghi, C. (2019). Water-based and inkjet printable inks made by electrochemically exfoliated graphene. *Carbon, 149*, 213–221.

10. Zhong, Q., Li, Y., & Zhang, G. (2021). Two-dimensional MXene-based and MXene-derived photocatalysts: Recent developments and perspectives. *Chemical Engineering Journal, 409*, 128099.

11. Zhang, C. J., McKeon, L., Kremer, M. P., Park, S. H., Ronan, O., Seral-Ascaso, A.,... & Nicolosi, V. (2019). Additive-free MXene inks and direct printing of micro-supercapacitors. *Nature Communications, 10*(1), 1–9.

12. Abdel Maksoud, M. I. A., Bedir, A. G., Bekhit, M., Abouelela, M. M., Fahim, R. A., Awed, A. S., ... & Rooney, D. W. (2021). MoS2-based nanocomposites: Synthesis, structure, and applications in water remediation and energy storage: A review. *Environmental Chemistry Letters, 19*(5), 3645–3681.

13. Mintz, L. (2017). *Molybdenum Disulphide (MoS2) Nanosheet Inks Evaluated for Printed Electronics and Application to Thin-Film Transistors* (Master's thesis, University of Waterloo).

14. Tan, W. C., Wang, L., Feng, X., Chen, L., Huang, L., Huang, X., & Ang, K. W. (2019). Recent Advances in Black Phosphorus-Based Electronic Devices. *Advanced Electronic Materials, 5*(2), 1800666.

15. Hu, G., Albrow-Owen, T., Jin, X., Ali, A., Hu, Y., Howe, R. C., ... & Hasan, T. (2017). Black phosphorus ink formulation for inkjet printing of optoelectronics and photonics. *Nature Communications, 8*(1), 1–10.

16. Zhang, K., Feng, Y., Wang, F., Yang, Z., & Wang, J. (2017). Two dimensional hexagonal boron nitride (2D-hBN): synthesis, properties and applications. *Journal of Materials Chemistry C, 5*(46), 11992–12022.

17. Petrone, N., Chari, T., Meric, I., Wang, L., Shepard, K. L., & Hone, J. (2015). Flexible graphene field-effect transistors encapsulated in hexagonal boron nitride. *ACS Nano, 9*(9), 8953–8959.

18. Goh, G. L., Zhang, H., Chong, T. H., & Yeong, W. Y. (2021). 3D printing of multilayered and multimaterial electronics: A review. *Advanced Electronic Materials, 7*(10), 2100445.

19. Xu, W., Jambhulkar, S., Zhu, Y., Ravichandran, D., Kakarla, M., Vernon, B., Lott, D. G., Cornella, J. L., Shefi, O., Miquelard-Garnier, G. and Yang, Y., 2021. 3D printing for polymer/particle-based processing: A review. *Composites Part B: Engineering, 223*, p.109102.

20. Zhou, S., Usman, I., Wang, Y., & Pan, A. (2021). 3D printing for rechargeable lithium metal batteries. *Energy Storage Materials. 38*(2021), 141–156, https://doi.org/10.1016/j.ensm.2021.02.041

21. Bekas, D. G., Hou, Y., Liu, Y., & Panesar, A. (2019). 3D printing to enable multifunctionality in polymer-based composites: A review. *Composites Part B: Engineering, 179*, 107540.

22. Siddiqui, G. U., Rehman, M. M., Yang, Y. J., & Choi, K. H. (2017). A two-dimensional hexagonal boron nitride/polymer nanocomposite for flexible resistive switching devices. *Journal of Materials Chemistry C, 5*(4), 862–871.

23. Hossain, R. F., & Kaul, A. B. (2020). Inkjet-printed MoS2-based field-effect transistors with graphene and hexagonal boron nitride inks. *Journal of Vacuum Science & Technology B, Nanotechnology and Microelectronics: Materials, Processing, Measurement, and Phenomena, 38*(4), 042206.

24. Sun, B., & Huang, X. (2021). Seeking advanced thermal management for stretchable electronics. *npj Flexible Electronics, 5*(1), 1–5.

25. Lei, I. M., Jiang, C., Lei, C. L., de Rijk, S. R., Tam, Y. C., Swords, C., ... & Huang, Y.Y.S. (2021). 3D printed biomimetic cochleae and machine learning

co-modelling provides clinical informatics for cochlear implant patients. *Nature Communications*, *12*(1), 1–12.

26. Guiney, L. M., Mansukhani, N. D., Jakus, A. E., Wallace, S. G., Shah, R. N., & Hersam, M. C. (2018). Three-dimensional printing of cytocompatible, thermally conductive hexagonal boron nitride nanocomposites. *Nano Letters*, *18*(6), 3488–3493.

27. Yang, M., Liu, M., Cheng, J., & Wang, H. (2021). A movable type bioelectronics printing technology for modular fabrication of biosensors. *Scientific Reports*, *11*(1), 1–9.

28. Orangi, J., Hamade, F., Davis, V. A., & Beidaghi, M. (2019). 3D printing of additive-free 2D Ti3C2T x (MXene) ink for fabrication of micro-supercapacitors with ultra-high energy densities. *ACS Nano*, *14*(1), 640–650.

29. Brown, E., Yan, P., Tekik, H., Elangovan, A., Wang, J., Lin, D., & Li, J. (2019). 3D printing of hybrid MoS2-graphene aerogels as highly porous electrode materials for sodium ion battery anodes. *Materials & Design*, *170*, 107689.

30. Liu, J., Li, W., Guo, Y., Zhang, H., & Zhang, Z. (2019). Improved thermal conductivity of thermoplastic polyurethane via aligned boron nitride platelets assisted by 3D printing. *Composites Part A: Applied Science and Manufacturing*, *120*, 140–146.

31. Liang, Z., Pei, Y., Chen, C., Jiang, B., Yao, Y., Xie, H., ... & Hu, L. (2019). General, vertical, three-dimensional printing of two-dimensional materials with multiscale alignment. *ACS Nano*, *13*(11), 12653–12661.

32. Chandrasekaran, S., Cerón, M. R., & Worsley, M. A. (2021). Engineering the Architecture of 3D Graphene-based Macrostructures. In Graphene-based 3D Macrostructures for Clean Energy and Environmental Applications, 2021, pp. 1–40. DOI: 10.1039/9781839162480-00001

33. de la Osa, G., Perez-Coll, D., Miranzo, P., Osendi, M. I., & Belmonte, M. (2016). Printing of graphene nanoplatelets into highly electrically conductive three-dimensional porous macrostructures. *Chemistry of Materials*, *28*(17), 6321–6328.

34. García-Tuñon, E., Barg, S., Franco, J., Bell, R., Eslava, S., D'Elia, E., ... & Saiz, E. (2015). Printing in three dimensions with graphene. *Advanced Materials*, *27*(10), 1688–1693.

35. Marzo, A. M. L., Mayorga-Martinez, C. C., & Pumera, M. (2020). 3D-printed graphene direct electron transfer enzyme biosensors. *Biosensors and Bioelectronics*, *151*, 111980.

36. Aki, D., Ulag, S., Unal, S., Sengor, M., Ekren, N., Lin, C. C., ... & Gunduz, O. (2020). 3D printing of PVA/hexagonal boron nitride/bacterial cellulose composite scaffolds for bone tissue engineering. *Materials & Design*, *196*, 109094.

37. Mensing, J. P., Lomas, T., & Tuantranont, A. (2020). 2D and 3D printing for graphene based supercapacitors and batteries: A review. *Sustainable Materials and Technologies*, *25*, e00190.

38. Yang, Y., Yuan, W., Zhang, X., Yuan, Y., Wang, C., Ye, Y., ... & Tang,Y. (2020). Overview on the applications of three-dimensional printing for rechargeable lithium-ion batteries. *Applied Energy*, *257*, 114002.

39. McOwen, D. W., Xu, S., Gong, Y., Wen, Y., Godbey, G. L., Gritton, J. E., ... & Wachsman, E. D. (2018). 3D-printing electrolytes for solid-state batteries. *Advanced Materials*, *30*(18), 1707132.

40. Li, J., Leu, M. C., Panat, R., & Park, J. (2017). A hybrid three-dimensionally structured electrode for lithium-ion batteries via 3D printing. *Materials & Design*, *119*, 417–424.

41. Fu, K., Wang, Y., Yan, C., Yao, Y., Chen, Y., Dai, J., ... & Hu, L. (2016). Graphene oxide-based electrode inks for 3D-printed lithium-ion batteries. *Advanced Materials*, *28*(13), 2587–2594.

42. Ganesh, V., Pitchumani, S., & Lakshminarayanan, V. (2006). New symmetric and asymmetric supercapacitors based on high surface area porous nickel and activated carbon. *Journal of Power Sources*, *158*(2), 1523–1532.

43. Zhang, J., Jiang, J., Li, H., & Zhao, X. S. (2011). A high-performance asymmetric supercapacitor fabricated with graphene-based electrodes. *Energy & Environmental Science*, *4*(10), 4009–4015.

44. Sun, G., An, J., Chua, C. K., Pang, H., Zhang, J., & Chen, P. (2015). Layer-by-layer printing of laminated graphene-based interdigitated microelectrodes for flexible planar micro-supercapacitors. *Electrochemistry Communications*, *51*, 33–36.

45. Jiang, Y., Shao, H., Li, C., Xu, T., Zhao, Y., Shi, G., ... & Qu, L. (2016). Versatile graphene oxide putty-like material. *Advanced Materials*, *28*(46), 10287–10292.

46. Choi, H. W., Zhou, T., Singh, M., & Jabbour, G. E. (2015). Recent developments and directions in printed nanomaterials. *Nanoscale*, *7*(8), 3338–3355.

47. Wang, Z., Gao, W., Zhang, Q., Zheng, K., Xu, J., Xu, W., ... & Liu, Y. (2018). 3D-printed graphene/polydimethylsiloxane composites for stretchable and strain-insensitive temperature sensors. *ACS Applied Materials & Interfaces*, *11*(1), 1344–1352.

48. Kim, J. H., Chang, W. S., Kim, D., Yang, J. R., Han, J. T., Lee, G. W., ... & Seol, S. K. (2015). 3D printing of reduced graphene oxide nanowires. *Advanced Materials*, *27*(1), 157–161.

49. An, B., Ma, Y., Li, W., Su, M., Li, F., & Song, Y. (2016). Three-dimensional multi-recognition flexible wearable sensor via graphene aerogel printing. *Chemical Communications*, *52*(73), 10948–10951.

50. Zhang, Q., Zhang, F., Medarametla, S. P., Li, H., Zhou, C., & Lin, D. (2016). 3D printing of graphene aerogels. *Small*, *12*(13), 1702–1708.

6 SMART Nano-Sensors via 3D Printing Technology

S. Deepak Kumar, G. Arun Manohar, and P. S. V. Ramana Rao
Department of Mechanical Engineering, Centurion
University of Technology and Management,
Vizianagaram, India

A. Mandal
School of Minerals, Metallurgical and Materials Engineering,
Indian Institute of Technology, Bhubaneswar, India

CONTENTS

DOI: 10.1201/9781003189404-6

6.1 INTRODUCTION TO SENSORS

A sensor is basically an electronic device which consists of a recognition element, a transduction element, and a signal processor that may provide chemical or physical information continuously and reversibly [1,2].

A modern definition of sensors includes the following:

- Convert non-electric information into electrical signals
- Have direct contact with the subject matter under investigation
- Respond fast and accurately
- Operate in repeated cycles or continuously
- Be small and cheap

A smart sensor is a device that receives input from the physical world and has built-in or computed capabilities to perform predetermined functions upon detection of specific input, as well as process data before sending it on.

An intelligent sensor is one that performs a specified action when it detects an appropriate input. The two key factors of an Internet of Things (IoT) ecosystem include the internet and the physical devices, such as sensors and actuators. Thus, the primary function of sensors is to collect data from their surroundings.

Sensor is thus defined as, "an input device that produces an output (signal) in relation to a specified physical quantity (input). It's a gadget that transfers signals from one energy domain to another" [3,4].

6.1.1 CLASSIFICATION OF SENSORS

There are a variety of classifications of sensors made by different experts. Some are very simple, and some are very complex [5,6].

A. Classification of sensor on the basis of their location:
   (a) Internal sensors and (b) External sensors
   a. **Internal sensors:** employ feedback information to determine the current state.
      These sensors are used to examine and control the positions and velocities of various joints of the robot [7,8].
      Example: The location of the robot arm is controlled using potentiometers and by means of optical encoders. Tachometers are employed to control the speed of a robot arm.
   b. **External sensors:** are frequently visible since they are physically mounted on the robot or on process equipment in the robot cell. External sensors are utilized to control the robot's interactions with the other equipment in the work cell.
      Examples: micro-switches; touch and tactile sensors; photoelectric devices, etc.
B. Classification of sensor on the basis of their physical activation.
   (a) Contact sensors and (b) Non-contact sensors [9,10]
   a. **Contact (or Tactile) sensors:** are devices used in robotics to capture data on physical contact between a manipulator hand and objects in the workspace.
      The two main types are (i) touch sensors and (ii) force sensors.
      i. *Touch sensors:* also called binary sensors, are used to indicate whether contact has been made between two objects or not.
         Examples: Limit switches, micro-switches, etc.
      ii. *Force sensors:* are utilized to signal not only whetherany contact with has been formed with the object or not. They also identify the size of the contact force.
   b. **Non-contact sensors:** They measure the state of an object/thing without physically touching it. Non-contact sensors rely on the response of a detector in audio or electro-magnetic radiation.
      Examples: Range sensors; proximity sensors; acoustic sensors, and vision sensors.

6.1.2 NANO-SENSORS

"Nano-Sensors are tiny sensors in the size of a few nanometers about 10 to 100 nm." One can detect the existence of a nanomaterial/ molecules within that size or even smaller [11]. The main properties of nano materials that are utilised to make a viable sensor are: physical and optical properties; properties from high surface to volume ratio. Nowadays, nano-sensors have become more important in several parts of human life as a result of technology advancements. The sensors were manufactured using a variety of production techniques. In several sectors, "additive manufacturing (AM) has recently become a typical technology for fabricating a large variety of technical sensor components" [12]. "This manufacturing route, also known as 3D printing, is based on melting and solidification, resulting in the manufacture of a component with high dimensional accuracy and a homogeneous surface finish." Since accuracy and well-designed procedures are required in the manufacture of sensors, AM has been used in the creation of sensor parts in recent years [13,14].

In this chapter, we summarised and categorised the uses of several AM processes in the production of sensor components. Further, a comparison of AM processes and classification of 3D-printed sensors depending on their uses and applications is demonstrated. Furthermore, the production of sensors using AM is presented in depth, and the challenges and future prospects of this technology are elaborated. The findings of the current investigation reveals that printing quality with high resolution and faster speed with good efficiency are achieved, which leads to significant advancement in the manufacturing of 3D-printed sensors [15,16]. The information supplied here can be used not only for comparing, improving, and optimising production methods, but also for further research studies in the creation of very sensitive sensors.

6.2 OVERVIEW OF ADDITIVE MANUFACTURING

AM has grown into a multibillion-dollar industry involving a variety of materials and processes all around the world. "It is also known as Rapid Prototyping (RP) or 3D printing and it is described as the process of connecting materials to generate components from 3D model data, generally layer by layer, as opposed to subtractive and formative manufacturing techniques" [17,18]. "International Organization for Standardization (ISO)/American Society for Testing and Materials (ASTM) 52900:2015 standards" AM entail a number of stages, beginning with a virtual CAD model and ending with a real component, and the general AM process is illustrated in Figure 6.1.

1. Create a concept: In most cases, AM procedure begins with the creation of a CAD model in a computer. The majority of 3D parts use solid models with surfacing features.
2. Conversion to STL, "Stereo-lithography": It works by removing all structural data and model history from the surfaces before reducing them to a collection of triangular facets. As a result, STL is essentially a model's surface picture.
3. Upload to a 3D printer and start printing: Once generated, the STL file is delivered immediately to the AM machine.
4. 3D printing machine setup: Most AM machines have some built-in settings that are specific to a given process or machine. Complex situations have default parameters to speed up the AM machine setup procedure and reduce errors.

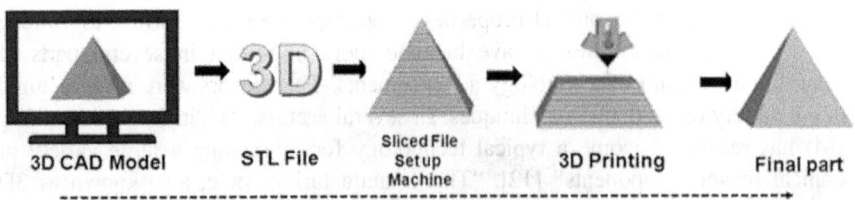

3D CAD Model   STL File   Sliced File Setup Machine   3D Printing   Final part

FIGURE 6.1 Steps in 3D printing process [18–20].

5. Build: The phases of the AM process are semi-automated and need a lot of manual control and decision-making. Until the build-up model is generated, the machines require a follow-up in organization of controlling the layer thickness, deposition of material, followed by creating complex cross-sections.
6. Parts removal and cleaning: The AM-output machines should be ready for usage. The components are separated from the excess materials that surround them.
7. Post-processing: This includes steps of finishing parts, sandpapering, polishing or coating.
8. Usage: AM components are ready for use in a variety of industrial applications.

6.2.1 ADDITIVE MANUFACTURING'S (3D PRINTING) POTENTIAL APPLICATIONS

Rapid prototyping is being extensively adopted as a strong tool for product development across a wide range of sectors, with substantial advances made [21].
The following are some of the most common AM applications:

- *Design:* A prototype is required throughout the product design and development phase. Design changes are substantially aided by RP components, which save development time, lower costs, and improve product quality.
- *Manufacturing:* RP models are used to create sand casting and investment casting dies and moulds.
- *Tool and mould production* includes consistent pocket tool holders, die casting moulds, and injection mould tooling.
- *Automobile:* Components for airplanes, luxury sports cars, vintage motorcycles, and other vehicles.
- *Electronics:* Embedded systems in solid materials, such as radio frequency Identification (RFID). Three-dimensional micro-electromechanical structures.
- *Art:* Creates art pieces and incorporates lifelike objects onto the screens.
- *Medical:* RP has just added a new dimension to the medical business. Patients can get customized medical equipment and implants thanks to 3D printed medical items like hearing aids. RP products have also gained popularity for dental applications.
- *Bio-engineering:* Rapid prototyping holds fresh potential for tissue engineering applications and provides a more efficient method of fabrication [22,23].

Originally designed as a rapid prototype/3D printing process, additive manufacturing (AM) today covers a wide variety of production applications in virtually every sector, ranging from tooling to mass customisation.

6.2.2 METHODS OF ADDITIVE MANUFACTURING FOR NANO-SENSOR FABRICATION

Based on ASTM standards, AM processes are classified into seven groups, which are demonstrated in Figure 6.2. The seven AM families are briefly described below [24,25]

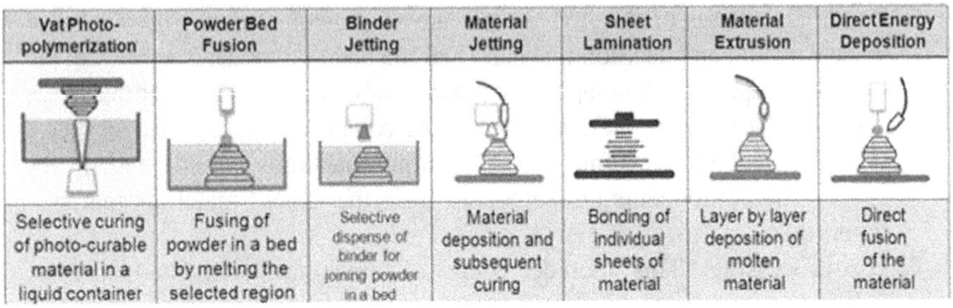

Vat Photo-polymerization	Powder Bed Fusion	Binder Jetting	Material Jetting	Sheet Lamination	Material Extrusion	Direct Energy Deposition
Selective curing of photo-curable material in a liquid container	Fusing of powder in a bed by melting the selected region	Selective disperse of binder for joining powder in a bed	Material deposition and subsequent curing	Bonding of individual sheets of material	Layer by layer deposition of molten material	Direct fusion of the material

FIGURE 6.2 Families of additive manufacturing technologies [5,19].

- Vat photo polymerization (VP): This technology employs a liquid resin that solidifies when exposed to ultraviolet light. The laser cures photosensitive resin nearly point-by-point and fabricates the desired item from a CAD file.
- Powder bed fusion (PBF): A thermal laser source is employed to promote fusion between the powder particles. After scanning the layer, powder material is dispersed over it in this fabrication process.
- Binder jetting (BJ): This process involves applying an adhesive to a powder bed to create the layers of a component. As a result, a fresh powder layer is distributed. Finally, the constructed part is post-processed.
- Material jetting (MJ): Here, the raw material is blasted onto the platform in order to solidify and manufacture the model layer upon layer. Finally, the part and the support material is removed from the building platform.
- Sheet lamination (SL): In this approach, a 3D item is created from a thin sheet of raw material. Thin sheets of material are bonded together layer by layer to build a part, which is then cut into the desired shape.
- Material extrusion (ME): This is a technique in which raw material (for example, poly-lactide (PLA)) is forced out through a nozzle by the application of pressure. Layers are formed by extruding molten material onto a substrate.
- Directed energy deposition (DED): In this technique, in order to melt the raw materials, a concentrated energy source is used. After the first layer is deposited, the nozzle and energy source are adjusted to deposit the second layer [26–28].

3D-printed sensors are comprised of both printed and non-printed components. Table 6.1 summarises the some information on 3D-printed sensors manufactured using various AM methods.

6.3 CLASSIFICATION OF 3D-PRINTED NANO-SENSORS

3D-printed sensors are classified based on their engineering and medical applications. The three broad categories of 3D-printed nano-sensors include: physical nano-sensors, bio-nano-sensors, and chemicalnanosensors [29]. In the current work,

TABLE 6.1

Sensors with 3D Printed and Non-printed Parts

AM Process	Parameter Measured	Materials	3D Printed Parts	Non-printed Parts	Reference
VP	Viscosity	Transparent resin	Micro viscometer	Container pump	[27, 28]
	Gas	Acrylic resin	Gas cell		
PBF	Gas	Metallic particle	Housing	Electrode housing	
	pH	ceramic	Electrode		
BJ	Strain	Nano particle silver	Sensor patches	Silicon layer	
	Force	ceramic	Sensing part	Fiber cladding	
MJ	Pressure	Steel foil	Light travels	Accelerometers	
	pH	Photo resin	Platform	Micro controller	
SL	Temperature	Copper	Sensing part	Housing	
	Strain	ceramic	Sensing part	Ceramic shield	
ME	Liquid	Polyethylene	Platform	Micro pipette	
	pH	ABS	Fluidic device	Electrode	

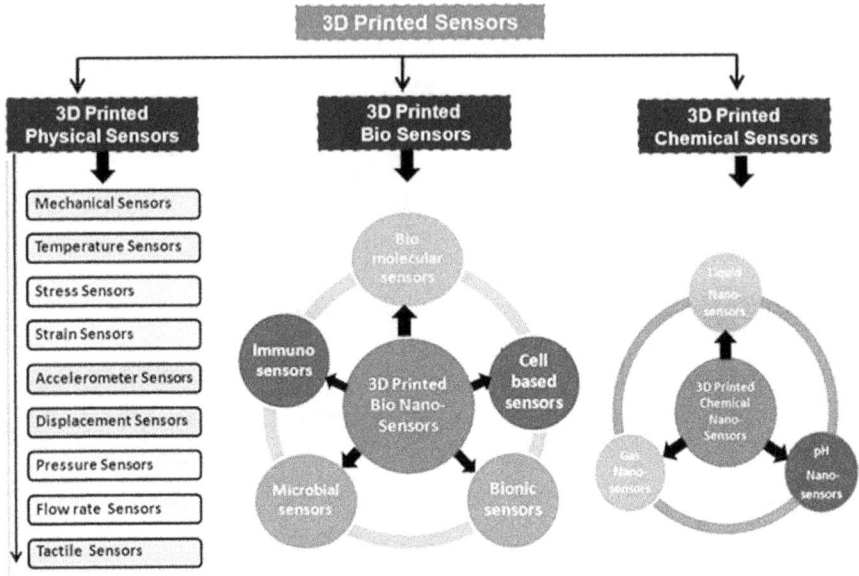

FIGURE 6.3 Classification of 3D-printed sensors [30–32].

various 3D-printed sensors are classified based on their applications and are illustrated in Figure 6.3.

6.3.1 3D-PRINTED PHYSICAL NANO-SENSORS

Physical sensors detect variations induced by physical effects and transform them into electrical signals like capacitance, resistance, inductance, current, or voltage.

3D printing physical sensors offers various advantages, such as unique shapes for fitting with commercial items like electrodes, textiles, or built with human parts like hand or knee joints. Physical sensors can attain high resolution or detecting sensitivity [30–32]. The applications of physical sensors that sense the mechanical quantities such as temperature, radio frequency (RF) harvesting sensing, particle, tactile sensing, gas concentration, etc., are illustrated in Figure 6.4.

6.3.1.1 Mechanical Nano-Sensor

Mechanical quantity sensors are capable of detecting mechanical quantities such as stress, strain, displacement, acceleration, flowrate, liquid viscosity, scale or angular position, among others. These mechanical quantities can subsequently be transformed into variations in resistance or capacitance. Many varieties/types of sensors have been created using 3D printing techniques, depending on various mechanical characteristics.

6.3.1.2 Strain Sensors

Strain sensors (strain gauges) convert tensile or compressive strains into electrical signals. They have been employed in a variety of scientific activities. Several materials, like nano particles, grapheme, and nanotubes, are being used as building

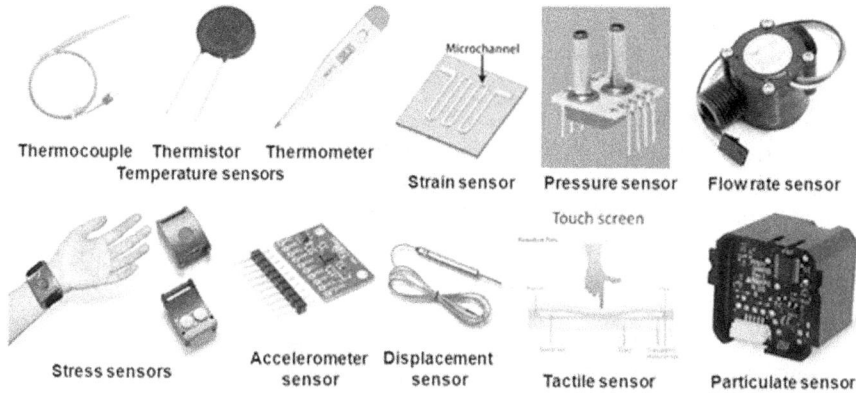

FIGURE 6.4 3D-printed physical sensors.

blocks in 3D-printed strain sensor inks. In short, using a smaller nozzle and a faster printing speed results in a thinner 3D-printed strain sensor.

6.3.1.3 Accelerometer Sensor

An accelerometer sensor is a device that measures the acceleration of any body or object while it is at rest. It does not indicate a coordinate acceleration. Accelerometer sensors are used in a wide range of applications, such as various electrical appliances, smart phones, and wearable devices.

Accelerometer sensors are employed in biomedical applications and are mostly used in step counting and activity monitoring and suppression.

6.3.1.4 Displacement Sensor

This category, an inkjet printing technique was applied, which falls under the category of material jetting. Displacement sensors thus created were used to analyze the displacement of a huge target as well as a little steel ball.

6.3.1.5 Force Sensors

Force sensors have been used in a variety of applications. They have been utilised in medical instruments, logistics, automobile industry, robotics, and other applications. Force sensors transform applied forces into electrical signals. This type of sensor typically consists of three major components: a transducer, flexure, and packaging. The shape, size, and mechanical qualities of a flexure influence the accuracy, sensitivity, and directional response of a force sensor. Because 3D-printed force sensors can be integrated into the system, a solid stand contact between the sensor and the system may not be required. Various pressure sensors can measure and record pressure fluctuations, structural stresses, and a desired pressure.

6.3.1.6 Stress Sensors

Customized 3D-printed main parts paired with mass-produced electrodes for stress monitoring can reduce production costs while boosting production flexibility. The change of three capacitors, which were formed by each paired electrode, can be employed to detect the stress of each direction.

6.3.1.7 Flow Rate Sensors

In comparison to traditional manufacturing processes, 3D printing technology used in the manufacture of flow sensors has the advantage of not only putting the least amount of material where it is particularly needed, but also creating a novel shape device required for small batch research. The average rotation difference between the 3D-printed flow sensor and the commercial flow sensor is 2.2 Hz.

6.3.1.8 Temperature Sensors

3D printing technology is used to build a glove-shaped structure for fitting a hand as a sensing device. Fabrication of a personalised glove is coupled with a liquid-state temperature sensor as a wearable warming device that can be adjusted to the patient's hand and comfortably warmed for an extended period [33]. This 3D-printed glove may fully contain solid-state components such as integrated circuit (IC) chips and others, liquid-state circuit components such as an inbuilt programmable heater and a temperature sensor, and liquid metal interconnects. In A flexible temperature sensor was created using inkjet-based 3D printing. The printed sensor was made with composed silver and then placed on a polymeric substrate.

6.3.1.9 Particle Sensors

Particle pollution, particularly micrometre particles, may cause harmful diseases. Hence, detecting airborne particulate in the atmosphere is essential. The sophisticated construction, which incorporates fine channels and integrates sensors, may be constructed using 3D printing technology for detecting fine particles. Particulate matter sensors detect and count particles in a particular environment using the light-scattering method. 3D printing was used to create a small system for detecting airborne particulate particles. The virtual impact or has two primary flow channels and one minor flow channel for separating particles into two groups: small and large particle streams. Particles with diameters of 2.5 μm or less were injected into two principal flow channels and absorbed by the surface of the QCM resonant sensor.

6.3.1.10 Tactile Sensors

These sensors are used as human interface devices (HIDs) to detect touched and pressed materials. This class of sensors simulates the sense of touch and can be utilised as a human interface device, such as a touch screen [34]. Fused deposition modelling (FDM) technology can be utilised to create a fully enclosed capacitive sensor capable of differentiating between three metallic materials, or acting as HIDs. The performance is determined by the qualities of the sensing materials. 3D printed materials of multi-walled carbon nanotubes (MWNTs) exhibit strong electrical qualities and can be used to make touch sensors.

6.3.2 3D-Printed Bio-NanoSensors

Bio sensors utilize biochemical mechanism for the recognition system. To address the demands of the health care sector, many sensors have been used in biology research. Furthermore, biosensors can be used to replace human body parts by incorporating living cells and electronic components as a bionic sensor [35]. Biosensors also find applications in food industry, medical research, and the marine sector. Biosensors detect biological mechanisms by combining classical recognition elements such as enzymes, antibodies, nucleic acids, and whole cells with recognition elements, such as aptamers, MIPs, phages, and affibodies. 3D-printed biosensors have several advantages in bio sensing, such as increased sensing sensitivity and production of tiny structures. Different AM processes are used in the manufacturing of biomedical sensors. The main domains of bio-nanosensor sare discussed below.

6.3.2.1   Biomolecular Sensors

Bio molecule sensors are categorised as enzyme sensors and DNA sensors. DNA sensors were made using 3D printing techniques. In this case, the VP approach was employed to create DNA-coated graphite electrodes [36]. Lactate and glucose are appealing indicators that provide information on energy metabolism; thus, the development of bio molecule based sensors to assess lactate and glucose levels has progressed in biomedical applications.

6.3.2.2   Immunosensors

Immunosensors are based on antigen and antibody responses that are very selective and sensitive. In 3D-printed immunosensors, biomarker proteins, pathogenic microorganisms, or viruses can be employed as antibodies. The intensity of immunological response can be detected by luminescence intensity, optical density, or fluorescence. A sensitive ECL immunosensor powered by a super capacitor, for example, can be used to detect cancer biomarker proteins on low-cost carbon sensors.

6.3.2.3   Microbial Sensors

In medical applications, an electrochemical device and immobilised microorganisms are being used to create microbial sensors that can detect organic molecules. 3D-printed microfluidic chips are created to investigate the gliding movement of cyanobacteria. 3D-printed microfluidic chips were developed and utilised to detect influenza virus. PLA material was used by the researchers to print such micro chip shaving to an accuracy of 0.1 mm in x and y and 0.08 mm in Z respectively. For an example, a 3D-printed micro fluidic chip functions as a microbial sensor to detect hemagglutinin (HA) in a quick, sensitive, and specific manner. This device is notable for its ability to be integrated with a magnetic pad for the isolation of para-magnetic particles mixed HA, and three electrode system for the electrochemical detection of coding sequence (CdS) quantum dots-labelled vaccine HA. The acquired results demonstrated that 3D printing is a very powerful and resilient tool for the development of sensors in the detection of influenza antigens. Another example is the use of integrated 3D printed fluidic devices equipped with LED sensors and photodiodes to monitor bacterial growth in response to variations in absorption intensity. Micro fluidic devices can be used to reduce observation periods and limit water evaporation.

6.3.2.4   Cell-Based Sensors

Living cells create signals are utilised in medical diagnostics and tailored medication.

Living cells are now used in 3D-printed cell-based sensors, and these form of sensors can be used for a variety of biological detections like imaging of cell density and cytotoxicity.

Because living cells may be used to detect a large range of substances, nowadays, cell-based sensing is one of the upcoming bio-sensor applications.

6.3.2.5   Bionic Sensors

3D printing technology has recently been utilised in the production of bionic sensors. As a result, living cells can be nurtured with electronic components and grown into functional organs. Mannoor *et al.* [37] developed a bionic sensor with improved auditory detecting capabilities for RF reception. As a result, this bionic ear, like normal human ears, has the capability to listen to stereo/audio music, which is depicted in Figure 6.5. Researchers also demonstrated a bionic ear that responds to frequency changes, which is similar to creating bionic ears using 3D printing technology.

6.3.3   3D-PRINTED CHEMICAL SENSORS

6.3.3.1   Liquid Sensors

Nowadays, 3D printing technology can be used to detect fluidic solvents using optical or electrochemical approaches. In comparison to the straight line sensor, a freestanding feature helical sensor with exceptional sensitivity for liquid trapping has been created using 3D printing technology. Liquid intent in a helical shape can sustain for 100 to 360 seconds, whereas liquid embedded in a straight structure can survive for 10 to 40 seconds. It demonstrates that more liquid can be sustained with high sensitivity in a 3D helical structure at the same amount of time. Figure 6.6 depicts the chemical nano-sensors. Accordingly, FDM technology was used to

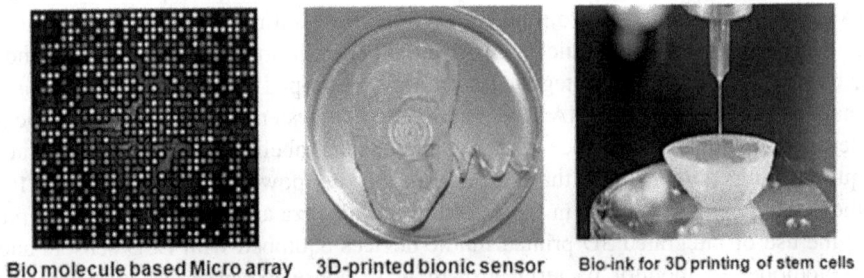

Bio molecule based Micro array   3D-printed bionic sensor   Bio-ink for 3D printing of stem cells

FIGURE 6.5   3D-printed bio-nanosensors [37].

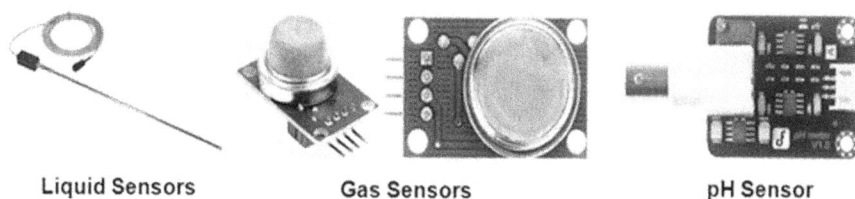

FIGURE 6.6   3D-printed chemical nano-sensors.

develop a fluidic device for detecting H_2O_2. Peak current signals were observed to respond linearly to H_2O_2 concentrations ranging from 100 nm to 20 μm.

6.3.3.2   Gas Sensors

Three electrodes implanted in 3D-printed nanostructures can be used to detect not just liquid concentration or flow rate, but also gas sensing via electrical (current or voltage) changes caused by chemical reaction. A 3D-printed sensor device, for example, is used to measure nitric oxide concentrations with strong linear correlation (7.6–190 μm), and another to measure oxygen tension in a stream of red blood cells. A commercial gas sensor paired with a 3D-printed gas-phase photo catalytic reactor for air pollution monitoring has also been developed [38].

6.3.3.3   pH Sensors

In general, the optical method for detecting pH fluctuation requires some optical components that are not suited for on-site examination. As a result, the camera on a smartphone can serve as an optical detector for simple experimentation and rapid colorimetric analysis. As 3D printing technology is used in pH monitoring, pH fluctuation can be monitored by optical methods, electrochemical methods, or Plasmon rulers. For example, a sample chamber that may be used in conjunction with a smartphone to create a portable fluorometer for measuring the pH of environmental water chemical reaction wares, as we all know, can be employed to monitor the reaction situation. 3D-printed chemical reaction products offer benefits such as custom monolithic features with no liquid leakage, high resolution structure with accurate details, and rapid quantitative detection, making them suitable for monitoring organic and inorganic synthesis.

6.3.4   NANOMATERIALS FOR NANO-SENSORS

Among the nanostructured materials reported for use in nano-sensors were metal, metal oxide, carbon nanotubes, polymers, graphene and biomaterials [39,40].

*   Gold nanoparticles: Researchers have made several efforts to develop gold (Au) based nano-sensors for ecological applications. Au nanoparticles are regarded as excellent electrocatalysts in a variety of electrochemical reactions because of their high stability. Further, gold nanoparticles have

interesting chemical, optical, and catalytic capabilities and are widely used in biological and chemical sensors.

- Silver nanoparticles: Because of their outstanding surface enhanced Raman scattering (SERS) and catalytic characteristics, silver nanoparticles with adjustable dimension and size distribution have picked up the interest of researchers as a typical nanoparticle utilised in biosensors.
- Platinum nanoparticles: They offer excellent catalytic characteristics and have been employed in electrochemical studies.
- Palladium nanoparticles: They are characterized by a wide range of sensor and catalytic applications to biomolecules, gases, and dangerous poisonous compounds.
- Copper: Because of its exceptional stability, excellent electrical conductivity, electrocatalytic capabilities, and inexpensive cost when compared to noble metals such as gold, silver and platinum, copper is an ideal sensing material.
- Metal oxides nanoparticles (MOX): have a diverse set of chemical, electrical, and physical properties that are sensitive to changes in the chemical environment.
- Tin oxides: SnO_2 nanoparticles are one of the most commonly used materials in gas sensors.
- ZnO nanostructures can extract the hidden electrochemical capacity of biomolecules and promote their direct electrochemistry due to their outstanding electron transfer rate.
- NiO nanostructures are p-type semiconductors and are widely used in applications, such as catalysis, battery electrodes, and gas sensors.
- TiO2 nanostructures may be employed on electrochemical sensors for medicinal and pharmaceutical purposes.
- Carbon-based nanomaterials: They have exceptional characteristics, such as high conductivity, stability, and low cost. Carbon nanotubes (CNTs) are one of the most important nanomaterials as nano-sensors because of their unique properties. Many CNTs based environmental nano-sensors serve as scaffolds for immobilization of biomolecules at their surface, and they combine several exceptional characteristics.
- Graphene is a one-of-a-kind two-dimensional nanostructure that allows for fast electron transport. It could be employed in electrochemical sensors and biosensors.
- Polymer nanomaterials: Many efforts have done into developing polymeric nanoparticles technology for the detection of food and environmental pollutants.
- Bio-nanomaterials: The combination of biomolecules' catalytic function and specific properties of nanoscalematerials for developing single and multi-walled carbon nanotubes (MWNTs) of nano-biocomposites provides several opportunities for nano-sensors produced via 3D printing [41–43].
- In comparison to bulk planar devices, nanomaterials in one dimension such as nanowires and nanotubes can be employed as nano-sensors. They can serve as

both transducers and cables for signal transmission. Because of their compact size, they can multiplex individual sensor units in a small device.

6.3.5 APPLICATIONS OF NANO-SENSORS IN ROBOTICS

Sensors are a robot's sensory system (similar to the human system's five senses of touch, sight, hearing, smell, and taste), and they measure external data (such as touch, distance, light, sound, strain, rotation, magnetism, smell, temperature, inclination, pressure, altitude, etc.). Sensors are used in robotics for both internal feedback control and exterior contact with the environment. [6]

"A robot sensor is a device or transducer that detects information about the robot and its surroundings, and transmits it to the robot's controller" [3]. Nano-Sensors have a wide range of environmental applications. One of the primary concerns of environmental authorities is the ability to detect dangerous chemicals and biological organisms in the air and water. Because of its small size, thickness, and measurement accuracy, nano-sensors will revolutionize how air and water quality is measured.

The major advantages of sensors in industrial robots are to:

  i. Identify the position and orientation of component/parts;
 ii. Maintain consistency in product/part quality;
iii. Determine variations in the shape and size of components;
 iv. Identify unknown objects; and
  v. Determine and analyze system malfunctions.

Sensors may be manufactured to fit practically anyplace in consumer devices that detect any motion or applications in robotics, autos, and even human bodies using micro and nano technology. Intelligent sensors are also increasingly being used in counter-terrorism, cargo tracking, and biometrics, among other applications. The most advanced sensors are used in automobiles to detect oncoming collisions and identify the type of airbags to be deployed, as well as the force and speed with which they deploy. MEMS are increasingly being used in medical applications, including as implanted devices and handheld devices for diagnostics and monitoring systems. Looking ahead, with technological breakthroughs, a new generation of sensors, including IoTs and wearables, will revolutionise the electronics sector in the next years.

6.4 SUMMARY AND CONCLUSIONS

In this chapter, comprehensive information on the available nano-sensors is described. Advancements in development of several nano-sensors with numerous applications have demonstrated that 3D printing process plays a major role in the creation of tiny, low-cost, and sensitive and effective nano-sensors. Theoretical understanding and experimental methodologies, as well as recent research, have been examined. In summary, this study demonstrated the fascinating 3D printing routes for the production of various sensors by way of emphasis on applications in engineering and medical fields. Industrial and academic case studies have been addressed that incorporated 3D printing techniques to fabricate various nano-sensors. Based on the studies conducted and case studies examined, the material limitations for each 3D printing process, as well as installation and assembly strategies for 3D-printed nano-sensors, are briefly demonstrated. 3D-printing applications have progressed beyond prototype and into real-world digital manufacturing. Indeed, a wide spectrum of 3D-printed sensors, which includes polymer-based materials to metallic components that meet engineering and medical applications. Optimization of the manufacturing processes can result in faster reaction, superior flexibility, and excellent sensitivity in 3D-printed sensors. Furthermore, the challenges and opportunities that forecast the road ahead is addressed, as well as some suggestions are recommended.

The key challenges and future prospects are outlined below:

6.4.1 CHALLENGES

- Despite the fact that multiple AM techniques are underway in the fabrication of various nano-sensors, there is still scope to improve the technology in sensors. Design methods and printing processes of electronic devices must be coupled in the creation of 3D-printed nano-sensors. This combination necessitates knowledge and competence in electronics and mechanics, as well as material science.
- Many factors play critical roles in the sensor development. Nano-material combination and biometric design, for example, demonstrated their major roles in sensor printing. The major limitations in the manufacture of a 3D-printed sensor should be the material's biocompatibility. The creation of printed implanted sensor devices in the current study by using 3D-printed nano-sensors as an implantable element is currently under research investigation.
- Material limits in any 3D printing method could be a hindrance or challenge in this quick prototyping process. However, available materials for each AM process of different sensor applications should be thoroughly addressed.
- Drift, generating reproducible calibration procedures, and applying pre-concentration and separation processes are all issues for nano-sensors. Integrating the nano-sensor with other components of a sensor package in a reliable manner is also a biggest challenge.
- Product life cycle: In the creation of 3D-printed sensors, it is essential to have knowledge on the engineering components' life cycle. Because 3D-printed nano-sensors are new and have not been studied for many years, there is little information on durability difficulties. Thus, every production stage of a 3D-printed nano-sensor must be examined to assess the product life cycle.
- In this context, an intriguing research project is the material selection and development of innovative materials for the printing of diverse nano-sensors

with high sensitivity and efficiency. Because the qualities of 3D-printed sensors can be varied and optimised by adjusting and optimising printing parameters, future study should thoroughly investigate the potential uses of these 3D printed nano-sensors and their process parameters.

- Variations in printing speed, for example, can cause damage in the line width of the 3D-printed sensor. The employment of advanced technology should be avoided in this regard, as it is a disadvantage in the rapid prototyping process.
- There is a large scope for carrying out research and development in both the technical and medical domains; future applications of 3D-printed sensors have high prospects. Biodegradable sensors materials development, novel ways of printing with greater resolutions and improved adhesive attachment of 3D-printed nano-sensors are examples of study issues that can be addressed as a scope of future research and development.
- Furthermore, nanowaste recycling must be addressed in future studies because nanomaterials are widely used in 3D-printed nano-sensors.

6.4.2 FUTURE PROSPECTS

The following are some suggestions for improving the applications of 3D printing methods in the manufacturing of nano-sensors.

- Design and build portable 3D printers for manufacturing of complex-geometry sensor parts. Further, modification or updating of software, hardware, standards, and tools will lead to substantial saving in time and increase efficiency and productivity.
- Developing and integrating smart software for 3D printing of sensors into complex shapes and surfaces. The smart nano-sensors can be employed in a variety of infrastructures, such as mining equipment, where process planning and project management is facilitated and maintained automatically by the smart nano-sensors.
- Standardization: International standards for the selection of nano materials for 3D printing of nano-sensors especially for medical and bio-medical applications. Efforts should be made to make complete standards and documentation procedures should be implemented for 3D-printed sensors developed for bio-field applications.
- Much research is underway in the development of sensor-based closed-loop control systems for metal 3D printing, which finds a variety of advantages and will lead to fine tune the process parameters in real time.
- Future research should focus on recycling of 3D printed nano-sensors. Indeed, research into the recyclability of discarded 3D-printed nano-sensors is much required to reduce environmental waste and implications, electronic waste, and reduction of manufacturing costs.
- Integrating of artificial intelligence (AI) technology in 3D-printed sensors. Intelligent sensing can be created with AI in various 3D-printed sensors. The collected data can assist AI in making human-like decisions.

- Nano materials-based sensors provide significant advantages over standard materials in terms of sensitivity and specificity, as well as cost and reaction time. Since nano-sensors function on the same scale as normal biological processes, they can have more specificity.
- Nano-Sensors are suitable for high-throughput applications due to their low cost.

Overall, many AM methods continue to be of significant interest. The rapid increase in the number of published research articles in this scientific and emerging domain, as well as the list of topics of interest in today's research programs, demonstrate that 3D printing technology will become more intelligent and flexible in the near future, making it one of the most robust methods for fabricating nano-sensors at various scales. Thus, in order to sustain in today's competitive world, industrial organisations should adopt smart nano-senors and connected technologies to meet the challenges of Industry 4.0.

ACKNOWLEDGEMENT

The authors would like to thank the National Project Implementation Unit (NPIU) India and TEQIP-III for their financial support as part of the Collaborative Research Scheme Project grant [CRS ID: 1-5732031971]. Conflict of Interest: The authors declare that they have no conflicts of interest.

REFERENCES

1. Behera A., Rajak D.K., Hussain P.B. (2021) 3D Printing and Nano-Sensors. In: Thomas S., Nguyen T.A., Ahmadi M., Farmani A., Yasin G. (eds) *Nano-Sensors for Smart Manufacturing: Micro and Nano Technologies*. Elsevier, pp. 183–198. ISBN: 9780128233580. 10.1016/B978-0-12-823358-0.00010-1
2. Kalsoom T., Ramzan N., Ahmed S., Ur Rehman M. (2020) Advances in Sensor Technologies in the Era of Smart Factory and Industry 4.0. *Sensors* 20(22): 6783. 10.3390/s20236783
3. Vamos T. (1988) Automation Production Systems and Computer Integrated Manufacturing: Mikell P. Groover. *Automatica* 24(4): 587. 10.1016/0005-1098(88)90106-9
4. Radhakrishnan P., Subramanyan S., Raju V. (2016) CAD/CAM/CIM, New Age International, 3rd (ed.), p. 673. New Age International.
5. Deepak Kumar S., Dewangan S., Jha S.K., Parida S.K., Behera A. (2021) 3D and 4D Printing in Industry 4.0: Trends, Challenges, and Opportunities. In: Proc of RDMPMC-2020, NIT Jamshedpur, 26–27th August 2020. Bag S., Paul C.P., Baruah M. (eds) *Next Generation Materials and Processing Technologies*. Springer, Singapore 9, pp. 579–587. 10.1007/978-981-16-0182-8_43
6. Trisha, Kumar SD (2020) Design and Development of IoT-Based Robot. In: *International Conference for Emerging Technology (INCET-2020)*, Belgaum, India, IEEE Xplore, pp. 1–4. 10.1109/INCET49848.2020.9154175
7. Abubakr M., Abbas A.T., Tomaz I., Soliman M.S., Luqman M., Hegab H. (2020) Sustainable and Smart Manufacturing: An Integrated Approach. *Sustainability* 12(6): 2280. pp. 1–19. 10.3390/su12062280

8. Kusiak A. (2019) Fundamentals of Smart Manufacturing: A Multi-Thread Perspective. *Annual Reviews in Control* 47: 214–220. 10.1016/j.arcontrol.2019.02.001

9. Poletti A., Treville A. (2016) Nano and Microsensors: Real Time Monitoring for the Smart and Sustainable City. *Chemical Engineering Transactions* 47: 1–6. 10.3303/CET1647001

10. Kumar R. (2018) Smart Micro/Nano Sensors and Their Applications in Intelligent Sensory Network System. *Int J SensNetw Data Commun* 7(1): 1–2. 10.4172/2090-4 886.1000e113

11. Nazari A. (2020) Nano-Sensors for Smart Cities: An Introduction.In: Han B., Tomer V.K., Nguyen T.A., Farmani A., Singh P.K. (eds) *Nano-Sensors for Smart cities: Micro and Nano Technologies.* Elsevier, p. 6. 10.1016/B978-0-12-819870-4.00001-3

12. Ni Y., Ji R., Long K., Bu T., Chen K., Zhuang S. (2017) A Review Of 3d-printed Sensors. *Applied Spectroscopy Reviews* 52(7): 623–652. 10.1080/05704928.2017.12 87082

13. Das S., Arvind P., Chakraborty S., Kumari R., Deepak Kumar S. (2021) IoT Based Solar Smart Tackle Free AGVs for Industry 4.0. In: Misra R., Kesswani N., Rajarajan M., Bharadwaj V., Patel A. (eds) *Proc of Internet of Things and Connected Technologies. (ICIoTCT 2020). Advances in Intelligent Systems and Computing.* Springer, 1382, pp. 1–7. 10.1007/978-3-030-76736-5_1

14. Das S., Kumari R., Deepak Kumar S. (2021) A Review on Applications of Simultaneous Localization and Mapping Method in Autonomous Vehicles. In: Kumar N. Tibor S., Sindhwani R., Lee J., Srivastava P. (eds) *Advances in Interdisciplinary Engineering. Lecture Notes in Mechanical Engineering.* Springer, Singapore, pp. 367–375. 10.1007/978-981-15-9956-9_37

15. Kusiak A. (2018) Fundamentals of Smart Manufacturing. *Int. J. Production Research* 56(1–2): 508–517. 10.1080/00207543.2017.1351644

16. Machado C.G., Winroth M.P., Ribeiro da Silva E.H.D. (2020) Sustainable Manufacturing in Industry 4.0: An Emerging Agenda. *International Journal of Production Research* 58(5): 1462–1484. 10.1080/00207543.2019.1652777

17. Gao W., Zhang Y., Ramanujan D. *et al.* (2015) The status, challenges, and Future of Additive Manufacturing in Engineering. *Computer Aided Design* 69: 65–89. 10.1016/j.cad.2015.04.001

18. Gibson I., Rosen D.W., Stucker B. (2010) *Additive Manufacturing Technologies: Rapid Prototyping to Direct Digital Manufacturing.* Springer, p. 459. 10.1007/978-1-4419-1120-9

19. Sahini D.K., Ghose J., Jha S.K., Behera A., Mandal A. (2020) Optimization and Simulation of Additive Manufacturing Processes: Challenges and Opportunities – A Review. In: Balasubramanian K., Senthilkumar V. (eds) *Additive Manufacturing Applications for Metals and Composites.* IGI Global, pp. 187–209. 10.4018/978-1-7998-4054-1.ch010

20. Ngo T.D., Kashani A., Imbalzano G., Nguyen K.T.Q., Hui D. (2018) Additive Manufacturing (3d Printing): A Review of Materials, Methods, Applications and Challenges. *Compos B* 143: 172–196. 10.1016/j.compositesb.2018.02.012

21. Deepak Kumar S., Ghose J., Mandal A. (2019) Thixoforming of Light-weight Alloys and Composites: An Approach toward sUstainable Manufacturing. In Kumar, K. , Davim, P. & Zindani, D. (Eds.), *Sustainable Engineering Products and Manufacturing Technologies.* Academic Press, Elsevier, pp. 25–43. 10.1016/B978-0-12-816564-5.00002-5

22. Deepak Kumar S., Karthik D., Mandal A., Pavan kumar, J.S.R. (2017) Optimization of Thixoforging Process Parameters of A356 Alloy Using Taguchi's Experimental Design and DEFORM Simulation. *Materials Today: Proceedings* 4(9): 9987–9991. 10.1016/j.matpr.2017.06.307

23. Xu Y., Wu X., Guo X. *et al.* (2017) The Boom in 3D-Printed Sensor Technology. *Sensors* 17: 1166. p. 37. 10.3390/s17051166
24. Chen D., Pei Q. (2017) Electronic Muscles and Skins: A Review of Soft Sensors and Actuators. *Chem. Rev.* 117: 11239–11268. 10.1021/acs.chemrev.7b00019
25. Middelhoek S., Noorlag D.W.J. (1981–1982) Three-Dimensional Representation of Input and Output Transducers. *Sensors and Actuators* 2: 29–41. 10.1016/0250-6874(81)80026-1
26. Landaluce H., Arjona L., Perallos, A. *et al.* (2020) A Review of IoT Sensing Applications and Challenges Using RFID and Wireless Sensor Networks. *Sensors* 20(9): 2495. p.18. 10.3390/s20092495
27. Khosravani M.R., Reinicke T. (2020) 3D-Printed Sensors: Current Progress and Future Challenges. *Sensors and Actuators A* 305: 111916. p. 17.
28. Imanzadeh H., Bakirhan N.K., Sınag A., Ozkan S.A. (2020) Methods for Design and Fabrication of Nano-Sensors and Their Electrochemical Applications on Pharmaceutical Compounds. In: Han B., Tomer V.K., Nguyen T.A., Farmani A., Singh P.K. (eds) *Nano-Sensors for Smart Cities: Micro and Nano Technologies.* Elsevier, pp. 31–61. 10.1016/B978-0-12-819870-4.00003-7
29. Leal-Junior A., Casas J., Marques C., Pontes M.J., Frizera A. (2018) Application of Additive Layer Manufacturing Technique on the Development of High Sensitive Fiber Bragg Grating Temperature Sensors. *Sensors* 18(12): 4120 p. 15. 10.3390/s18124120
30. Nehra M., Dilbaghi N., Hassan A.A., Kumar S. (2019) Carbon-Based Nanomaterials for the Development of Sensitive Nanosensor Platforms. In: Deep A., Kumar S. (eds) *Advances in Nano-Sensors for Biological and Environmental Analysis.* Elsevier, pp. 1–25. 10.1016/B978-0-12-817456-2.00001-2
31. Kang H.K., Lee J.Y., Choi S. *et al.* (2016) Smart Manufacturing: Past Research, Present Findings, and Future Directions. *Int. J. Precis. Eng. and Manuf.-Green Tech.* 3(1): 111–128. 10.1007/s40684-016-0015-5
32. Fraden J. (2016) Temperature Sensors. In: *Handbook of Modern Sensors.* Springer, Cham, 2016. ISBN: 978-3-319-19302-1. 10.1007/978-3-319-19303-8_17
33. Behera A., Pan J., Behera A. (2021) Temperature Nano-Sensors for Smart Manufacturing. In: Thomas S., Nguyen T.A., Ahmadi M., Farmani A., Yasin G. (eds) *Nano-Sensors for Smart Manufacturing: Micro and Nano Technologies.* Elsevier, pp. 249–272. 10.1016/B978-0-12-823358-0.00013-7
34. Liu C., Huang N., Xu F. *et al.* (2018) 3D Printing Technologies for Flexible Tactile Sensors toward Wearable Electronics and Electronic Skin. *Polymers* 10: 629, p. 31. 10.3390/polym10060629
35. Khorsandi D., Nodehi M., Waqar T. *et al.* (2021) Manufacturing of Microfluidic Sensors Utilizing 3D Printing Technologies: A Production System. *Journal of Nanomaterials* 2021: 5537074, p.16. 10.1155/2021/5537074
36. Han T., Kundu S., Nag A., Xu Y. (2019) 3D Printed Sensors for Biomedical Applications: A Review. *Sensors* 19: 1706 p. 22 10.3390/s19071706
37. Mannoor M.S., Jiang Z.W., James T. *et al.* (2013) 3D Printed Bionic Ears. *Nano Lett* 13(6): 2634–2639. 10.1021/nl4007744
38. Deepak kumar S., Chattree A., Jha S.K., Singh N.K., Mandal A. (2019) Deformation Behavior of Semi-solid forged A356-5TiB$_2$ Nano in-situ Composites. In: Shanker K., Shankar R., Sindhwani R. (eds) *Advances in Industrial & Production Engineering. Lecture Notes in Mechanical Engineering.* Springer, Singapore, pp. 77–84. ISBN: 978-981-13-6411-2. 10.1007/978-981-13-6412-9_7
39. Behera A., Sahini D., Pardhi D. (2021) Procedures for Re-cycling of Nano-Materials: A Sustainable Approach. In: Rai M., Nguyen, T.A. (eds) *Nanomaterials Recycling.* Elsevier, pp. 175–207. 10.1016/B978-0-323-90982-2.00009-3
40. Abdel-Karim R., Reda Y., Abdel-Fattah A. (2020) Review-Nanostructured Materials-

Based Nano-Sensors. *Journal of the Electrochemical Society* 167: 037554, p. 10. 10.1149/1945-7111/ab67aa

41. Asim Kumar (2018) Nano-Sensors: Applications and Challenges. *International Journal of Science and Research* 8(7): 1472–1474. 10.21275/ART20199881

42. Deshpande A.H., Weldode J.M., Pise J.S. (2018) Applications of Nano-Sensors in Various Fields: A Review. *International of Management Engineering and Technology* 8(XI): 1403–1408. ISSN NO: 2249-7455.

43. Deepak Kumar S., Arun Manohar G., SuryaTeja R. (2022) The State of the Art 3D Printing: A Case Study of Ganesh Idol. *Materials Today Proceedings*, 1–9. https://doi.org/10.1016/j.matpr.2022.01.418

7 3D Nanoprinting in the Biomedical Industries

Vaibhavi Srivastava and Mayank Handa
Department of Pharmaceutics, National Institute of
Pharmaceutical Education and Research, Raebareli,
Lucknow, India

Rahul Shukla
Department of Pharmaceutics, National Institute of
Pharmaceutical Education and Research, Raebareli,
Lucknow, India

CONTENTS

7.1   INTRODUCTION

Three-dimensional printing (3D printing) was invented by Emanuel Sachs and his colleagues at Massachusetts Institute of Technology, USA in the late 1980s (US patent number 5204055) [1]. 3D printing is a technique that involves fabrication of stereoscopic, real, three-dimensional objects using a printer controlled by CAD softwares or CT scanners [1]. 3D printing is a rapid, creative, and admissible technique, thus attracting medical researchers' attention after its successful utilisation in the fields of automobiles, architecture, medical devices, aerospace industry, and electronics. It is an iconic example for transformation of analogue processes to digital manufacturing processes. 3D printing involves layer-by-layer addition of raw materials to create a physical three-dimensional object, thus known as layered manufacturing (see Figure 7.1) [2]. Fabricating objects traditionally requires removal of material by means of milling, carving, or shaping, whereas 3D printing deposits or adds material over the previous layer to create a desirable shape; therefore, the technology is widely known as additive manufacturing. Several techniques lie under the umbrella of 3D printing, e.g., stereolithography apparatus (SLA), fused deposition modelling (FDM), selective laser sintering

DOI: 10.1201/9781003189404-7

Physical mixture   3D manufacturing device

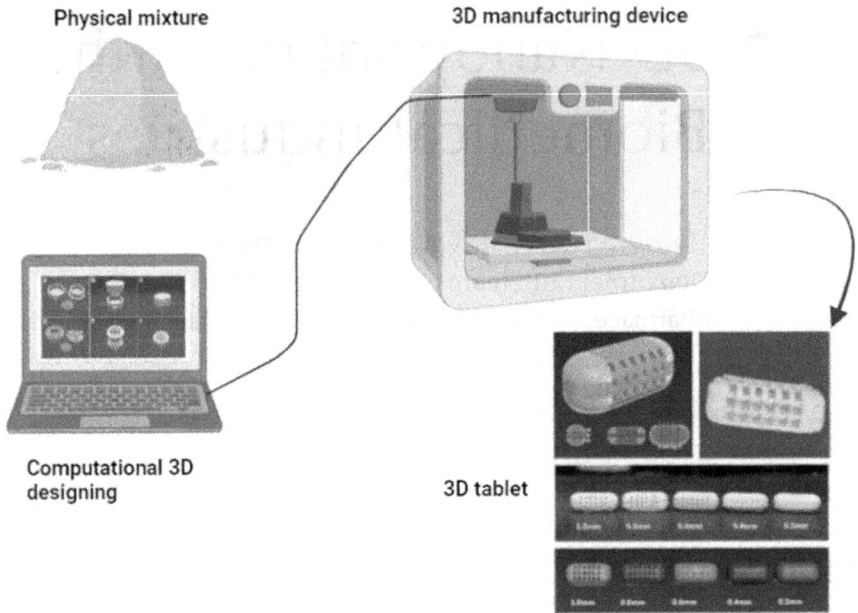

Computational 3D
designing   3D tablet

FIGURE 7.1   An illustrative representation of 3D based manufacturing.

(SLS), digital light processing (DLP), solid state extrusion (SSE), hot melt extrusion (HME), binder jetting, and vat polymerisation [1].

The first record of applying 3D printing was Hideo Kodama, a Japanese scientist, who utilised UV lights to harden polymers and developed an object in 1981. This was said to be the foundation of stereolithography (SLA). Further, in 1986, Charles Hull invented stereolithography (patent 4575330), a process bearing similarity with 3D printing. SLA is a 3D printing process that exploits CAD software monitored laser beams for the development of concept models or physical prototypes for testing purpose [3]. Selective laser sintering (SLS) uses an additive process to fix powder layer-by-layer and transform it in solid geometry; thus, it is also known as powder bed fusion. SLS was developed by Joseph Beaman and Carl Deckard (patent 4863538). FDM is the most common type of 3D printing and was developed by Scott Crump (patent 5121329). This technique involves melting of thermopolymer in liquid and extruding it layer upon layer. Digital light processing is similar to SLA, with the only difference in light source. DLP requires arc lamp with liquid crystal panel. This technique primarily works with photopolymer resins that alter its properties in the presence of light. Binder jetting or binder jet printing (BJP), the pioneer technique developed by Emanuel Sachs, involves a liquid layer over a powder layer for geometrical object manufacturing. Every technique holds its own disadvantages and advantages, and selected on the basis of area of application. Biomedical and pharmaceutical sector generally require strength, biocompatibility, and safety for human use [1].

3D printing is very popular in industries because of its ability to fabricate complex models, objects, less demand of raw materials, rapid and high-speed production. In spite of all advantages, additive manufacturing is still in its cradle stage in the pharmaceutical field. Additive manufacturing is capable of creating pristine tissues, dental and other medical prostheses, blood vessels, artificial organs, etc. This organ printing technology brings revolutionary progress in the biomedical field and pharmaceutical market. Application of 3D printing in designing novel drug-delivery systems, personalised medications, implants and inserts manufacturing, and many more make it a fast growing and expanding technology in the healthcare profession [3].

In this chapter, we will learn about restrictions of conventional medications and how 3D printing overcomes them, application of 3D nanoprinting in nanotechnology and the healthcare sector, and regulatory considerations in 3D nanoprinting.

7.2   CHALLENGES IN CONVENTIONAL MEDICATION

Traditional medication is more like fitting all shapes in one groove, but in reality, there may be significant differences among individuals against drug responses, even at the same dose [4,5]. Personalized medicine can be the solution for this problem as it is able to reduce the risk of drug adverse effects. When the medical treatment provided to the patient suits the characteristics of the individual, it is said to be personalized medicine. This depends on the fact that every individual has its own molecular and genetic profile, and so every individual will react differently to the same dose of the same drug. Personalized medicines are more satisfactory to the patient as they adhere to individual pharmacokinetic properties [6]. Manufacturing of conventional dosage forms occur in bulk, thus personalized dosing is expensive and impractical. In order to provide personalized medication, often available dosage forms like tablets or capsules could be modify by breaking or crushing, but this may lead to incorrect dosing [7]. 3D printing or additive manufacturing is a peculiar technique for rapid modelling, and prototyping involves the fabrication of geometrical solid objects by depositing raw materials in a layer-by-layer fashion. 3D printing is procuring the attention of pharmaceutical researchers and formulation scientists as an effective approach to overcome the above mentioned challenges. Additive manufacturing technology can design drugs and dosage forms immediately, on demand. Thus, the application of 3D nanoprinting in pharmaceutical industries pushes it a step closer to customized medicine [8]. Oral dosage forms are the most popular and compliant form. 3D printing can fabricate tablets of different geometry and drug release patterns. Polypill, created by a 3D printer, contains multiple drugs in a single tablet with different release time, and thus is beneficial for geriatric patients as they are generally on more than one therapy. A 3D printer can fabricate solid geometries, oral dosage forms, numerous types of tablets, and medical devices. Additive manufacturing is slow compared to traditional mass production, even though it possesses its own advantages, like personalization, cost effectiveness, applicability for small batches, and more compatibility to the patient. Few 3D printing techniques are best fit for the manufacturing of nano and micro-scale delivery systems, e.g., DLP and SLA. The conventional production strategy is unable to produce high-dose tablets due to the limitation of blending and punching steps, while Khaled *et al.* published an article about 3D printing of high-dose paracetamol tablets [9]. Every coin has two faces. In spite of all, additive manufacturing possesses a few limitations; for instance, it is not suitable for bulk production, it possesses low loading capacity, and it requires skilled personnel.

7.3   NANOTECHNOLOGY BASED 3D PRINTING

Nanotechnology and 3D printing are two very fascinating words belonging to two novel fields that, on combining, create interesting outcomes. The established manufacturing process involves the removal of materials to obtain the final product, e.g., carving, or cutting, or milling. Additive manufacturing acts opposite to this and deposits raw materials layer-by-layer to yield the final product. This process is called the bottom-up method since the process initiates with a tiny particle or a single layer and moves the 3D printer, layer over layer, to build a geometrical three-dimensional object [10].

Nanotechnology deals with the formation and development of tiny structures of size 0.1 to 100 nm. For potential delivery of drugs, nanotechnology is being applied in drug-delivery systems. Utilisation of nanotechnology in delivering drugs improves poor water-soluble drug bioavailability, improves absorption, targets drug delivery, provides sustainability to highly soluble drugs, enhances permeation through barriers, and amplifies the therapeutic efficiency of the drug. Nanotechnology helps to explore new probabilities like dual targeting, theranostic applications, and stimuli responsive release. Examples of nanotechnology applied in drug-delivery systems are polymeric nanoparticles, metallic nanoparticles, nanocrystals, micelles, dendrimers, nanogels, nanodiamonds, carbon nanotubes, quantum dots, nanoemulsions, etc.

Pal P. and co-workers fabricated a composite bilayer scaffold encompassing nanofibrous layer and 3D printed layer-for-tissue regeneration. They reported that two layers had different traits in terms of pore diameter, structural attributes, composition, and hydrophobicity, which ultimately gives the scaffold a tensile strength of 6.12 ± 1.26 MPa and a water-retaining capacity of 95%. These parameters confirmed that the abovementioned composite was potentially applicable for tissue regeneration [11]. Ceylan and his team used a 3D printer to design a biodegradable microswimmer for pathological markers detection. These hydrogels-based microswimmers are magnetically monitored and composed of two photons polymerized from suspension of gelatine methacryloyl and superparamagnetic iron oxide nanoparticles. It was reported that matrix metalloproteinase-2 (MMP-2) degrades microswimmer completely and thus testing it in SKBR3 breast cancer showed its potential as a theranostic agent [12]. Another example of using 3D printing as a nanotechnology is dual-photon lithography. Unlike stereo-lithography, dual-photon lithography absorbs two photons at a time. This technique has a resolution of less than 50 nm and can build scaffolds for living cells.

a. **Nanoparticles:** Macroparticles have a number of flaws and defects that can be avoided using nanoscopic structures. Application of nanoparticles in drug delivery offers numerous advantages like high loading capacity, excellent absorption, bioavailability of drugs, good permeation ability, better mechanical strength, etc. Considering the benefits of nanoparticles, researchers strive to print using a 3D printer. Recently, researchers from recognised foreign universities designed a nanoparticles assembly using additive manufacturing. Lee *et al.* present an example of using 3D printing to generate nanoparticles for nerve regeneration. The research team designed a three-dimensional biocompatible porous assembly encompassing nanoparticles for neurogenic factor delivery. They used SLA and co-axial electrospray technique for this purpose. It was reported that assembly containing nanoparticles had better proliferation and showed longer neuritis generation [13]. Another example involves magnesium oxide nanoparticles for bone regeneration. 3D printer-generated magnesium oxide nanoparticles with PCL and HAP tablet can enhance proliferation of osteoblast cells and thus proved beneficial in bone regeneration [14]. Similar effort was taken by Abdal-hay *et al.* who integrated magnesium hydroxide nanoparticles with polycaprolactam and designed a composite fabricated by additive manufacturing. They claimed that nanoparticles incorporated composite showed positive result compare to blank [15]. On the basis of the abovementioned examples, it is clear that application of 3D printing for nanoparticles can result in excellent customized medication and delivery systems. Additive manufacturing can produce personalized delivery system with predetermined structure, shape, size, porosity, and release pattern.

b. **Nanocapsules:** According to IUPAC, hollow nanoparticles consist of a solid polymeric shell that encapsulate a drug present at core of the sphere are known as nanocapsules. The size of nanocapsules varies from 10–1000 nm. Nanocapsules can be used as drug-delivery system, nutraceuticals, additive food supplements, etc. Nanocapsules offer several advantages like being smart carriers of drugs, improved bioavailability, enhanced drug efficacy, protected drug from adverse environment, sustainability and precised targeting. Beck *et al.* strived to deliver deflazacort by oral route. For this effort, they utilised 3D printer to print a tablet containing nanocapsules of deflazacort using the FDM technique. The filaments of polycaprolactam and Eudragit$^®$ RL100 were prepared using 3D printing and channelled with nanocapsules. The 3D printed nanocapsules showed PDI of 0.1, size of 0.28 μm, and 65% of *in vitro* drug release within 24 h. Another research team utilised computer-aided design to fabricate a two-step 3D printed nanocapsule in which polycaprolactam, a thermoplastic polymer, sandwiched different oils, such as, linalool, farnesyl limonene, and trivalent alkyne. The oil was loaded by inkjet print head method, and the nanocapsule obtained was of size 200–800 μm [16].

The first marketed nanocapsule product was vitamin E containing antiwrinkle lotion [17]. Layered manufacturing proved beneficial in the production of nanoformulations like nanocapsules and nanoparticles as it reduces total time consumption and wastage of materials. SLA, FDM, SLS, and pressure-assisted microsyringes are 3D printing techniques that are widely used in nanocapsule preparation.

c. **Hydrogels:** Hydrogel is a highly cross-linked, three-dimensional network of hydrophilic polymers that can swell in the presence of water and retain a large amount of water (at least 10% of the total weight). High water-holding capacity is responsible for its structural integrity. Hydrogels were first reported by Wichterle and Lim in 1960 [18]. Hydrogels can exist in continuous phase as colloidal state with water. They are widely applicable in different scientific fields due to their peculiar properties, like excellent absorbability, swell ability, flexibility, and self-healing ability. Noticeably, one can utilise only one property of a hydrogel at a time; i.e., the existing form is applicable for a single purpose only. Thus, research is still going on to form an existing form of hydrogel with more than one property in order to utilise it more. Recently, several research articles explained the utilisation of 3D printing for hydrogel to expand its utilisation. Common 3D technologies applicable in hydrogels are SLA, extrusion-based printing, i.e., solid state extrusion and liquid state extrusion, inkjet printing, laser-assisted forward transfer, microvalve-based printing, etc. Additive manufacturing brings revolutionary change to the field of hydrogel.

Ma *et al.* used extrusion-based printing to formulate cellulose nano-crystal (CNC) hydrogel with excellent viscoelasticity. The concentration of CNC varied from 0.5%–25% w/w in order to achieve good rheological properties. At 20 wt.% of CNC concentration, desirable characteristics were shown; i.e., good printability, better rheological traits, and fidelity [19]. Cheng *et al.* utilised extrusion-based 3D printing to develop a HPMC K4M or E4M hydrogel with storage modulus, high-yield stress, and hardness. They strived to develop semi-solid tablets of theophylline having drug-loading capacity of 75 to 125 mg [20]. Another example involves extrusion-based 3D-printed hydrogel consisting of a mixture of alginate and gelatine at different ratios of 1:2, 1:1, and 2:1, respectively. The resulting hydrogel was sufficiently porous and possesses the capability to encapsulate and deliver therapeutics like proteins, enzymes, vitamins, antioxidants, etc. [21]. Abouzeid *et al.* applied 3D printing for preparation of hydrogel for bone tissue regeneration. They designed an aqueous hydrogel prepared by mixing PVA cellulose nanofibers with hydroxyapatite (HAP) and sodium alginate [22]. Zhang *et al.* implemented the concept of combining electrostatic interaction with hydrogen bonding. This helped in the formation of double-networked hydrogel due to the physical interaction of poly (sulfobetaine-co-acrylic acid)/chitosan-citrate. The transparent hydrogel formed was good in its self-healing ability (95.4%), highly sensitive, and showed electrical conductivity of 0.11 S/m [23].

Thermoresponsive hydrogels (TH) are another novel modification of hydrogels, which are sensitive to temperature change. These TH easily convert from solid to gel state, along with change in temperature, and 3D-printed TH showed good biocompatibility; thus, it can be used in drug-delivery systems. Kesti *et al.* developed a thermoresponsive polymer consists of PNIPAM decorated hyaluronan crosslinked by light source having fast gelation ability and quick structure fidelity after printing with methacrylatedhyaluranon (MAH). MAH provides a long-lasting mechanical stability and thus can be printed as a high-resolution structure. The PNIPAM decorated hyaluronan cross linked polymer possessed low critical temperature of 25.7°C–29.7°C. Results showed that the gelation temperature of thermoresponsive hydrogel was influenced by concentration and charge of polymer and cells, all these leading to a highly printable polymer [24]. Designing of three-dimensional geometric tissues that have natural organ functionality using 3D printing is catching attention globally for regenerating damaged tissue. Heishand and his team developed a thermoresponsive cell-printing ink to print neuro stem cells using polyurethane dispersion system. The polyurethane has desirable modulus and chemical constituents that favor neuro stem cells printing. They tested it on a zebrafish embryo model by inducing a neural injury to it and found that polyurethane hydrogel strengthens healing of the nervous system. This can be proved as a promising treatment for neurodegenerative disorders [25]. In another study, Tsukamoto *et al.* utilised hydroxybutyl chitosan, a thermoresponsive hydrogel, for developing a 3D tissue printed by a robotic dispensing printer. The cell orientation and design of tissue was precisely monitored and controlled using a layer-by-layer technique [26]. Another research team developed a thermoresponsive polymer from methacrylated chondroitin sulfate and a thermosensitive poly(N-(2-hydroxypropyl) methacrylamidemono/dilactate)-polyethylene glycol triblock copolymer (M15P10) to fabricate a biocompatible polymer for artificial tissue generation. 3D printed hydrogel possessed excellent porosity and mechanical strength. It was reported that 3D-printed chondrogenic cells from the same polymer have good viability and proliferating ability [27].

d. **Nanofibers:** As the name suggests, nanofibers are a novel type of material consisting of fibers having diameter in nano range. These are highly porous with large surface-to-volume ratio. Numerous polymers, like chitosan, poly caprolactone, poly-lactic acid, cellulose, copolymers of PLGA, etc., can be exploited for nanofiber generation. In 1934, the idea of fine nanofiber was explained and patented by Formhals. Because of its unique characteristics, nowadays, it is widely applicable in tissue nanoengineering, drug-delivery systems, wound healing, nanocomposites, etc. Fabrication of nanofibers using 3D printing is quite beneficial and economic; the layer-by-layer technique is used to design electrospun nanofibers. An electrospinning system involves a printer with nozzles that spray polymeric solution, and in presence of an electric field, the polymeric fluid turns in to nanofibers.

Yu *et al.* thought about fabrication of 3D-printed bone tissue regeneration. To turn this idea in reality, Yu infused PCL/gelatine dispersed nanofibers into PCL printing mesh. This porous PCL scaffold produces a good cellular response [28]. Similarly Huang *et al.* employed PCL nanofibers for damaged bone repairing [29]. Another example involves fabrication of PCL nanofibers composed of nanohydroxyapatite (nHAP) and laponite. nHAP helps in osteogenesis, i.e., new bone tissue generation and laponite supports bone formation, proliferation, and adhesion because it is a smectic clay-type material. The PCL nanofibers was 3D printed using an extrusion technique, and it was reported that the resulting nano filament was nontoxic and exhibited a promising ability for bone tissue regeneration [30]. An idea for utilisation of nanofibers in drug-delivery systems was explored further by Ambrus and his team. They designed a 3D-printed nanofiber by electrospun technology and loaded it with the less aqueous, soluble drug loratidine with an objective to investigate the influence of nanofiber on the solubility of drug. Noticeably, results showed that nanofibers enhanced aqueous solubility of drug by 26 fold and 60% of drug releases from the delivery system, which is 15 times higher than the pure drug release profile [31].

7.4 APPLICATION OF 3D-BASED NANOPRINTING IN HEALTHCARE

3D printing is a quick, creative, and admissible technique that offers several advantages to almost every aspect of production and industries. Combining additive manufacturing with a drug-delivery system provides customization of treatment in no time and has proved to be more beneficial. Additionally, the technique is cost-effective and provides a scalable production opportunity. Besides the pharma industry, the healthcare sector is widely utilising 3D printing for numerous applications. 3D printing proved itself beneficial in biomedical engineering of medical aids in diagnostic or therapeutic regions. 3D printing is also involved in several advanced areas, such as tissue regeneration, tissue and organ models, replica designing, prosthetics and implants fabrication, etc. [3].

In the following section, we discuss these findings, the development required for translational modifications, and the advantages and limitations.

a. **Implants:** Medical devices make life much easier; they play a crucial role in the healthcare sector. Applying 3D printing in the production of medical devices like implants, prosthetics, orthodontics, etc., would provide an opportunity to form peculiar and personalized items with the best fit to the individual. Implants and prosthetics need to meet patient's requirement and fit their anatomy and pathology. Traditional casting methods possess limitations of inappropriate joint fixation and reconstruction, even after the implication of extra tools and kits. Thus, the application of 3D printing for customized implant fabrication is considerable. This can produce a specific computer-aided design model using an individual's biometric scans. A standard size is not suitable for all patients, and this creates problems in movement (in the case of prosthetics); thus, personalized implants are more acceptable. Another advantage of 3D-printed implants is the implication of biocompatible materials as raw material.

3D-printed microswimmer devices showed various applications in nanomedicines. Microswimmer device mechanisms involve three stages: loading of the drug, transportation to targeted site, and release of the drug. Loading is similar to all other delivery systems, but the transportation mainly focuses on modification to control movement of microswimmers to target tissues. Microswimmers can achieve desirable kinetics by sensitization against stimulus like, change in shape, magnetic actuation, etc. Zhu *et al.* developed a microimplant-by-material extrusion method having sustained a drug-release profile [32]. Another research team designed a 3D-printed intervertebral implant with minimal error and cost-effectiveness compared to traditional implants [33]. 3D printing is worth exploring in the field of spine surgery or vertebral skeleton model as it is simple and cost-effective. Gbureck *et al.* developed a bioceramic implant using inkjet printing for angiogenesis [34]. Another interesting study involves incorporation of multiple drugs in a single implant manufactured by 3D printing to obtain dual-pulsed release system for bone tuberculosis treatment [35].

b. **Anticancer drug delivery:** Cancer is a major cause of death globally, recording about 10 million deaths in the last year alone. The uncontrolled growth of cells anywhere inside the body is termed as cancer. To mitigate the disease, numerous drug therapies are being used, but most of the anticancer drugs have the limitation of poor aqueous solubility and thus face problems in reaching the desirable site. Again, the cumulative effect of drug results in toxicity to healthy tissues that would worsen the situation. Thus, use of traditional delivery systems like oral dosage forms or parenteral dosage forms causes difficulty to patients.

Recently, the use of 3D-printed polymeric scaffolds have been considered as a good alternative for anticancer therapy. Complexes consisting of polymers like PCL and PLGA have been fabricated by 3D printing and can be used to deliver antibiotics or antitumour agents. 3D-printed delivery systems can be in form of patches or scaffolds that can deliver a long-lasting release of drugs with pre-defined kinetics. This helps in patient compatibility and acceptability.

Maher and co-workers developed a 3D-printed titanium implant that involves microparticles and tubular arrays on the surface fabricated by 3D printing. These topographical microparticles and nano-tubular arrays could deliver doxorubicin and apoptosis-inducing ligand (Apo2L/TRAIL) directly to targeted sites. It was reported that these nanosurfaces could be worthy of exploration for localized delivery of drugs in bone cancer and fracture support [36]. Another example involves the manufacturing of a microfluidic chip using additive manufacturing for the delivery of combinational chemotherapeutics. These chips have tiny channels that allow inter-mixing of therapeutic solutions quickly. This delivery system proved efficient in cytotoxicity study [37]. Similar efforts were observed in one more study where the SLA technique was used to manufacture microneedles of biodegradable resins. These needles showed noticeable suppression of the tumour [38]. Sarkar N. and Bose S. concluded a liposome-encapsulated curcumin with the help of porous calcium phosphate podium. It was believed that curcumin will release from 3D-printed porous complex and will promote apoptosis of osteosarcoma cells and proliferation of osteoblast cells [39]. Treatment of cancer requires personalized and targeted delivery systems since there is always a risk of harming healthy cells along with cancerous cells. Thus, for the treatment of cancer, a smart therapeutic system could only be possible by utilising additive manufacturing.

c. **Peptide delivery:** Peptides are the building blocks of proteins having a significant role in bio-organisms. Recently, peptides owned noticeable utilisation in drug delivery and nanomedicine because they offer a great chemical diversity and can be obtained from various unicellular and multicellular organisms. However, using peptides as a drug is associated with several challenges.

Tiny needle structures present on the surface of the delivery system enhance the skin penetration of peptides and proteins. Recently, 3D printers were used to fabricate microneedles on the surface of system, which proved more efficient than patches. Microneedles can fabricate by other simple techniques as well, but the use of 3D printing enables the design of more complex microneedles. For the delivery of insulin, Pere *et al.* designed a transdermal patch containing a polymeric microneedle on its surface using the SLA technique. Biocompatible resins are allowed to be crosslinked in the presence of light for fabrication of microneedles having conical and pyramidal shape; further inkjet printing was used to coat these needles with insulin. The idea behind the fabrication of needle-based insulin formulation was excellent skin penetration capacity, thus good delivery ability. It was reported that insulin was delivered within half an hour by 3D-printed microeedles [40]. Lu and his colleagues performed similar work by designing microneedles of poly(propylene fumarate) using the SLA technique for skin cancer treatment. They added diethyl fumarate as an adjuvant to improve mechanical strength of needles. They reported that microneedles released dacarbazine for up to 5 weeks [41].

Small sized peptides can be used as bioinks in 3D printing. Short peptides bear resemblance to collagen because of their nanofibrous topographical aggregation, being the same as hydrogels, and hence, they possess similar architecture as extracellular tissue space. For instance, ultrashort peptide hydrogel showed biocompatibility if employed in stem cell printing or intestinal epithelium cells culturing. In a cell potency study, numerous markers of pluripotency, like Tra-I-60, Tra-I-81, Oct4, and Nanog, showed that ultrashort peptides possess embryonic stem cell potency [42].

d. **Wound healing:** The occurrence of wounds is very common in day-to-day life, and this is because there is great demand for customized novel materials to heal wounds. Several nanotechnology-based systems proved more beneficial than currently available conventional wound healing methods, but still safety is a big concern. For wound healing, uses of nanoparticles of antibiotic agents are becoming more popular, although their manufacturing is difficult at such a large scale. However, recently, several innovations can be observed in dressing materials and wound healing systems, like incorporation of antimicrobial agents in bandages and other dressing materials or the use of silver and zinc in sterilized polymeric cotton gauges. Polymeric filaments coated with metals are manufactured using the extrusion method, and these filaments show a sustained release of metals and antimicrobial activity.

The 3D-printed hybrid matrix of pericardium and poly (ethylene glycol) was successfully utilised in healing injured blood vessels post surgery. Often wounds are observed during the replacement of damaged blood vessels strengthened by vascular grafts after surgery. Abovementioned hybrid matrix can suppress inflammatory signals and thus proved beneficial [43]. Another advanced innovation observed in the field of wound healing is the generation of new human skin at the wound site. Lee and his colleagues successfully generated a human skin using a layer-by-layer 3D printing technique. The 3D-printed skin had high similarity index with natural human skin in terms of shape, structure, texture, and, architecture [44]. Similar work was done by Xu C. *et al.* by designing bio platters and encapsulating cellulose nanoparticles associated with customized tissues and engineered at every inch. The whole system was optimized with hydrogel of methylcellulose and alginate. It was reported that these bio platters could be used for cartilage and bone regeneration [45]. The major risk associated with wounds is microbial infections. Considering the problem, researchers developed a 3D-printed alginate-based system encompassing nanoparticles of zinc oxide (ZnO) because it possesses good antibacterial activity. The 3D-printed system proved more advantageous compared to conventional delivery systems due to its large pore size and stability. ZnO incorporated alginate hydrogel showed significant inhibition activity over *Staphylococci* without affecting the fibroblast cells [46].

e. **Regenerative medicine:** With the advancement in technology, now it is possible to replace damaged tissues with new ones or to regenerate the new, healthy tissue. Engineering tissue is also playing a critical role in implants insertion. Tissue engineering and regeneration involves live cells, biocompatible polymers, and other growth factors to design an implant that aids in the growth of healthy cells. 3D printing proved to be the best technique for designing systems similar to connective tissues, and thus, it is helpful in bone regeneration. Dealing with small size bone damage is more challenging than dealing with large bones. Recent medical options for bone damage is bone graft vascularisation, but it is more applicable for later ones. Additive manufacturing dispenses effective solutions that will easily insert inside the bone and promote bone growth. A research article discussed the regeneration of articular surface of rabbit synovial joint, and to achieve this outcome, they developed a HAP/PCL implant system by additive manufacturing and inserted it in a rabbit. The system was doped with transforming growth factor β3. Results showed a successful regrowth of humeral joints of a rabbit [47]. Another study showed that a fibrin and mesenchymal stem cells coated 3D-printed trachea consists of PCL would regenerate cartilage [48]. Research is still going on to obtain a 3D-printed nano system consist of nanoparticles of calcium phosphosilicate of size 20–100 nm. Utilisation of additive manufacturing in nano technology could yield numerous advance systems of nano- and micro-size range with desirable unique properties such as osteointegration, as required in bone regeneration. By the whole discussion, one can come to the point that combined utilisation of 3D printing and nanotechnology can provide un-rivalled results in terms of designing regeneration medicines.

f. **COVID-19 implications:** COVID-19 created havoc globally, presenting new challenges and difficulties every day all over the whole world. The situation demands new techniques and delivery systems to overcome medical emergencies that occur due to the disruption in supplies and transportation. Additive manufacturing or three-dimensional printing was exploited well to fulfill the requirements of medical devices, diagnostic kits, emergency dwellings, and other medical accessories, etc. Masks, sanitizers, and PPE kits are few armours against COVID-19, but providing a personal protective equipment (PPE) kit in such a large amount was a big challenge. 3D printing was used for the mass production of PPE kits and masks in no time. 3D-printed adapters were the major attention. Erickson *et al.* developed a 3D-printed adapter for Flyte Helmet, and this was the innovative way to convert the helmet into PPE [49]. The best use of 3D printing was done by developing real-size human body dummies and body organs to train medical staff for diagnostics, swab tests, ways of dealing COVID patients, and other COVID-compatible behaviours.

g. **Diagnosis and detections:** Diagnosis of disease is prioritized over treatment. This is the reason why medical imaging has been come to the fore in the health sector. A complete inspection of the internal body allows health practitioners to accurately detect the disease. 3D diagnosis is done in three steps: proper image navigation followed by image post processing and 3D printing. Like all other fields, layered manufacturing proved beneficial in field of diagnosis also. In recent past, 3D printing exploited well for diagnosis since it involves computers and automation provides 3D-printed images [37]. Diagnostic tools and equipments can be prepared using cost-effective and fast three-dimensional printing. One very familiar but interesting example of the application of 3D printing in diagnosis is microfluidic chips and diag-nostic kits. These are gaining in popularity because of their compatibility, easy-to-use kits, and need for a very small volume of sample.

3D printing can take the already existing diagnostic methods to another level; for example, 3D-printed assistive devices can make advancement in polymerase chain reaction and can sense an individual's pressure through mobile phones. 3D bio printing is one modification in 3D printing; it involves the incorporation of bio-molecules and biological structures like cells, growth factors, proteins, etc., in printable bio-ink solutions.

7.5 REGULATORY CONSTRAINTS

In the last decade, the tremendous use of additive manufacturing is being observed in almost every industry. The reason behind such popularity of 3D printing technology is the quick and customized production of any kind of device. 3D-printed complex geometry meets the desired criteria. Therefore, to regulate the use of 3D printing in manufacturing devices, the FDA issued a draft in May 2016 entitled, "Technical Considerations for

Additive Manufactured Devices." As 3D printing can produce customized anatomical implants, prostheses, and artificial organs, thus the FDA guidelines asked for a clear explanation of manufacturing steps and details of device. Guideline also involves an explanation of factors that affect the 3D-printed devices [50].

In the FDA guidelines, it is clearly mentioned that raw materials used for additive manufacturing of any article should be from a recognized source. There must be a description about all the properties of materials if affecting the finished product since 3D printing is widely applicable in the medical sector; thus, the regulation of raw materials is very important. 3D-printed products are regulated for the problems associated with software that is employed for finished product fabrication.

7.6   CONCLUSION

This chapter is all about three-dimensional printing, its application in the biomedical and pharmaceutical industry, and the regulatory aspects of 3D printing. The chapter also describes the limitations of conventional drug-delivery systems and how utilization of a 3D-based approach in manufacturing of delivery systems creates advancement in novel drug-delivery systems. Employing 3D printing in the pharmaceutical industry is in its cradle stage, but because of its peculiar advantages, like being economical, being fast and rapid to produce, requiring less raw material, yielding effective finished products, and many more, will surely lead to achieve high popularity and acceptance.

Additive manufacturing helps to deliver personalized medication, which is one of the essential goals of advanced medication. This chapter provides several examples where nanomaterials were processed by additive manufacturing and yields different types of biocompatible scaffolds, implants, prostheses, and tissues. 3D printing has envisaged its ability in tissue engineering and regeneration. 3D-based approaches are under limited FDA regulation and are still tested to preclinical studies. It can be said that 3D printing will be worth further exploration for the establishment of 3D printing for human use.

ACKNOWLEDGMENT

The authors acknowledge the Department of Pharmaceuticals, Ministry of Chemical and Fertilizers, Government of India. The NIPER-R communication number for this article is NIPER-R/Communication/242.Conflict of Interest: The authors declare no conflict of interest among themselves.

REFERENCES

1. Beg S, Almalki WH, Malik A, Farhan M, Aatif M, Rahman Z, et al. 3D printing for drug delivery and biomedical applications. *Drug Discov Today [Internet]*. 2020 September; 25(9): 1668–1681. Available from: https://linkinghub.elsevier.com/retrieve/pii/S1359644620302841
2. Kumar M, Sharma A, Mohanty UK, Kumar SS. Additive manufacturing with welding. *Advances in Welding Technologies for Process Development*. 2019;77–100. https://www.taylorfrancis.com/chapters/edit/10.1201/9781351234825-5/additive-manufacturing-welding-manish-kumar-abhay-sharma-uttam-kumar-mohanty-surya-kumar

3. Jain K, Shukla R, Yadav A, Ujjwal RR, Flora SJS. 3D printing in development of nanomedicines. *Nanomaterials*. 2021.

4. Cui M, Pan H, Su Y, Fang D, Qiao S, Ding P, et al. Opportunities and challenges of three-dimensional printing technology in pharmaceutical formulation development. *Acta Pharm Sin B*. 2021; 11(8): 2488–2504.

5. Shukla R, Kumar J, Dwivedi P, Gatla P, Mishra PR. Microparticles of diethylcarbamazine citrate for the treatment of lymphatic filariasis. *Asian Journal of Chemistry*. 2013.

6. Okwuosa TC, Soares C, Gollwitzer V, Habashy R, Timmins P, Alhnan MA. On demand manufacturing of patient-specific liquid capsules via co-ordinated 3D printing and liquid dispensing. *Eur J Pharm Sci*. 2018;118: 134–143.

7. Sharma V, Shaik KM, Choudhury A, Kumar P, Kala P, Sultana Y, et al. Investigations of process parameters during dissolution studies of drug loaded 3D printed tablets. *Proc Inst Mech Eng Part H J Eng Med*. 2021; 235(5): 523–529.

8. Chen G, Xu Y, Kwok PCL, Kang L. Pharmaceutical applications of 3D printing. *Addit Manuf*. 2020;34(November 2019): 101209.

9. Khaled SA, Alexander MR, Wildman RD, Wallace MJ, Sharpe S, Yoo J, et al. 3D extrusion printing of high drug loading immediate release paracetamol tablets. *Int J Pharm*. 2018;538(1–2): 223–230.

10. Handa M, Tiwari S, Yadav AK, Almalki WH, Alghamdi S, Alharbi KS, et al. Therapeutic potential of nanoemulsions as feasible wagons for targeting Alzheimer's disease. *Drug Discov Today [Internet]*. 2021 July; Available from: https://linkinghub.elsevier.com/retrieve/pii/S1359644621003251

11. Selvaraj S, Fathima NN. Fenugreek incorporated silk fibroin nanofibers – A potential antioxidant scaffold for enhanced wound healing. *ACS Appl Mater Interfaces*. 2017;9(7): 5916–5926.

12. Ceylan H, Yasa IC, Yasa O, Tabak AF, Giltinan J, Sitti M. 3D-printed biodegradable microswimmer for theranostic cargo delivery and release. *ACS Nano*. 2019;13(3): 3353–3362.

13. Lee SJ, Zhu W, Heyburn L, Nowicki M, Harris B, Zhang LG. Development of novel 3-D printed scaffolds with core-shell nanoparticles for nerve regeneration. *IEEE Trans Biomed Eng*. 2017;64(2): 408–418.

14. Roh HS, Lee CM, Hwang YH, Kook MS, Yang SW, Lee D, et al. Addition of MgO nanoparticles and plasma surface treatment of three-dimensional printed polycaprolactone/hydroxyapatite scaffolds for improving bone regeneration. *Mater Sci Eng C*. 2017;74: 525–535.

15. Kim CG, Han KS, Lee S, Kim MC, Kim SY, Nah J. Fabrication of biocompatible polycaprolactone–hydroxyapatite composite filaments for the FDM 3D printing of bone scaffolds. *Appl Sci*. 2021;11(14): 6351.

16. Rupp H, Binder WH. 3D printing of core–shell capsule composites for post-reactive and damage sensing applications. *Adv Mater Technol*. 2020;5(11): 1–8.

17. Katz LM, Dewan K, Bronaugh RL. Nanotechnology in cosmetics. *Food Chem Toxicol*. 2015;85: 127–137.

18. W.L.N. T. Hydrogels. Nature. 1960;185:112–118.

19. Ma T, Lv L, Ouyang C, Hu X, Liao X, Song Y, et al. Rheological behavior and particle alignment of cellulose nanocrystal and its composite hydrogels during 3D printing. *Carbohydr Polym*. 2021;253(October): 117217.

20. Cheng Y, Qin H, Acevedo NC, Jiang X, Shi X. 3D printing of extended-release tablets of theophylline using hydroxypropyl methylcellulose (HPMC) hydrogels. *Int J Pharm*. 2020;591(October): 119983.

21. Kuo CC, Qin H, Cheng Y, Jiang X, Shi X. An integrated manufacturing strategy to fabricate delivery system using gelatin/alginate hybrid hydrogels: 3D printing and freeze-drying. *Food Hydrocoll*. 2021;111(June 2020): 106262.

22. Abouzeid RE, Khiari R, Salama A, Diab M, Beneventi D, Dufresne A. In situ mineralization of nano-hydroxyapatite on bifunctional cellulose nanofiber/polyvinyl alcohol/sodium alginate hydrogel using 3D printing. *Int J Biol Macromol.* 2020;160: 538–547.

23. Zhang J, Chen L, Shen B, Wang Y, Peng P, Tang F, et al. Highly transparent, self-healing, injectable and self-adhesive chitosan/polyzwitterion-based double network hydrogel for potential 3D printing wearable strain sensor. *Mater Sci Eng C.* 2020;117: 111298.

24. Kesti M, Müller M, Becher J, Schnabelrauch M, D'Este M, Eglin D, et al. A versatile bioink for three-dimensional printing of cellular scaffolds based on thermally and photo-triggered tandem gelation. *Acta Biomater.* 2015;11(1): 162–172.

25. Hsieh FY, Lin HH, Hsu S Hui. 3D bioprinting of neural stem cell-laden thermo-responsive biodegradable polyurethane hydrogel and potential in central nervous system repair. *Biomaterials.* 2015;71: 48–57.

26. Tsukamoto Y, Akagi T, Shima F, Akashi M. Fabrication of orientation-controlled 3D tissues using a layer-by-layer technique and 3D printed a thermoresponsive gel frame. *Tissue Eng Part C Methods.* 2017;23(6): 357–366.

27. Abbadessa A, Blokzijl MM, Mouser VHM, Marica P, Malda J, Hennink WE, et al. A thermo-responsive and photo-polymerizable chondroitin sulfate-based hydrogel for 3D printing applications. *Carbohydr Polym.* 2016;149: 163–174.

28. Yu Y, Hua S, Yang M, Fu Z, Teng S, Niu K, et al. Fabrication and characterization of electrospinning/3D printing bone tissue engineering scaffold. *RSC Adv.* 2016;6(112): 110557–110565.

29. Huang B, Aslan E, Jiang Z, Daskalakis E, Jiao M, Aldalbahi A, et al. Engineered dual-scale poly (ε-caprolactone) scaffolds using 3D printing and rotational electro-spinning for bone tissue regeneration. *Addit Manuf.* 2020;36(February): 101452.

30. Liakos IL, Mondini A, Del Dottore E, Filippeschi C, Pignatelli F, Mazzolai B. 3D printed composites from heat extruded polycaprolactone/sodium alginate filaments and their heavy metal adsorption properties. *Mater Chem Front.* 2020;4(8): 2472–2483.

31. Cubo-Mateo N. Design of thermoplastic 3D-printed scaffolds for bone tissue engineering: Influence of parameters of "Hidden" importance in the physical properties of scaffolds. 2020; 12(7): 1–14.

32. Min Z, Kun L, Yufang Z, Jianhua Z, Xiaojian Y. 3D-printed hierarchical scaffold for localized isoniazid/rifampin drug delivery and osteoarticular tuberculosis therapy. *Acta Biomater.* 2015;16(1): 145–155.

33. Domanski J, Skalski K, Grygoruk R, Mróz A. Rapid prototyping in the intervertebral implant design process. *Rapid Prototyp J.* 2015;21(6): 735–746.

34. Gbureck U, Hölzel T, Doillon CJ, Müller FA, Barralet JE. Direct printing of bio-ceramic implants with spatially localized angiogenic factors. *Adv Mater.* 2007;19(6): 795–800.

35. Wu W, Zheng Q, Guo X, Sun J, Liu Y. A programmed release multi-drug implant fabricated by three-dimensional printing technology for bone tuberculosis therapy. *Biomed Mater.* 2009;4(6): 1–10.

36. Maher S, Kaur G, Lima-Marques L, Evdokiou A, Losic D. Engineering of micro- to nanostructured 3d-printed drug-releasing titanium implants for enhanced osseointegration and localized delivery of anticancer drugs. *ACS Appl Mater Interfaces.* 2017;9(35): 29562–29570.

37. Waheed S, Cabot JM, Macdonald NP, Lewis T, Guijt RM, Paull B, et al. 3D printed microfluidic devices: Enablers and barriers. *Lab Chip.* 2016;16(11): 1993–2013.

38. Uddin MJ, Scoutaris N, Economidou SN, Giraud C, Chowdhry BZ, Donnelly RF, et al. 3D printed microneedles for anticancer therapy of skin tumours. *Mater Sci Eng C.* 2020;107: 110248.

39. Sarkar N, Bose S. Liposome-encapsulated curcumin-loaded 3D printed scaffold for bone tissue engineering. *ACS Appl Mater Interfaces.* 2019;11(19): 17184–17192.

40. Pere CPP, Economidou SN, Lall G, Ziraud C, Boateng JS, Alexander BD, et al. 3D printed microneedles for insulin skin delivery. *Int J Pharm.* 2018;544(2): 425–432.

41. Lu Y, Mantha SN, Crowder DC, Chinchilla S, Shah KN, Yun YH, et al. Microstereolithography and characterization of poly(propylene fumarate)-based drug-loaded microneedle arrays. *Biofabrication.* 2015;7(4): 1–13.

42. Sundaramurthi D, Rauf S, Hauser CAE. 3D bioprinting technology for regenerative medicine applications. *Int jjournal Bioprinting.* 2016;2(2): 9–26.

43. Jamróz W, Szafraniec J, Kurek M, Jachowicz R. 3D printing in pharmaceutical and medical applications – Recent achievements and challenges. *Pharm Res.* 2018;35(9): 1–22.

44. Lee W, Debasitis JC, Lee VK, Lee JH, Fischer K, Edminster K, et al. Multi-layered culture of human skin fibroblasts and keratinocytes through three-dimensional free-form fabrication. *Biomaterials.* 2009;30(8): 1587–1595.

45. Xu C, Zhang Molino B, Wang X, Cheng F, Xu W, Molino P, et al. 3D printing of nanocellulose hydrogel scaffolds with tunable mechanical strength towards wound healing application. *J Mater Chem B.* 2018;6(43): 7066–7075.

46. Cleetus CM, Primo FA, Fregoso G, Raveendran NL, Noveron JC, Spencer CT, et al. Alginate hydrogels with embedded zno nanoparticles for wound healing therapy. *Int J Nanomedicine.* 2020;15: 5097–5111.

47. Lee CH, Cook JL, Mendelson A, Moioli EK, Yao H, Mao JJ. Regeneration of the articular surface of the rabbit synovial joint by cell homing: A proof of concept study. *Lancet.* 2010;376(9739): 440–448.

48. Chang JW, Park SA, Park JK, Choi JW, Kim YS, Shin YS, et al. Tissue-engineered tracheal reconstruction using three-dimensionally printed artificial tracheal graft: Preliminary report. *Artif Organs.* 2014;38(6): E95–E105.

49. Erickson MM, Richardson ES, Hernandez NM, Bobbert DW, Gall K, Fearis P. Helmet modification to PPE with 3D printing during the COVID-19 pandemic at Duke university medical center: A novel technique. *J Arthroplasty [Internet].* 2020;35(7): S23–S27. Available from: 10.1016/j.arth.2020.04.035

50. Targhotra M, Aggarwal G, Popli H, Gupta M. Regulatory aspects of medical devices in India. *Int J Drug Deliv.* 2017;9(2): 18.

8 3D Printing of Nanocomposites

Vigneshwaran Shanmugam
Department of Mechanical Engineering, Saveetha School of Engineering, Chennai, India

Rajkumar Velu
Department of Mechanical Engineering, Indian Institute of Technology Jammu, Jammu, India

CONTENTS

8.1 INTRODUCTION

Additive manufacturing (AM) is a bottom-up approach that provides robust control over the dimension of the final product. AM process is a promising technology for a wide variety of applications ranging from sturdy mechanical devices like gears to delicate body implants [1]. Correspondingly, composites, are an ever-evolving field, and a wide range of advanced composites are propelling toward high potential applications [2]. An inadequate variation in the composition and quality of the raw material enhances the property of the end-use product. In consequence of development in AM technology and composite materials, nanocomposites are born from such an idea where the secondary additives are in the nano dimension, resulting in improved characteristics viz. mechanical, electrical, corrosion, and thermal properties. The current research era is developing nanocomposites for 3D printing processes, such as fused deposition modelling (FDM), selective laser sintering (SLS), [3], and stereolithography (SLA) nanocomposites. However, injection molding is a conventional manufacturing process that is still widely utilized for the fabrication of various components. Although 3D printers are extensively helpful for low volume production, the injection-molding process is highly endorsed for fabrication to attain reasonable product quality for specific applications. The choice of the manufacturing process is relative as it depends upon the number of components to be printed along with the requisite quality. Hence, this battle between a choice of injection molding and 3DP technology will always exist. This chapter focuses on the properties and performance of the 3D-printed nanocomposites.

8.2 3D PRINTING OF NANOCOMPOSITES

Processing nanocomposites through advanced techniques like 3D printing improves the possibilities of material applicability in various sectors and has significant potential for producing a new class of multifunctional nanocomposites [4]. Therefore, 3D printing has been a significant technique fabricating complex-shaped components in unique or small series production. Stereolithography (SLA), powder bed technologies (i.e., selective laser sintering and binder jetting), and fused deposition modelling (FDM) are well-known techniques in the era of 3D printing. Among these, FDM is the most commonly practiced method due to its simplicity and design flexibility. The FDM was first developed by Scott Crump mainly as a rapid prototyping (RPT) tool; it is the most admired 3D printing technique that transforms thermoplastic filaments into functional components through melting and extrusion [5]. In the past decade, the FDM is one of the most accepted techniques in 3D printing and is the base of most commercial 3D printers at an affordable costs that use different thermoplastic filaments as feeding materials. The FDM permits the printing of complex-shaped 3D parts via the extrusion of molten thermoplastic filaments deposited layer upon layer and solidified upon exposure to atmospheric air. The price of an FDM printer may be as low as a thousand dollars. Due to its low cost and variety of compatible materials, FDM is favourable for research communities, industries, and domestic users. Therefore, development on both techniques and feedstock materials is progressing significantly. The noteworthy thermoplastics such as polylactide (PLA), acrylonitrile butadiene styrene (ABS), or polyamide (PA) have been used as feedstock raw materials in the FDM technique, which makes the users adapt quickly and comprehend the conventional process. High thermal-resistance polymeric materials such as polyetherimide [6] and polyether ether ketone (PEEK) [7] are also being adapted for the FDM printing process. Filaments of various thermoplastics reinforced with nanomaterials, such as carbon nanotubes, graphene, and nanoclay, have been commercialised, leading to the fabrication of polymer nanocomposites. The maximum resolution of the FDM-printed parts is around 40 mm [8]. Recently, some new FDM 3D printers have dual nozzle heads through which a matrix and reinforcement can be fed separately and combined while printing to form nanocomposites. FDM utilise a polymer filament as a feeding material, and those polymers used must have definite properties (thermos-softening, photopolymerization, etc.) depending on the printing technique. However, reinforcing appropriate fillers to the matrix enhances the required property of nanocomposite-based functional components. Therefore, for most applications, it is essential to feed the polymer with a filler material specifically selected in view of the final product properties. In this perspective, nanomaterials or nanofillers are attractive fillers. The fabricated nanocomposite must satisfy the required properties of the printed object while complying with the printer specifications. The design and development of these nanocomposites is difficult and requires intensive investigation due to numerous practical limitations. Commonly, the use of nanoparticulates, nanofibres, or nanomaterials as reinforcement in polymeric matrix allows the fabrication of polymeric nanocomposites. These nanocomposites have been strongly marked; their footprint is present in various applications, such as structural, biomedical, and so on, due to their high mechanical performance and excellent functionality. In today's scientific practices, the literature is advocating that carbon-based nanoparticles, nano-ceramic based nanoparticles, and natural nanoparticles are the three most extensively used fillers for the fabrication of nanocomposites. Still, there is a considerable scope for these particles, reinforced nanocomposites printed through the 3D printing process. In the past, research has reported the use of polymer/filler nanocomposites manufactured using a conventional moulding process because of their benefits, specifically improved mechanical, thermal, and tribological properties. However, selected authors have reported using polymer/fillers nanocomposites manufactured through the FDM 3D printing, and mainly, there is limited research on PLA/clay nanocomposites. There are many opportunities to study the enhancement compatibility of filler and matrix, mechanical properties, and process improvement.

In addition, processing polymer nanocomposites to fabricate complex parts using a desktop 3D printer offers excellent multifunctional properties; these include good electrical conductivity, thermal conductivity, mechanical strength, and stiffness at a relatively low cost with a relatively low filler concentration of layered silicate fillers or conductive nanofillers, such as carbon nanotubes (CNT), graphene, and metal particles. Consequently, the material utilisation in various advanced applications like microelectronics and device packaging. Figure 8.1 shows various emerging applications of 3D-printed nanocomposites.

8.3 MATRIX MATERIAL FOR 3D PRINTING OF NANOCOMPOSITES

Additive manufacturing techniques involve a variety of fusion mechanisms such as thermal fusion, chemical fusion, and optical/photo-polymerisation fusion, as shown in Figure 8.2. For example, initially, for the FDM process, the polymer materials are formed as filament to extrude during the printing process, and noteworthy polymers such as ABS, PLA, polyvinyl alcohol (PVA), polycaprolactone (PCL), polymethylmethacrylate (PMMA) and polyamide (nylon) [9]. In addition, recent advances in this technology have also permitted the use of high-performance engineering polymers such as PEEK and polyetherimide (PEI) for the FDM process [10]. Subsequently, polymers, namely PEEK and PEI filaments, are commercially available 3D4Makers [11] and 3DXTech [12], respectively. High-performance engineering polymers provide excellent mechanical and thermal properties, but they are relatively expensive and capable for high temperature operating conditions.

Moreover, to print the high-performance thermoplastics, it requires a high temperature extruder, and the commercial FDM machine is not suitable, though their extrude temperature limits up to 250 °C. In this section, predominant polymer

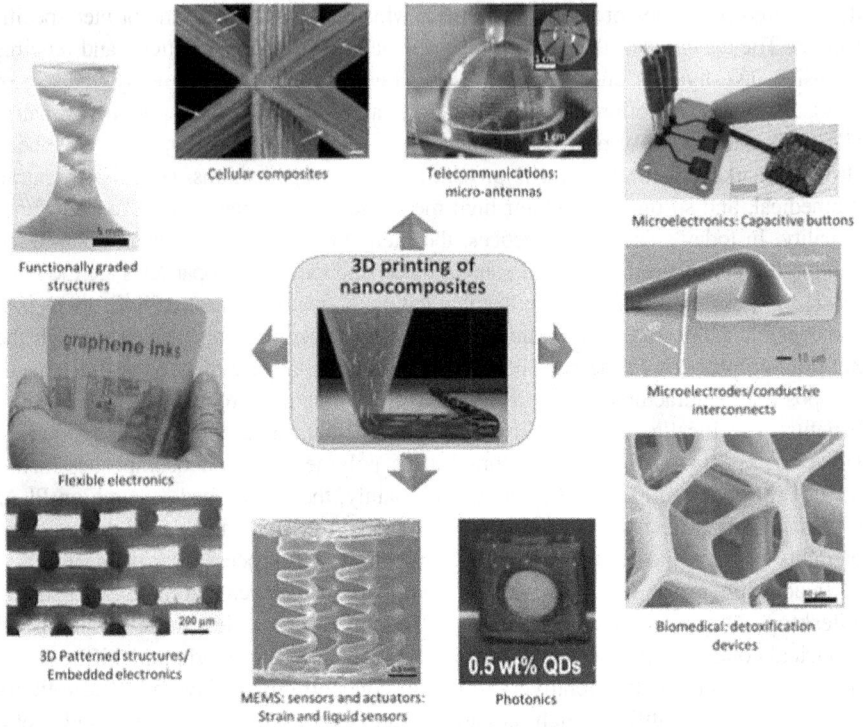

FIGURE 8.1 Macro and microstructures of different 3D-printed nanocomposite for a wide range of applications, such as MEMS, photonics, microfluidics, biomedical, microelectronics, and telecommunication sectors [8].

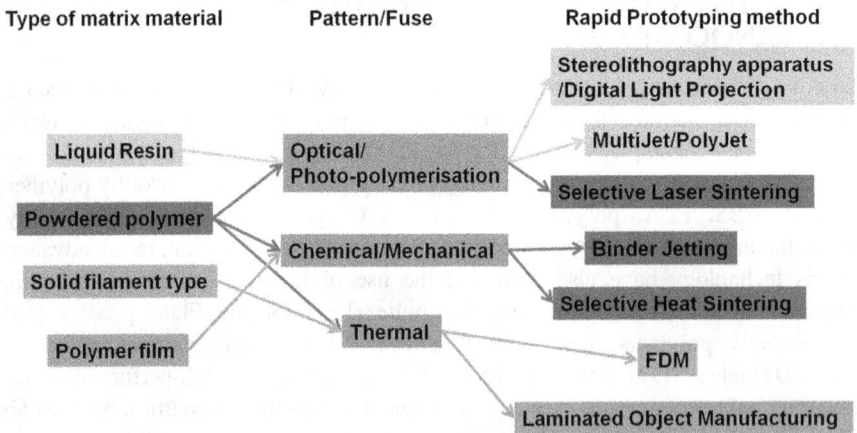

FIGURE 8.2 Outline of monomer/polymeric matrix materials used with specific layered building techniques in rapid prototyping [13].

matrixes used in 3D printing are discussed with respect to high potential applications like biomedical, flexible electronics, etc.

It is essential to understand the properties requisite of polymer matrix used for a 3D printing process. For direct ink writing (DIW) printable polymers, the viscosity is a significant property, and it should be in an appropriate range. In FDM process during extrusion, the viscosity of a polymer provides a material to be extrudable and be able to provide a seamless flow of filaments [14]. Once extruded, it desires to thicken to provide good mechanical properties to support the printed structures and prevent delamination. In addition, the polymer used in 3D printing requires the following properties: high thixotropic and shear-thinning performance, optimised viscosity, rapid solidification after extrusion to attain the printed shape, and sufficient mechanical strength, as shown in Figure 8.3.

Larraza et al. developed a flexible and printable polyurethane filament parting from biocompatible waterborne polyurethane for biomedical applications. The synergistic effect of cellulose nanofibers and graphene was utilised to strengthen the filament properties, and printability and those filaments was employed to prepare polyurethane-based nanocomposites. Fabricated nanocomposite filaments displayed altered properties, which has a direct influence on material printability. Graphene-added nanocomposites disclosed better thermal and mechanical properties for a good printing process. Furthermore, these filaments were utilised in FDM and fabricated 3D-printed parts, which displayed better shape fidelity. Properties revealed by polyurethane and graphene filaments are evident for their potential in the biomedical domain [16]. Concerning the modern environment, an eco-friendlier choice to conventional polyurethane is waterborne polyurethanes (WBPU) and waterborne polyurethane-ureas (WBPUU). These materials can be used as matrix material in 3D printing.

ABS-based nanocomposites were manufactured through the FDM process. These ABS samples were reinforced with multi-walled carbon nanotubes (MWCNT) and carbon black (CB) fillers [17]. The electromagnetic interference-shielding efficiency

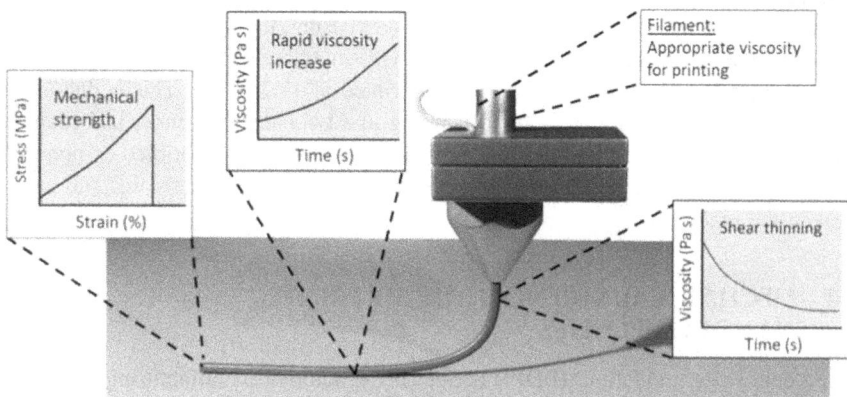

FIGURE 8.3 Important properties of polymers required for extrusion-based 3D printing [15].

(EMI SE) of produced FDM samples was assessed. Various concentrations of the WMCNT, CB, and MWCNT/CB fillers were added in an ABS matrix by a melt-compounding method using an internal mixer. Based on the rheological behaviour of the nanocomposites, the 3 wt.% of filler concentration was selected as the optimum quantity for the production of the filament. The ABS filaments were produced using the twin-screw extruder and fed into a conventional FDM printer for 3D-printed specimen fabrication along with three different growing directions. The effect of filler and increasing directions of the electrical conductivity, the EMI SE, and the mechanical properties of extruded filaments, as well as the 3D-printed samples, were tested. In general, the conductivity, EMI SE and mechanical properties of 3D-printed specimens were significantly dependent on the growing direction. The experimental findings of this research showed that an optimised EMI SE and mechanical properties were obtained through an appropriate choice of a polymer nanocomposite formulation alongside the 3D printing parameters [17]. In another research, Ahmed et al. [18] prepared nanocomposite, which contains poly (ionic liquid) (PIL), PMMA [19], and MWCNTs as fillers and an ionic liquid (IL) that acts as a plasticizer and dopant for the MWCNTs. The nanocomposites showed a wide range of mechanical properties (elongation at break varied from 50% to 250%) and conductive properties depending on the filler concentration level. It was noticed that the maximum conductivity of 520 Sm^{-1} was attained with 15 wt.% MWCNT. The dispersion was observed at this level of MWCNT filler concentration, which was detected in SEM micrographs. The thermal properties of the PMMA nanocomposites were assessed through thermogravimetric analysis (TGA), differential scanning calorimetry (DSC), and dynamic mechanical analysis (DMA). The high thermal stability of the nanocomposites was noticed up to 340 °C, regardless of the composition and a variable transition temperature dependent on the MWCNT, IL, and polymeric concentrations. Ultimately, the parameters for 3D printing were optimized. As a proof of concept, the authors demonstrated the fabrication of a flexible, 3D-printed circuit, which can be bent and twisted without any damage in the circuit [18].

Other than particle reinforced 3D printing, fibre reinforcement also gained significant attention among the research communities. For instance, Torrado et al. [20] used ABS and jute fibre as matrix filament and reinforcement, respectively, and fabricated ABS/jute fibre-reinforced composites using the FDM process. It was measured that the ultimate tensile strength of ABS/Jute fibre composite was 8.63 MPa, which is significantly lower than pure ABS printed of 17.73 MPa. This is due to the applied high temperature during melt blending in a twin-screw extruder and filament extrusion in nozzle result into thermal degradation of jute fibre in printed composites. Therefore, the thermal degradation of jute fibres also leads to voids that ultimately weaken the material.

8.4 MECHANICAL STRENGTH OF 3D PRINTED NANOCOMPOSITES

Fused deposition modelling (FDM) is one of the additive manufacturing technique utilized for the 3D printing of nanocomposites. The main advantages of FDM are the low cost, high printing speed, and process simplicity [21]. Even though the

FDM process has many benefits when compared to other additive manufacturing techniques, it has many disadvantages; these include poor surface finishing, the mechanical properties being mainly dependent on process parameters, and FDM is mainly limited to thermoplastic materials because this technique is primarily reliant on the thermoplasticity of the material to be printed [22,23]. Here, we will discuss the various mechanical properties, such as tensile strength, flexural strength, and hardness of 3D-printed nanocomposites manufactured by FDM technique.

8.4.1 Tensile Strength

The tensile properties of the 3D-printed nanocomposites are the most analysed and most discussed characteristics in comparison to the other mechanical characteristics of the 3D-printed specimens. Many factors clearly influence the tensile properties of the nanocomposites. The thickness of the layer deposited by the nozzle and raster width (width of the deposition path related to tooltip size) plays a vital role to enhance the tensile properties. S Rajpurohit and H Dave analysed tensile strength of a fused filament fabricated PLA part [24]. From their studies, it was clear that raster width, raster angle (angle made by the raster concerning the x-axis of the build table), and the interaction of raster width and layer thickness had a significant influence on the tensile properties of the PLA part. Razavi-Nouri, Mohammad, et al. analysed poly (acrylonitrile-butadiene-styrene) ABS/multi-walled carbon nanotubes nanocomposite specimens by considering the raster angles, such as 45/−45° and 0/90° along with the layer thickness, such as 0.05 mm, 0.1 mm and 0.2 mm [25]. From Figure 8.4(a) and (b), it is clear that the tensile strength decreases with the increase in layer thickness for a particular nanofiller content. This is mainly because as the layer thickness decreases, the nozzle tip will be much closer to the deposited layers, and the high temperature of the nozzle tip keeps the deposited layers in a molten state for a longer time. So this helps the layers to diffuse each other for a extended period. This increases the bonding strength between the adjacent layers and leads to an increase in the tensile strength of the nanocomposites [26–28].

From Figure 8.4, it is also clear that the tensile strength of 45/−45° raster angle specimens is slightly higher than the tensile strength of 0/90° raster angle specimens. This is mainly because the number of voids formed in 0/90° specimens is higher when compared to 45/−45° specimens. It is also attributed to the fact that the path travelled by nozzle is shorter for the printing of samples at 45/−45° compared to the printing of 0/90° samples. When the nozzle travels for a shorter distance, the heat transfer rate will be higher and the interdiffusion between the layers will be more efficient [29].

Another important factor that influences the tensile properties is the filler loading in nanocomposites. Zixiang Weng et al. analysed the 3D-printed ABS/montmorillonite nanocomposites manufactured through FDM and observed that tensile strength of the nanocomposite increased due to the 5 wt.% loading of organically modified montmorillonite (OMMT) on ABS when compared to the loading of 1 wt.% and 3 wt.% of the same filler [30]. The presence of graphene nanoplatelets on PLA will increase the tensile strength up to 27% and elastic modulus up to 30% of the PLA matrix when compared to a FDM-processed PLA without any fillers [31]. This

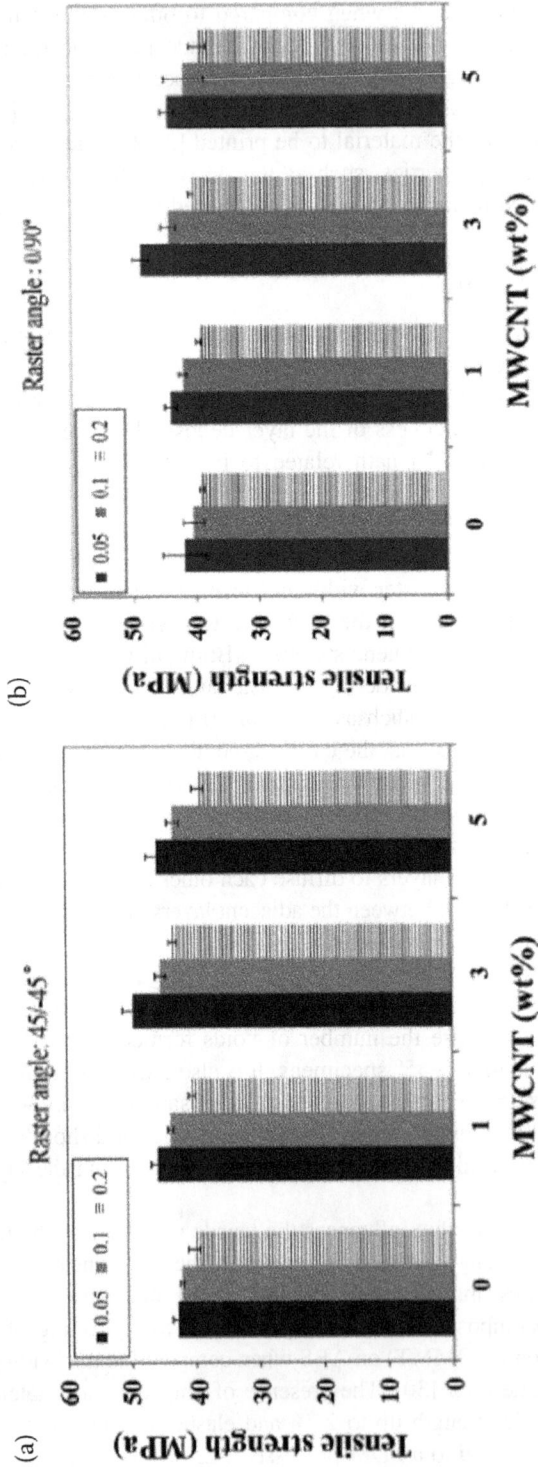

FIGURE 8.4 (a) and (b) Tensile strength Vs MWCNT (wt.%) for three different layer thicknesses at raster angles 45/–45° and 0/90°, respectively.

increase in tensile strength is mainly due to the agglomeration of the nanoparticles and its interaction with the polymer matrix in nanocomposites [32]. Here, the effective size of the filler plays a remarkable role: as the filler size decreases, its surface area increases, which leads to better interaction with the polymer matrix. However, the higher filler loading causes the polymer chains in immobilized conditions due the concentration of stress at the agglomeration points, which leads to the formation of fracture points and thus reduces the tensile strength of nanocomposites [25].

Printing pattern is another important factor that influences the tensile characterization of nanocomposites. According to Lu Yang et al., the parallel filling pattern (0 × 0) of the glycerol-modified water-treated PVA (W-mPVA)/graphene nanoplates (GNP) had the best tensile strength and Young's Modulus, whereas the perpendicular filling pattern (90 × 90) had the worst tensile properties [33]. For the parallel pattern, the loading direction aligns in the same direction with the filament deposition direction during the tensile testing, causing the stress to be predominantly preserved by the filaments.

8.4.2 Compressive Strength

The research on the various parameters influenced on the compressive strength of 3D-printed nanocomposites is still going on, and it should be thoroughly analysed for various applications of nanocomposites. From the different studies conducted on this specific property, it was concluded that the layer thickness and higher infill percentage plays a vital role for the enhancement of compressive strength of 3D-printed parts. The compressive strength of Hydroxyapatite/graphene nanocomposite with 0.4 wt.% graphene and with 125 μm layer thickness was 70 times greater than the hydroxyapatite sample with 0% graphene [34]. This enhanced mechanical property was mainly due to the high surface area of graphene flakes and its interaction with the matrix. The graphene flakes tend to withstand longitudinal as well as lateral loads, which makes it easier to properly orient with the matrix. Its random distribution in the matrix prevents the initiation of crack propagation. Build direction is also another parameter that clearly affects the compressive strength of 3D-printed nanocomposites. The studies of Lee et al. clearly showed that the compressive strength of axial FDM specimen was 11.6% higher than the transverse FDM specimen [35].

8.4.3 Flexural Strength

When compared to tensile and compressive strength, flexural or bending strength is the least analysed characteristics of 3D-printed nanocomposites. It is not easy to analyse the optimum parameter for the maximum bending strength because the 3D-printed part experiences tensile as well as compressive strength in different layers, when bending stress is applied to it [21]. N Vidakis et al. made a comparison study of the flexural strength between the ABS/graphene nanoplatelets powder and ABS/CNT nanocomposites [36]. By adding graphene nanoplatelets in the ABS matrix, there is a drastic reduction in flexural strength of nanocomposite, which is mainly due to the presence of carbon-based filler without having any functionality characteristics, and this leads to the weak bonding between the graphene nano-platelet and ABS matrix. However, the addition of CNT to the ABS matrix leads to an increase in flexural strength of around 50% compared to pure ABS is mainly due to the alignment of carbon nanotubes in the 3D printing direction. The addition of 5 wt.% of carbon fibre in the ABS matrix enhanced the flexural properties so that flexural toughness, flexural modulus, and flexural stress were increased by 21.86%, 16.82% and 11.82%, respectively, as compared to the pure ABS [37]. So, from the above studies, it can be concluded that the flexural strength will be maximum when the fibre orientation is 0 ° with the matrix. As per the literature survey, it is clear that the other process variables were not appropriately investigated. So a thorough analysis is required to determine an optimum combination of process variables to improve the flexural properties.

8.5 THERMAL PROPERTIES OF 3D-PRINTED NANOCOMPOSITES

Differential scanning calorimetry and thermal gravimetric analysis, along with thermal conductivity and thermal expansion, are some of the parameters widely adopted for characterizing the thermal behaviour [38,39]. The addition of nano-materials in the FDM polymers exhibit enhanced thermal properties. For instance, Wang et al. [40] printed PLA-based nanocomposite with graphite nanoplatelets (GNP) as the dispersed phase with a GNP modifier to enhance the properties of the nanocomposites. Higher dispersion of GNP in the PLA matrix was observed, which resulted in a strong interfacial binding interaction. A reduction in the melting point of the nanocomposite was reported with increasing concentration of the GNP additives, which was attributed to the physical space provided by the additive for the movement of the PLA molecular chain. The printed nanocomposite showed increased thermal stability, 60 °C increase in degradation temperature, and a 7% increase in the remaining residual weight. In another work by Ivanov et al. [41], the thermal conductivity of FDM-printed nanocomposite showed increased thermal conductivity by almost 181% compared to plain matrix polymer. The authors reported that this improvement in thermal values is because of an improved interface between the matrix and dispersed phase. Better adhesion between matrix and secondary addition material resulted in suppressed phonon scattering, resulting in improved thermal conductivity. This can be further attributed to the utilization of the FDM technique for the fabrication of nanocomposites. Tambrallimatha et al. [42] investigated the thermal properties of the ABS and PC-based graphene nanocomposites. Composites were prepared using the ABS and PC reinforced with 0.2, 0.4, 0.6 and 0.8 wt.% graphene. Addition of graphene increased the composite thermal stability. Compared to ABS and PC, the glass transition temperature of the 0.8 wt.% graphene added composites ca.6% and ca.8% higher, respectively. The result of this investigation was evidence of the effectiveness of the FDM-printed nanocomposites' thermal stability. Similar results were reported by Weng et al. [30] on the thermal stability of the ABS/montmorillonite nanocomposites prepared through FDM.

8.6   3D-PRINTED NANOCOMPOSITES AND BIOMIMICRY

Composites are a new generation of materials composed of a polymer matrix and a reinforcement. They have high specific strength and modulus and are primarily used in aircraft applications [43]. New manufacturing methods for developing composite materials, such as 3D printing, are being introduced to enhance the quality of composites and facilitate their manufacturing process. 3D printing is a manufacturing technique that has garnered much interest in recent years [44]. It is estimated that the revenue for the sale of 3D-printed parts will be around US$31 million by 2025 [45]. It can manufacture complex geometrically structured components that cannot be produced by traditional methods [46]. Due to the relatively high production cost, it is desirable for the manufacturing of low volume products. Recently, bioinspired composites that adopt biomimicry techniques are being fabricated [47]. Biomimicry is a rotational 3D printing technique that uses a rotating nozzle to adjust fibre alignment in composites by controlling the rotational speed relative to the printing speed [48] [47]. This method enhances the mechanical properties, including stiffness, toughness, strength, and shear resistance of the manufactured composites compared to the traditionally manufactured ones. In line with this, Chew et al. [49] investigated the mechanical performance of natural fibre-reinforced plastics (NFRP) laminates using a helicoidal flax fibre-epoxy laminates with cross-ply quasi-isotropic laminates as control samples. The samples were produced from flax-epoxy prepreg and tested under out-of-plane and impact loads. According to the study, the helicoidal NFRP laminates absorbed more energy under impact load. The 9° inter-ply angle configuration enhanced the peak load by 72% and 52% compared to the cross-ply and quasi-isotropic laminate samples, respectively.

The bio-mimicked composite concept can be used in conjunction with other manufacturing methods, such as direct ink writing, fused filament fabrication (FFF), and thermoplastic additive manufacturing. Biomimicry has been applied in the construction industry to produce biologically inspired building materials [50]. Fibre reinforcement for bio-mimicked has also been extensively studied. According to a study by Summe et al. [51] on hemp fibre bio-mimicked composites, it was shown that hemp fibre reinforcement had the capability of enhancing the tensile and bending strength compared to the composites with fibre-glass reinforcement. In addition, the fracture toughness of the hemp-fibre reinforced bio-mimicked composites was higher than that of monolithic structural materials but similar to fibre-glass reinforced composites. In another work, Alghahtani et al. [52] observed that the inclusion of fibre reinforcements is a requirement in bio-mimicked composites if they serve as a substitute for the traditional load-bearing structural composites. The authors conducted mechanical tests to elucidate the influence of volume fraction and fibre length (chopped and continuous fibres) on the fracture toughness of bio-mimicked composites. It was seen in the study that the constant carbon fibre increased the maximum bending load by 25% and the fracture energy by four times, while the chopped carbon fibre showed a 50% increment in maximum bending load and a 20-fold increment in fracture energy. The authors also noticed that fibre and material types, such as rigid ceramic and soft polymer, had a significant influence on the mechanical properties of bio-mimicked composites. Additionally, it has been envisaged that bio-mimicked composites reduce the risk in connection with a building's resistance to earthquakes [53]. It is quite challenging to determine the parameters to consider to increase the toughness of composites for residential buildings. Allameh et al. realised that increasing the length of fibres in bio-mimicked composites made from cement, polymers, and carbon fibre increases the strength, fracture energy, and toughness. Biologically inspired composites have become a mainstay in recent years due to their unique mechanical properties. More work should be done on fire behaviour of these composites to increase their relevancy and believe it more valuable, which is currently a grey area. It is undeniable that the biomimicry concept is very attractive, and its development could lead to the betterment of technological advancement in material research.

8.7   3D-PRINTED NANOCOMPOSITES IN BIOMEDICAL APPLICATION

The growth in 3D printing technology enabled its application in the field of biomedical [54]. The application includes drug-delivery systems, tissue engineering, tissue and organ models, prosthetics and replica fabrication, and implants [55]. The advantage of 3D printing technology enabled the printing of biomedical scaffolds; recent research in this area focused on testing the 3D-printed nanocomposite biomedical properties. For instance, Ranjan et al. [56] studied the biomedical properties of PLA nanocomposite developed with hydroxyapatite and chitosan. This investigation concluded the PLA-based nanocomposite with hydroxyapatite and chitosan was a suitable material for biomedical application due to improved mechanical properties. In another work, Alam et al. [57] developed PLA nanocomposite scaffolds through fused filament fabrication (FFF) by using metal/alloy. Three different composites were fabricated by using three reinforcing particles, namely, copper, bronze, and silver. The copper particle was reinforced at 70 wt.% having the particle size of 50–70 μm, bronze (60–100 μm) at 75 wt.%, and silver (90 μm) at 4 wt.% The developed composites exhibited excellent anti-bacterial property and mechanical strength. PLA-Br has a higher elastic modulus of 1.70 GPa when compared to PLA-Cu (1.65 GPA) and PLA-Ag (1.59 Gpa) samples. Higher mechanical strength was noted for the composites printed at 0° printing direction when the printing direction is considered. The anti-bacterial performance of the composites against E.coli and S. aureus was comparably better than the neat PLA. This investigation proved that through the treatment in the composites, it is possible to increase the anti-bacterial activity. Notably, acidic acid treatment on the PLA composites reported enhanced anti-bacterial performance against E.coli (up to 18% less viability). Overall, this investigation shows the performance of the metal particle reinforced nanocomposites for biomedical applications. Thus, the authors recommended these composites for bone scaffold applications.

Similarly, Naghieh et al. [58] investigated the potential of FDM-printed PLA/gelatin-forsterite scaffolds. The elastic modulus of the printed nanocomposite was 52% higher than that of the neat PLA. The presence of calcium phosphate-like precipitation formation on the surface of the nanocomposite indicates that the composites have improved bioactivity. Based on these findings, the author recommended FDM printed PLA/gelatin-forsterite scaffolds for use in bone tissue regeneration. The potential of 3D-printed nanocomposites for biomedical engineering applications can be deduced from these three studies.

REFERENCES

1. Velu R, Jayashankar DK, Subburaj K. Chapter 20 – Additive processing of biopolymers for medical applications. In: Pou J, Riveiro A, Davim JPBT-AM, editors. *Handbooks Adv. Manuf.*, Elsevier; 2021, pp. 635–659. 10.1016/B978-0-12-818411-0.00019-7
2. Velu R, Raspall F, Singamneni S. 3D printing technologies and composite materials for structural applications. *Green Compos Automot Appl* 2019;171–196. https://www.sciencedirect.com/science/article/pii/B9780081021774000082?via%3Dihub
3. Velu R, Singamneni S. Evaluation of the influences of process parameters while selective laser sintering PMMA powders. *Proc Inst Mech Eng Part C J Mech Eng Sci* 2015;229:603–613. 10.1177/0954406214538012
4. Velu R, Calais T, Jayakumar A, Raspall F. A comprehensive review on bio-nanomaterials for medical implants and feasibility studies on fabrication of such implants by additive manufacturing technique. *Materials (Basel)* 2019;13:92. 10.3390/ma13010092
5. Crump SS. Apparatus and method for creating three-dimensional objects, Patent no: US5121329A, 1989.
6. PEI Ultem 1010 Filament | 3D4Makers | 3D printing n.d. https://www.3d4makers.com/products/pei-filament#:~:text=Polyether%20Imide%20(PEI)%20Ultem%201010,Ultem%201010%20for%20their%20filament.
7. Velu R, Vaheed N, Ramachandran MK, Raspall F. Correction to: Experimental investigation of robotic 3D printing of high-performance thermoplastics (PEEK): A critical perspective to support automated fibre placement process (The International Journal of Advanced Manufacturing Technology, (2019), 10.100. *Int J Adv Manuf Technol* 2019. 4/2020, 10.1007/s00170-019-04763-2
8. Farahani RD, Dubé M. Printing polymer nanocomposites and composites in three dimensions. *Adv Eng Mater* 2018;20:1700539. 10.1002/ADEM.201700539
9. Tian X, Liu T, Yang C, Wang Q, Li D. Interface and performance of 3D printed continuous carbon fiber reinforced PLA composites. *Compos Part A Appl Sci Manuf* 2016;88:198–205. 10.1016/j.compositesa.2016.05.032
10. Wu H, Fahy WP, Kim S, Kim H, Zhao N, Pilato L, et al. Recent developments in polymers/polymer nanocomposites for additive manufacturing. *Prog Mater Sci* 2020;111:100638. 10.1016/j.pmatsci.2020.100638
11. Acquah SFA, Leonhardt BE, Nowotarski MS, Magi JM, Chambliss KA, Venzel TES, et al. Carbon Nanotubes and Graphene as Additives in 3D Printing. In Berber MR & Hafez IH (Eds.), *Carbon Nanotubes - Current Progress of their Polymer Composites* IntechOpen. 2016. https://doi.org/10.5772/63419
12. Home – 3DXTECH n.d. 3DXTech, https://www.3dxtech.com/
13. Stansbury JW, Idacavage MJ. 3D printing with polymers: Challenges among expanding options and opportunities. *Dent Mater* 2016;32:54–64. 10.1016/J.DENTAL.2015.09.018
14. Velu R, Vaheed N, Ramachandran K. Experimental investigation of robotic 3D printing of high-performance thermoplastics (PEEK): A critical perspective to support automated fibre placement process n.d. https://www.springerprofessional.de/en/experimental-investigation-of-robotic-3d-printing-of-high-perfor/17427482

15. Jiang Z, Diggle B, Tan ML, Viktorova J, Bennett CW, Connal LA. Extrusion 3D printing of polymeric materials with advanced properties. *Adv Sci* 2020;7:2001379. 10.1002/ADVS.202001379

16. Larraza I, Vadillo J, Calvo-Correas T, Tejado A, Olza S, Peña-Rodríguez C, et al. Cellulose and graphene based polyurethane nanocomposites for fdm 3d printing: Filament properties and printability. *Polymers (Basel)* 2021;13: 1–19. 10.3390/POLYM13050839

17. Schmitz DP, Ecco LG, Dul S, Pereira ECL, Soares BG, Barra GMO, et al. Electromagnetic interference shielding effectiveness of ABS carbon-based composites manufactured via fused deposition modelling. *Mater Today Commun* 2018;15:70–80. 10.1016/J.MTCOMM.2018.02.034

18. Ahmed K, Kawakami M, Khosla A, Furukawa H. Soft, conductive nanocomposites based on ionic liquids/carbon nanotubes for 3D printing of flexible electronic devices. *Polym J* 2019 515;51:511–521. 10.1038/s41428-018-0166-z

19. Velu R, Singamneni S. Selective laser sintering of polymer biocomposites based on polymethyl methacrylate. *J Mater Res* 2014;29:1883–1892. 10.1557/jmr.2014.211

20. Torrado AR, Shemelya CM, English JD, Lin Y, Wicker RB, Roberson DA. Characterizing the effect of additives to ABS on the mechanical property anisotropy of specimens fabricated by material extrusion 3D printing. *Addit Manuf* 2015;6:16–29. 10.1016/j.addma.2015.02.001

21. Jaisingh Sheoran A, Kumar H. Fused deposition modeling process parameters optimization and effect on mechanical properties and part quality: Review and reflection on present research. *Mater Today Proc* 2020;21:1659–1672. 10.1016/j.matpr.2019.11.296

22. Mohamed OA, Masood SH, Bhowmik JL. Optimization of fused deposition modeling process parameters: A review of current research and future prospects. *Adv Manuf* 2015;3:42–53. 10.1007/s40436-014-0097-7

23. Stansbury JW, Idacavage MJ. 3D printing with polymers: Challenges among expanding options and opportunities. *Dent Mater* 2015;32:54–64. 10.1016/j.dental.2015.09.018

24. Rajpurohit SR, Dave HK. Flexural strength of fused filament fabricated (FFF) PLA parts on an open-source 3D printer. *Adv Manuf* 2018;6:430–441. 10.1007/s40436-018-0237-6

25. Razavi-Nouri M, Rezadoust AM, Soheilpour Z, Garoosi K, Ghaffarian SR. Morphology and mechanical properties of poly (acrylonitrile-butadiene-styrene)/multi-walled carbon nanotubes nanocomposite specimens prepared by fused deposition modeling. *Polym Compos* 2021;42:342–352. 10.1002/pc.25829

26. Liu Q, Chen D. Viscoelastic behaviors of poly (e -caprolactone)/attapulgite nanocomposites 2008;44:2046–2050. 10.1016/j.eurpolymj.2008.04.035

27. Kumar A, Ohdar RK, Mahapatra SS. Parametric appraisal of mechanical property of fused deposition modelling processed parts. *Mater Des* 2010;31:287–295. 10.1016/j.matdes.2009.06.016

28. Wu W, Geng P, Li G, Zhao D, Zhang H, Zhao J. Influence of layer thickness and raster angle on the mechanical properties of 3D-pinted PEEK and a comparative mechanical study between PEEK and ABS. *Materials (Basel)* 2015;8:5834–5846. 10.3390/ma8095271

29. Dul S, Fambri L, Pegoretti A. Fused deposition modeling with ABS-graphene nanocomposites. *Compos PART A* 2016. 10.1016/j.compositesa.2016.03.013

30. Weng Z, Wang J, Senthil T, Wu L. Mechanical and thermal properties of ABS/montmorillonite nanocomposites for fused deposition modeling 3D printing. *Mater Des* 2016;102:276–283. 10.1016/j.matdes.2016.04.045

31. Prashantha K, Roger F. Multifunctional properties of 3D printed poly(lactic acid)/graphene nanocomposites by fused deposition modeling. *J Macromol Sci Part A Pure Appl Chem* 2017;54:24–29. 10.1080/10601325.2017.1250311

32. Skorski MR, Esenther JM, Ahmed Z, Miller AE, Hartings MR, Skorski MR, et al. The chemical, mechanical, and physical properties of 3D printed materials composed of TiO 2 -ABS nanocomposites. *Sci Technol Adv Mater* 2016;17:89–97. 10.1080/14 686996.2016.1152879

33. Yang L, Chen Y, Wang M, Shi S, Jing J. Fused deposition modeling 3D printing of novel poly(vinyl alcohol)/graphene nanocomposite with enhanced mechanical and electromagnetic interference shielding properties. *Ind Eng Chem Res* 2020;59:8066–8077. 10.1021/acs.iecr.0c00074

34. Azhari A, Toyserkani E, Villain C. Additive manufacturing of graphene-hydroxyapatite nanocomposite structures. *Int J Appl Ceram Technol* 2015;12:8–17. 10.1111/ijac.12309

35. Lee CS, Kim SG, Kim HJ, Ahn SH. Measurement of anisotropic compressive strength of rapid prototyping parts. *J Mater Process Technol* 2007;187–188:627–630. 10.1016/j.jmatprotec.2006.11.095

36. Vidakis N, Maniadi A, Petousis M, Vamvakaki M, Kenanakis G, Koudoumas E. Mechanical and electrical properties investigation of 3D-printed acrylonitrile–butadiene–styrene graphene and carbon nanocomposites. *J Mater Eng Perform* 2020;29:1909–1918. 10.1007/s11665-020-04689-x

37. Ning F, Cong W, Qiu J, Wei J, Wang S. Additive manufacturing of carbon fiber reinforced thermoplastic composites using fused deposition modeling. *Compos Part B Eng* 2015;80:369–378. 10.1016/j.compositesb.2015.06.013

38. Singamneni S, Velu R, Behera MP, Scott S, Brorens P, Harland D, et al. Selective laser sintering responses of keratin-based bio-polymer composites. *Mater Des* 2019;183. 10.1016/j.matdes.2019.108087

39. Velu R, Singamneni S. Thermal aspects of selective laser sintering of PMMA+ β-TCP composites. Sff symposium Engr Utexas Edu n.d.

40. Wang Y, Lei M, Wei Q, Wang Y, Zhang J, Guo Y, et al. 3D printing biocompatible l-Arg/GNPs/PLA nanocomposites with enhanced mechanical property and thermal stability. *J Mater Sci* 2020;55:5064–5078. 10.1007/s10853-020-04353-8

41. Ivanov E, Kotsilkova R, Xia H, Chen Y, Donato RK, Donato K, et al. PLA/graphene/MWCNT composites with improved electrical and thermal properties suitable for FDM 3D printing applications. *Appl Sci* 2019;9:1209. 10.3390/APP9061209

42. Tambrallimath V, Keshavamurthy RDS, Koppad PG, Kumar GSP. Thermal behavior of PC-ABS based graphene filled polymer nanocomposite synthesized by FDM process. *Compos Commun* 2019;15:129–134. 10.1016/J.COCO.2019.07.009

43. Yeong WY, Goh GD. 3D printing of carbon fiber composite: The future of composite industry? *Matter* 2020;2:1361–1363. 10.1016/j.matt.2020.05.010

44. Babu K, Das O, Shanmugam V, Mensah RA, Försth M, Sas G, et al. Fire behavior of 3D-printed polymeric composites. *J Mater Eng Perform* 2021; 30:1–11. 10.1007/s11 665-021-05627-1

45. Sriram V, Shukla V, Biswas S. Metal powder based additive manufacturing technologies—Business forecast. *3D Print. Addit. Manuf. Technol.*, Singapore: Springer Singapore; 2019, pp. 105–118. 10.1007/978-981-13-0305-0_10

46. Korkees F, Allenby J, Dorrington P. 3D printing of composites: Design parameters and flexural performance. *Rapid Prototyp J* 2020;26:699–706. 10.1108/RPJ-07-2019-0188

47. Allameh SM, Miller R, Allameh H. Mechanical properties of 3D printed biomimicked composites. *ASME Int Mech Eng Congr Expo Proc* 2018;12:1–5. 10.1115/IMECE2 018-86309

48. Raney JR, Compton BG, Mueller J, Ober TJ, Shea K, Lewis JA. Rotational 3D printing of damage-tolerant composites with programmable mechanics. *Proc Natl Acad Sci* 2018;115:1198–1203. 10.1073/pnas.1715157115

49. Chew E, Liu JL, Tay TE, Tran LQN, Tan VBC. Improving the mechanical properties of natural fibre reinforced laminates composites through Biomimicry. *Compos Struct* 2021;258:113208. 10.1016/j.compstruct.2020.113208

50. Allameh SM, Ogonek T, Cooper PD. Fabrication and characterization of bio-inspired structural composites. *Mech. Solids, Struct. Fluids, ASMEDC* 2008;12:369–372. 10.1115/IMECE2008-66778

51. Summe M, Allameh S. Mechanical properties of hemp-reinforced biomimicked composites. *Process. Eng. Appl. Nov. Mater., ASMEDC* 2010;12:71–76. 10.1115/IMECE2010-38440

52. Alghahtani H, Allameh SM. Effect of Fiber Form and Volume Fraction on Fiber-Reinforced Biomimicked Composites. Vol. 2 *Biomed. Biotechnol.*, American Society of Mechanical Engineers; 2012, pp. 1–5. 10.1115/IMECE2012-85718

53. Allameh S. Effect of Reinforcement Fiber Length on the Mechanical Behavior of Biomimicked Composites. Vol. 14 *Emerg. Technol. Mater. Genet. to Struct. Saf. Eng. Risk Anal.*, American Society of Mechanical Engineers; 2016. 10.1115/IMECE2016-65202

54. Velu R, Kamarajan BP, Ananthasubramanian M, Ngo T, Singamneni S. Post-process composition and biological responses of laser sintered PMMA and β-TCP composites. *J Mater Res* 2018;33:1987–1998. 10.1557/jmr.2018.76

55. Al-Dulimi Z, Wallis M, Tan DK, Maniruzzaman M, Nokhodchi A. 3D printing technology as innovative solutions for biomedical applications. *Drug Discov Today* 2021;26:360–383. 10.1016/J.DRUDIS.2020.11.013

56. Ranjan N, Singh R, Ahuja I. Material processing of PLA-HAp-CS-based thermoplastic composite through fused deposition modeling for biomedical applications. *Biomanufacturing* 2019:123–136. 10.1007/978-3-030-13951-3_6

57. Alam F, Shukla VR, Varadarajan KM, Kumar S. Microarchitected 3D printed polylactic acid (PLA) nanocomposite scaffolds for biomedical applications. *J Mech Behav Biomed Mater* 2020;103:103576. 10.1016/J.JMBBM.2019.103576

58. Naghieh S, Foroozmehr E, Badrossamay M, Kharaziha M. Combinational processing of 3D printing and electrospinning of hierarchical poly(lactic acid)/gelatin-forsterite scaffolds as a biocomposite: Mechanical and biological assessment. *Mater Des* 2017;133:128–135. 10.1016/J.MATDES.2017.07.051

9 Nanomaterial Used in 3D Printing Technology

Waleed Ahmed
Engineering Requirements Unit, College of Engineering, United Arab Emirates University, Abu Dhabi, United Arab Emirates

Essam Zaneldin
Civil and Environmental Engineering Department, College of Engineering, United Arab Emirates University, Abu Dhabi, United Arab Emirates

Amged Al Hassan
Mechanical Engineering Department, College of Engineering, United Arab Emirates University, Abu Dhabi, United Arab Emirates

Ali H. Al-Marzouqi
Chemical and Petroleum Engineering Department, College of Engineering, United Arab Emirates University, Abu Dhabi, United Arab Emirates

CONTENTS

9.1 INTRODUCTION

The world is transforming, and the engineering world is focusing on developing more economcally feasible and sustainable engineering solutions; therefore, the nanotechnology-oriented 3D printing technology is emerging as a new field as a means of creating a wide range of cost-effective products. 3D printing technology has dominated several industries and medical fields within a short time. This unique technology has provided remarkable advantages that were not previously available or were very expensive compared with other manufacturing technologies. In general, 3D printing depends on two main factors: the technical aspects necessary for printing materials and the developed materials used in the 3D printing process, which are considered the pillars of the industry. Therefore, the materials utilised in 3D printing are the main elements in developing this industry due to their broad applications in various areas of life, which directly impact the progress of the applied technology. In this chapter, different types of nanomaterials used in 3D printing technology are presented. Their benefits and impact on the properties of the products manufactured using 3D printing technologies are then discussed. Since enhanced applications at the nano-scale must be evaluated to address any potential risks to humans and the environment, the chapter also focuses on the gaps in risk assessment associated with 3D printing and nanotechnology. Furthermore, different applications and nanomaterials already implemented in 3D printing technology at a commercial scale are discussed to highlight the industrial impact of this promising technology.

9.1.1 Nanocomposites in Conjunction to Additive Manufacturing Technology

Additive manufacturing technology that is commercially regarded as 3D printing has a wide range of applications, as well as the adopted materials like ceramics, biomass, biocompatible, and antimicrobial material [1]. With the availability of low-cost FDM (fused deposition modeling) commercial 3D printers [2], it becomes possible to manipulate and tailor the properties of the 3D-printed objects, especially recycling the polymeric waste of the 3D printing process. One of the main challenges in reinforcing material with particles is the stress concentation initiated around the particles that lead to the degradation and the failure of the composite due to interfacial stresses and the size of the reinfected particles [3]. Therefore, many attempts are carried out to investigate using finite-element method and the reasons behind the failure [4], especially in the 3D printing process. Due to a larger surface-to-volume ratio, nanomaterials have become very useful for many applications, like the construction sector [5], where the different materials can be used with 3D printing concrete.

Nanomaterials has been a center of focus for many researchers for linking with the 3D printing technology to manufacture diverse range of engineering solution products. Weng et al. [6] examined the reinforcement impact of various nanomaterials viz. stereolithographic resin-based nanocomposites composed from montmorillonite, nano SiO2, and attapulgite, etc. The study stressed the potential of nanocomposites for utilization at the desktop-level SLA 3D printing. Furthermore, it involved studying the viability of the nylon-type six-based nanocomposite material as a finite difference method-oriented filament material. Hence, the study concluded with the discussion on the potential of the said material (i.e., nylon-type six-based nanocomposite) as a replacement material for the ABS as a 3D printing material [7]. Despite the bright avenues for successful research, development and application of the nanocomposite and nanoparticles' aggregation and occurrence of light-scattering phenomenon as result of high nano-filler content still remain the major setback in the popularity of the novel material [8]. Zhong et al. [9] studied the nanostructure of a geopolymer (i.e., alkali-activated, Portland cement-free concrete) to deduce the effects of nano-graphene oxide for the rheological extrusion properties and predicted the direct relationship of compressive strength of the material with its electrical conductivity. Furthermore, Wu et al. [10] adopted an analytical approach to predict the trends of material density of the alumina particles as attributes of size of particles (i.e., micro, nano, and mono-sized distribution particles). Whereas, Kazemian et al.'s [11] study deduced that shape-stability of 3D-printed cement paste can be enhanced by the addition of the silica fume and nano-clay particles into the samples. The advent of tunability of the physiochemical properties of the nanomaterials and their composites has opened fields of application, e.g., agriculture to additive manufacturing (AM) [12]. Moreover, the addition of carbon nano fibers, carbon nano tubes, graphene oxide, metal oxide nanoparticles, and the metal nanoparticles in various polymer matrix has enhanced the tensile strength and shear strength of various materials for different engineering applications [13]. Literature reveals that ancient glass paintings were created by integrating the gold nanoparticles in the PVA matrix to create a unique sparkling effect. This PVA matrix integration was different from conventional approaches since the gold particles were embedded inside the polymer rather than in a top coat. The unique sparkle can be characterized by an optical phenomenon that displays different colors on variable reflection and transmission of the incident light rays [14]. In fact, there is limited data for the hardened material properties in the literature, even though there are quite well-recognized impacts of the nano-sized admixtures on the fresh properties of the 3D-printed cementitious composites. Nanomaterials have substantial consequences on speeding up the hydration process, which ultimately leads to quicker formation of the strength-giving phases. The incorporation of nanomaterials appears to be a desirable solution in the direction of enhancing the initial strength advancement of printable composites. Kruger et al. [15] studied the effects of NC (nanoclay) and the SiC nanoparticles on the mechanical properties (i.e., strength) of 3D-printed concrete. He nano-silica integrated concrete samples exhibited significant enhancement in the compressive and flexural strengths, while the scope of mechanical property enhancement in the SiC-incorporated specimens was limited. Cho et al. [16] reported that nanoclay had substantial impacts on improving compressive strength, bending strength, and the modulus of elasticity in the lightweight 3D-printed foamed concretes. Literature includes a study on the compressive strength property of the high-volume printable mortars (fly ash) incorporated with nano attapulgite clay, which exhibited a noticeable improvement in the compressive strength [17]. On the other hand, the attapulgite incorporation in the mortars reduced the tensile bond strength, especially with every successful time interval during printing. Furthermore, Chougan et al. [18] also extended the nano-material incorporation technique to integrate the nano-graphite (of variable dosages) on the 3D-printed geopolymer composites to evaluate the enhanced mechanical performance in comparison to the corresponding traditionally cast samples. The findings demonstrated that the addition of nano-graphite particles has a constructive effect on the mechanical properties of the samples after the curing process. The flexural-strength of the printed elements was presented to constantly improve with the nano-graphite particles, and the strength value increased. Furthermore, the compressive strength of the 3D-printed samples that are integrated with NGPs was elevated to the control sample. Zhong et al. [9] 3D printed the graphene oxide particles infused geopolymers. They studied the impact of graphene-oxide loading measures on the electrical conductivity and mechanical property of 3D-printed sections. Their findings revealed that boosting the graphene oxide produced enhancements in elastic modulus and compressive strength of the 3D-printed elements enhanced with nanomaterials. Nevertheless, because of the nanoparticle agglomeration observed problem, the inclusion of unnecessary graphene-oxide particles lowers together with the elastic modulus and compressive strength of the 3D-printed samples [19]. Zhou et al. [20] explored the results of graphene-oxide lateral size on the altered mechanical properties of the geopolymer nanocomposites. The findings disclosed that the addition of graphene-oxide particles might change the geopolymers' rheology and the composite's appropriateness for the 3D printing procedure. Moreover, the study reported that the nanoparticle infusion in the geopolymers is further favorable with reference to the mechanical properties viz. modulus of elasticity, ultimate strength, fracture ductility, and compressive strength compared to the use of big-size particles [20]. Mendoza et al. [21] carried out an experimental study to form a comparative account on the effects of two groups of micro-and nano-sized admixtures, where silica fume versus the nano silica and nanoclay versus the metakaolin seemed to affect a large number of findings. It was observed that nano-sized admixtures were extra efficient in boosting the yield stress along with the rate of thixotropic buildup. Based on the chemical composition, silica nanoparticles are better than the clay-based nanoparticles in controlling the rheological properties of the samples. In the study of Kruger et al. [22], the effects of nanosilica (NS) were assessed on the rheological characteristics of 3D-printed concrete samples. The research verified that NS boosts the re-flocculation rate. On the contrary, greater amounts of nanosilica were observed to substantially reduce re-flocculation rates, thus causing an adverse impact on thixotropic performance. Moreover, the large quantity of nanosilica caused a substantial increase in the dynamic shear stress of the aterials for extended periods. Thus, the mixture NS was deemed suitable for progression to the 3D printing phase. A Dung et al. research article [23] detailed the procedure of preparation of nanofilm (chitosan/nano-TiO2) with various contents of nanoparticles (TiO2) by using 3D printing technology. They also examined the ink solution for the purpose of 3D printing technology, which was prepared by dissolution of chitosan by using the ultrasonic method. It was observed from the obtained results that the chitosan films were the most promising adsorbents for the removal of coloring matter. The study conducted by Konuray et al. [24] utilized a mixture of cycloaliphatic epoxy along with an anhydride curing agent in order to enhance the mechanical strength and performance of the commercially available photocurable resin. They treated the photocured samples in the subsequent step for curing of the epoxy resin to obtain an interpenetrated polymer network. It was examined from the obtained results that mechanical property named as young modulus of resin enhanced by 900% over the neat resin.

9.1.2 DEVELOPMENTS OF ADDITIVE MANUFCTURING TECHNOLOGY: 3D PRINTING

Nowadays, 3D printing technology is growing all over the globe in the development of biomaterials. Ulrich et al. [25] studied modulus thiolene resin for ultrafast two-photon micro fabrication of complex objects and facile postmodification by utilizing the different series of donor-acceptor stenhouse adduct photoswitches. This study can be beneficial for engineers and researchers in the biomedical field. A numerical study was conducted to investigate the effective surface agglomeration ability of the surfactants during the formation of nanodroplets [25]. It was observed from the numerical research that adding surfactants may be the additional benefit of raising the total kinetic energy of the ink composition and decreasing the possibility of agglomeration within the ink. Currently, the reputation of traditional type bio inks (gelatin–alginate–montmorillonite (GT–AT–MMT) is limited in the application of bio composite materials because of low mechanical strength. 3D printing technology and nano composite materials' mechanical characteristics to develop hydrogel bio-ink were examined [26]. This type of bio-ink (gelatin–alginate–montmorillonite (GT–AT–MMT) can be observed as an essential composite in 3D printing technology. Ulrich et al. [27] proposed an improved responsiveness to the 3D printing technology by introducing a unique approach characterized by various molecularly engineered multicolored microfabrication of 3D photochromic materials. This research can be beneficial for engineers and researchers to study diverse custom microfabrication of the 3D photochromic objects with molecularly engineered multicolors and responsiveness.

The application of the 3D printing technology is a promising technology that is growing worldwide in the domain of biomedical science and engineering. The application of 3D printing technology in the wiring process, which is widely employed in electronic packaging, was investigated. It was observed from the Ahn et al. study [28] that 3D printing technology has the potential of replacing the wiring that is used for the application of electronics packaging. To increase cellular function in response to the implanted surfaces, various nano and microscale topography was suggested as an essential element. Furthermore, bone implants mainly based on 3D-printed nanoparticles with dual micro topgraphy and nano topography promote interactions [29]. In addition, the study also involved the analysis of the 3D-printed titanium alloy bone implants (Ti6Al4V), which featured a unique dual topography consisting of micron-sized spherical particles and vertically aligned titanium nanotubes. This type of study can be beneficial for engineers and researchers to develop and improve 3D printing technology in biomedical science. In the Eickelmann et al. investigation [30], a cost-effective setup of microarrays was generated. It was observed from the study that Blu-ray players accurately transferred a minute amounts of biomolecules surface. They performed all the steps for microarray generation in any chemistry labs all over the globe, at a lower cost, without any special experimental apparatus. This study may be helpful to the design and development of a cost-effective laser-based nano 3D polymer printing system. In another study by Eickelmann [30], a low-cost laser-based 3D printing setup was proposed to allow for the spot-wise pattering of the surface with specific polymer nano layers. They also conducted all the steps for the generation of microarrays without any defined tools at a low cost. The proposed system enabled the re-generation of array production and offered a flexible and more economically viable alternative to the costly spotting robot technology. Furthermore, the convenient pattering method exhibited in this approach is characterized by microarrays pro-duction for different biomolecules like glycans intended for antibody and lectic screen interactions. Henceforth, the study also involved laser-transferring and re-action of the amino acid-containing nanolayers and the functionalized acceptor surface (pre-defined pattern), and hence, the description of solid-phase synthesis of peptides at each individual layer (i.e., on microarray).

9.1.3 3D PRINT MATERIAL OPTIMIZATION

Additive manufacturing is the fastest growing technology that allows the building of automated 3D objects in a layered manner. Sikora [31] addressed the recent development in the application of nano and microparticles like graphene-based materials, nanosilica, and clay nanoparticles and chemical admixtures in the domain of 3D printing and summarized how they impact the performance of 3D printing construction materials. This review study may be helpful toward the vital influence of material on the 3D printing application. 3D printing applications in the domain of nanomaterials are growing all over the world to achieve higher electronic con-ductivities. The 3D printing of polypyrrole (ppy) into a nafion sheet was conducted to examine the conductivity (electronic) of the product microstructures formed by multiphoton sensitized polymerization. This kind of research can be very useful for engineers and research to examine the characteristics of nanoparticles. The devel-opment and design of 3D printing technology are most important to improve the performance of 3D printing materials in different research areas. The enhancement of carbon nanotubes-based filament was modeled and formulated at different concentrations [32].

Furthermore, the 3D printing of PPy into a nafion sheet was performed to de-termine the electronic conductivity of the product microstructures formed by the multi-photon sensitized polymerization process. In addition, the filament formulations were printed in the form of series of complex structures. The material used for 3D printing sheets for the production of micro-structure is vital in 3D printing tech-nology. Nowadays, photonic integrated circuits technology is growing all over the world. The photonic integrated circuit was manufactured by using 3D printing techniques in conjunction with the advanced inorganic (hybrid) materials [33]. The organic and inorganic materials are developed on the fast sol-gel process for the production of advanced optical materials. The advantage of the above-mentioned bonding for micro and nanostructures' production plays a vital role in 3D printing technology. In the investigation by Ramachandran and Rajeswari [19], the nano-silica or the polylactic acid (PLA) bio nanocomposite filaments were produced to carry out the additive manufactured process (FDM). In addition, the impact of nano-silica on mechanical characteristics of additive manufactured PLA biocomposite was investigated.

Furthermore, nano-silica with several weight percentages were also incorporated with PLA and prepared the filament by the extrusion process. This experimental study was helpful in the domain of biocomposite manufacturing for the applications of 3D printing. 3D printing technology is the most important and growing technology in the biomedical field. In this research study, evaluation of polylactides and the carbon nanocomposite-based filament materail for 3D printing was carried out [34]. In addition, the possible application of biocomposite materials for the manufacturing of electrical circuitry for 3D printed robots was also studied Moreover, the scope of the study also encompassed the investigation of the mechanical and electrical properties of the filaments produced at a 0%–15% weight ratio of carbon nanoparticles (NC) in dichloromethane (DCM) and polylactide (PLA) solvent. The significant advantages of this technique can be beneficial to examining the nanomaterials' characteristics in 3D printing technology. The 3D printing of nanocomposite materials was carried out for the application of the biomedical field [35]. In addition, Silva et al. also gave a comprehensive account on potential applications of the graphene derivatives (i.e., polymers reinforced with graphene) with reference to the enhanced electrical and mechanical properties, and optimum cell response in context to the biomedical field. This advancement of 3D printing of nanocomposite materials can be beneficial for engineers and researchers to develop biomaterials.

9.1.4 OTHER APPLICATIONS OF NANOPARTICLE-BASED MODIFIED MATERIALS

The usage of nanomaterials in 3D printing technology is growing all over the world. The utilization of nanomaterial in 3D printing technology for industrial application in human health and safety was studied [36]. Furthermore, the regulatory development and their effect on the nanomaterial applications were also investigated. Gao and Zhang [37] chose the SiC and SiO2 particles as the reinforcing media, and an FGM featuring a volume fraction growing from 0%–20% (with increment of 1) layer by layer was prepared. Moreover, the experimental results concluded that the thermal conductivity and the mechanical properties of the SiC reinforced PDMS-based FGM particles deviated with the changing layers. The selected composition of nanoparticles may have a good performance in 3D printing technology. Nano-metric hydroxyapatite plays an essential role as a constituent of a hybrid system of bone scaffold fabrication because of its biocompatibility and biomimicry. HA nano-rods measuring (40 nm–60 nm × 20 nm) were prepared by a hydro-thermal method, including the utilization of a fully characterized ammonium-based dispersing agent (Darvan 821-A). A novel type of nanoparticles was introduced to produce a composite mimicking the bone composition. The obtained results showed that these nanoparticles are more viable for extrusion printing applications [38]. Energy-storing devices perform an essential role in the area of electrical power-generation systems. A detailed procedure of energy devices preparation by using 3D printing was presented [39]. In addition, waste and recycled polymeric materials with nanoparticles were used to print dry cells as ESD. It was investigated that 3D printed ESD was suitable and reliable to produce energy storage devices. The design and development of biomimetic structures are growing worldwide in the domain of medical 3D printing. Micro and mano-hierarchical structure coatings were manufactured using 3D printing titanium [40]. In addition, several characteristics of nanoparticles have also been investigated, including the morphology, hydrophobicity, corrosion resistance, and chemical compositions. 3D printing technology in the domain of biomedical is growing day by day all over the world. Several biomaterials were studied and investigated to improve the characteristics of biomedical materials [41]. There is immense importance of biomaterials for the 3D bioprinting processes to cater against the structural limitations of the native hydrogel-based bio-inks. The considerable advantages of this nano-bio material are to enhance the 3D printing technology in the biomedical domain. The evolution of the 3D printing sector in the area of biomedical engineering was examined [42]. The twin-type screw compounding technique was employed to prepare feedstock filaments viz. PVDF, PVDF-2.5%PPy, PVDF-2.5%CNT, and PVDF-2.5% (CNT + PPy) to examine the viability of 3D printing for the application of the biomedical domain. It was also discussed from the obtained results that the addition of PPy and CNT in PVDF matrix had modified the specific heat and charge capacity of the samples. The main advantages of these nanomaterials are the enhancement of performance of 3D printing technology in the domain of the biomedical field. Kumar et al. [43] conducted an experimental study to examine the mechanical and thermal properties of the material used in 3D printing. The performance of materials used in 3D printing technology was improved by using a twin-screw compounder for the preparation of feedstock filaments. It was also investigated from the obtained results that including ZnO in the PLA result in the development of highly reactive composites with water acted as a stimulus at room temperature and pressure with a weak, porous, and mechanically soft surface, whereas the 2% ZnO incorporated PLA sample exhibited less porosity, a higher hardness number, and highly responsive composite stimulus to the 40°C warm water media. The benefit of the above-mentioned materials is to increase the mechanical and thermal performance in the biomedical domain. Bone tissue engineering plays a vital role in biomedical applications and is strongly dependent on the use of 3D scaffolds. The biomedical performance of scaffolds was investigated using different mechanical testing techniques [44]. This research study focusses on analyzing the effect of nHA and mHA particles (i.e., nano-hydroxyapatite and micro-hydroxyapatite respectively) on the in-vitro biomechanical performance of hydroxyapatite scaffolds. The main feature or advantages of these above methods or materials used to modify the 3D printing technology in the domain of the biomedical field.

There are various instances in the literature for the natural hierarchical structures, e.g., tree leaves and butterfly wings, in context to their purposed-based design and functionality rather than just the aesthetics; such exclusive functionalities include ultralight nature due to the high surface area-to-volume ratio and high chemical reaction performance factor [45]. Similarly, replicating the natural hierarchical structures, TiO2 nanoparticle embedded hierarchical gyroid structure can be 3D printed to evaluate its photocatalytic performance. Therefore, such a replicative approach to the natural hierarchical structures can be applied to material science technology for developing materials for various specialized applications. 3D printing technology plays a vital role in biomedical applications. A 3D printable radiopaque ink was developed and successfully printed a finished artifact [46]. In addition, a 3D printable ultraviolet (UV) curable resin including zirconium oxide (ZrO2) nanoparticles was established. Five wt.% ZrO2 was disseminated in a base resin by employing a high-shear mixer. This type of 3D printable bio-ink on the artifact plays an essential role in the biomedical domain. Beloshenko et al. [47] investigated the effects of technological conditions of such a process and the zirconia incorporation on the characteristics of fine-fibrous filter materials. It is important to note that the filter materials were obtained by matrix polymer extraction from a micro-fibrillar composite through a material extrusion machine. This type of study can be beneficial for researchers and engineers in biomedical applications. A susceptible and low-cost temperature and humidity-independent strain sensor was developed based on the microdispensing direct-write (MDDW) technique [48]. In addition, a composite of silver and carbon was utilized as an active sensor material where both materials in the composite have opposite resistance temperature coefficients. It was observed from the obtained results that a sensor response showed an excellent temperature and humidity compensation, reducing the relative effect of temperature on the resistance by 99.5% and humidity by 99%. This type of sensor is best suitable to measure strain in the body.

The BTO nanoparticles are modified and incorporated with acrylate surface groups to enhance the mechanical and electrical conversion efficiency of the composites through the direct covalent linkages with polymer matrix I bright light exposure [49]. The experimental results obtained from this study are crucial for the project since they provide the basis of 3D piezoelectric polymers' fabrication and also lay the basis for production of the highly effective and efficient piezoelectric polymer material through the nano-interfacial tuning. Tao et al. [50] further represent a more promising model of the nerve conduit (nano-particle enhanced) for the boosting and regeneration of the peripheral nerves. Tao's conduit model comprises gelatin-methacryloyl (GelMA) hydrogels matrix disseminated with the drug-loaded poly (ethylene glycol)- poly(3-caprolactone) (MPEG-PCL) nanoparticles. The manufacturing process is accomplished by continuous 3D printing technology. These conduits have the ability to induce the recovery of sciatic injuries in the morphology, histopathology, and all other potential clinical conditions associated with regeneration of the peripheral nerves. Furthermore, Thomas et al. [51] 3D printed an antiskating polyamide spacer through the process of membrane distillation; meanwhile, the top surface of the spacer was quoted with fluorinated silica nanoparticles that were artificially fabricated via the solgel procedure. The study further evaluated the performance of the FS-coated printed surface in comparison to the conventional top-coat materials with different chemical properties. This research study can be beneficial for engineers and scientists to develop 3D-printed feed spacers via facile nanoparticles. And it discusses the feasibility and efficiency of R-NPs within a 3D-printed nerve conduit in the regenerative process of the of nerve defects [52]. The generated nerve conduit can care for the survival and explosion of Schwann cells. In addition, they also investigated that the regenerative RGFP966 drug offers remyelination by targeting the PI3K–AKT–ERK signaling pathway. The functional biomaterials can inspire this research study in tissue engineering. Kunchala and Kappagantula [53] investigated the impact of nanoparticle densifiers added to the liquid (print) for the mechanical performance and the manufacturability of ceramics fabricated by the binder jetting technique. The green alumina sample was manufactured with filler particles of average size (i.e., 40 μm) entrenched with nanoparticles of the average size (i.e.,50 nm) suspended in between the printi liquid with a variable concentration of about 0 wt.%–15 wt.%. This type of research can be beneficial for engineers and researchers to examine the performance of ceramic. Kumat et al. [43] used Zno nanoparticles grafted on polylactic acid (PLA) by utilizing the twin-screw compounders to produce feedstock filaments. The experimental results fetched from this study revealed that inducing 1% concentration ZnO in PLA gives a highly responsive composite that uses water as a stimulus at r.t.p (i.e., 25°C) and is porous in nature, mechanically weak, and soft-surfaced, whereas the 2% concentration ZnO in PLA offers a less porous, harder, and more actively responsive composite at 40°C water media as the stimulus. This type of study can be beneficial for engineers and researchers in biomedical 3D printing technology. A process was established to manufacture and characterize an alumina-enhanced nanocomposite with optimised mechanical properties for the injecting printing process [54]. Furthermore, the ceramic nanoparticles that have an average primary particle size (i.e., APPS) 16 nm and 31 nm were also analysed through the high-resolution scanning electron microscopy. This kind of investigation can be extremely beneficial for researchers and engineers in 3D printing applications. A newly developed 3D printing ink was used to print the pre-designed optimum-quality, three-dimensional structures. A 3D printing ink is also called a bionanocomposite nanoparticles [55]. It was also observed from the obtained results that magnetic ink aids in the fabrication of multi-material structures viz. hydrogels produced with a magnetic nanoparticle gradient. This type of study can be beneficial for engineers and researchers to develop new technology in biomedical applications. Nowadays, size engineering plays an important role in adaptable methods to boost potassium ion storage's electrochemical properties. A series of MoP nanoparticles splotched nitrogen-doped carbon nanosheets (MoP@NC) was manufactured [56]. The obtained results showed that ultrafine MoP nanoparticles enabled high-performance 3D-printed potassium ion hybrid capacitors.

Calamak and Ermis et al. [57] employed the situ silver nanoparticles (AgNP) manufacturing approach to functionalize a variety of biomaterials through investigation of the ultraviolet irradiation. The scope of study also included employment of the UV irradiation method and a green physical approach to design an AgNP functionalized 3D-printed polylactic acid (PLA) composite scaffold, regardless of any heat treatment or reducing agent. Hence, this methodology and green approach can be extensively utilized for the diverse 3D printing applications. Meanwhile, the study by Dudukovic et al. [58] gave a comprehensive account on an approach based on overcoming the 3D printing issues by quantifying and controlling the viscoelasticity of inks through direct ink writing method for the multi-material print of silica-titania glassy. The authors of the study also developed simple silica and silica-titania inks by utilizing the suspension of fumed silica nanoparticles in an organic solvent along with the dissolved molecular titania precursor. The results of the study indicate that there is potential of extending this 3D printing approach to colloidal systems, thereby allowing the predictive ink formulation design for the acquired printability in the direct ink writing method as discussed earlier.

9.1.5 Scope of Nanoparticle-Based Materials in Construction Industry

In recent years, many researchers have focused on studying various uses of nanoparticles in construction applications because there is a huge potential of nanoparticle incorporated construction materials, e.g., concrete, mortar, bricks, etc., for immense and promisingly extraordinary physical and chemical properties [59]. As a consequence of the pandemic crisis and the slump in oil prices, the construction industry worldwide is under tremendous pressure, and contractors are doing everything possible to become more competitive locally and globally. Despite the current circumstances, the construction industry in the United Arab Emirates (UAE) remains resilient, with several planned projects ready to start. The UAE federal government invests every year to construct new projects and maintain existing facilities to improve its domestic infrastructure [60]. According to a recent report, the UAE construction sector is anticipated to grow at 3.8% through 2022 [60]. The federal government of the United Arab Emirates spends Billions of dollars every year on developing new facilities in order to improve the domestic and commercial infrastructure. Despite the subdued oil prices, huge sums of money are still being invested into such mega projects in the UAE. By 2030, the UAE government is expected to spend about a whopping $300 billion on infrastructure development projects [61]. The UAE World Expo 2020 is organized by Dubai and, due to the coronavirus pandemic, is postponed to October 1, 2021, to March 31, 2022 [62]. This event will occupy more than 400 ha. constructed midway between emirates Dubai and Abu Dhabi. The exposition is anticipated to bring together more that 20 million visitors from nations all over the world, where important investments and developments in the infrastructure had been accomplished effectively by the Expo 2020 mega event start date [63]. Local municipalities and contractors in UAE face massive tasks to maintain the existing facilities and construct new projects and infrastructure utilities. The majority of these tasks are mainly linked using construction material that is robust, durable, quickly produced, and cost-effective [64].

9.2 ADDITIVE MANUFACTURING TECHNOLOGY AND THE CONSTRUCTION INDUSTRY

The construction industry has one of the slowest automation rates in comparison to other industries, such as the manufacturing industry. As a result of this low automation level, the construction industry has been characterized as having high instability, leading to unsafe work conditions and resulting in dangerously high accident rates [65]. Therefore, various automation methods in construction have been introduced to boost up the pace of the construction processes and increase application of the safety precautions. Additive manufacturing, also described as 3D printing, is a technology that allows building physical elements of a three-dimensional piece at individual layers in a multiple-layered system. In recent years, it is believed to be one of the most rapidly booming fields in the civil engineering sector and one of the critical pillars of the concept of Industry 4.0 [66]. With 3D printing, innovative digital manufacturing technology is now readily available for printing 3D concrete objects and prefabricated building blocks with efficient use of materials [67]. It is pretty clear that there is a global need to increase the construction of homes and meet the required demand and, to achieve this objective, the construction industry needs to use this innovative technology. The increase in the number of projects using 3D printing for construction between the years 1997 and 2018 [68].

In recent years, 3D printing technology has contributed to advancing the construction industry and can extend it even more. It has various benefits as compared to conventional construction methods. This technology allows new structural opportunities that cannot be accomplished with conventional construction techniques [69]. Also, concrete structures with complex designs can still be produced without actually increasing the costs or decreasing the productivity whatsoever. Since 3D printing does not require formwork and because of the time constraints of projects nowadays, a sound fall in project costs can be achieved due to the continuous work of 3D printers and the smaller labor force required for 3D printing [70]. In addition, it is well established that 3D printing reduces the production of construction wastes.

Despite the many benefits of 3D printing, more steps need to be taken. Additional developments are still required to make this technology effective for producing sturdy construction material that can bear structural loads and be cost-effective [71]. For example, the reduction in strength of 3D-printed parts due to the weak interlayer bonding is unavoidable, which can limit the wide range of applications of this manufacturing technology in the construction industry [72]. When the subsequent layers of materials are printed, voids tend to form, which lead to an additional porosity during the manufacturing process. The formation of voids between the layers deteriorates the mechanical properties of the overall composite material since the adhesion between printed layers is mitigated. Zareiyan and Khoshnevis [73] reported that the weak interlayer bonding forces between the layers could occur due to the formation of chemical forces resulting in the micro-scale surface roughness in the material. Another challenge of implementing 3D printing in the construction industry is the negative environmental impact of ordinary Portland cement, which has restricted its use in 3D printing. A shift to using geopolymers as an eco-friendlier material took place, and geopolymers gathered considerable interest due to their high early-age strength gain, low setting time, freeze-thaw resistivity, anti-corrosive properties, and cost efficiency [74]. However, some of the disadvantages of the geopolymers, such as the low flexural strength, the need for high temperatures, and the high risk associated with the eco-toxicity generated during the production of sodium silicate, has tendency to compromise the feasibility of using geopolymers in large-scale construction production [75]. To resolve some of these issues, several additives, such as the nanoparticle and fiber reinforcements, have been introduced to improve the interlayer bonding of 3D-printed geopolymers and their low flexural strength [76]. Fiber reinforcement (such as polymer, carbon, glass, and steel) has also been used to improve the flexural strength of 3D-printed materials [77]. To evaluate the effectiveness of fiber reinforcement, Hambach and Volkmer [77] compared carbon, glass, and basalt, and glass fibers on the mechanical performance of ordinary Portland cement 3D-printed parts. The results of the study revealed that some fibers improved the flexural strength of the elements. In another study, Bos et al. [77] also noticed an improvement in the flexural strength of 3D-printed samples of mortar reinforced with steel fibers compared to pieces not reinforced with steel fibers. On the other hand, few research efforts focused on studying the mechanical performance and characteristics of the 3D-printed polymers. Nematollahi et al. [59], for example, investigated the use of polypropylene fiber-reinforced geopolymers in 3D-printed parts and noticed an improvement in the workability and mechanical performance of the printed part.

9.3 SUMMARY

This preliminary part of this study is focused on analyzing the literature for understanding the technological background of modifying the existing engineering materials by incorporating the nanoparticles to enhance and optimize the material's mechanical and thermal properties with respect to specific applications, e.g., cement used in the construction industry. Most of the references discussed in the study are centered around development and testing of nanocomposites, e.g., stereolithographic resin-based nanocomposites, nylon-type six-based nanocomposite, nano-graphene oxide-infused geopolymer composites, silica fume and nano-clay particle-based nanocomposites, ZnO, SiC, SiO incorporated PLA composites, and hydroxyapatite-based composites, etc.

Moreover, the study is carried out in conjunction with its application for the additive manufacturing technologies, i.e., 3D printing technology. Many researches present in this study are centered around replacing the ABS plastic as the conventional 3D printing material with the nano-structure modified materials. For instance, the research of Weng et al. [6] stressed the potential of nanocomposites for utilization at the desktop-level SLA 3D printing and involved studying the viability of the nylon-type six-based nanocomposite material as a replacement material for the ABS as a 3D printing material. Furthermore, the advent of tunability of the physiochemical properties of the nanomaterials and their composites has opened diverse fields of application, e.g., agriculture to additive manufacturing (AM) [12]; the addition of carbon nano fibers, carbon nano tubes, graphene oxide, metal oxide nanoparticles, and the metal nanoparticles in various polymer matrix has enhanced the tensile strength, shear strength, and thermal properties of various materials for different 3D-print-based engineering applications [13]. The mechanical property enhancement and responsiveness (with regard to 3D printing manufacturing technology) of the modified nanocomposites is promptly studied and advocated in various works of literature discussed in this study.

Hence, the entire study stresses the scope and potential application of products made from the nanom,aterial-infused materials and fabricated using the 3D printing technologies. Because the world is transforming and the engineering world is focusing on developing more economically feasible and sustainable engineering solutions, the nanotechnology-oriented 3D printing technology is deemed the most viable means of creating a wide range of cost-effective and effective products. Later parts of the literature review in his study stress the utilization of the said technology for developing optimum construction materials and practices e.g., nanoparticle-infused self-healing cements, mortars, scaffolding materials, viscoelasticity management of concrete-based structures, etc. However, it is worthy to note that the application of such materials and products is not only limited to construction or the civil or mechanical engineering sectors, but other diverse range of fields, too, e.g., aerospace engineering (e.g., nano-composite structures), biomedical engineering (e.g., bio inks and bone implants etc.), and electrical engineering (e.g., piezo-electric polymers) sectors, too.

REFERENCES

1. Ahmed W., Siraj S., Al-Marzouqi A. Embracing additive manufacturing technology through fused filament fabrication for antimicrobial with enhanced formulated materials. *Polymer* 2021; 13:1523.
2. Ahmed W., Alabdouli H., Alqaydi H., Mansour A., Khawaja H.A. Open Source 3D Printer: A Case Study. In Proceedings of the International Conference on Industrial Engineering and Operations Management, Dubai, United Arab Emirates, 10–12 March 2020, p. 10.
3. Ahmed W.K. Investigating impact of the interfacial debonding on the mechanical propertiesof nanofiber reinforced composites. *Journal of Nano- and Electronic Physics* 2 13;5(4), pp. 1–5.
4. Ahmed W., Al-Rifaie W., Zaneldin E. Chapter 4—Mathematical modeling and simulation of interfaces between fiber and its matrix. In *Micro and Nano Technologies, Fiber-Reinforced Nano-composites: Fundamentals and Applications*, Elsevier: Amsterdam, The Netherlands, 2020, pp. 91–99. https://essuir.sumdu.edu.ua/bitstream-download/123456789/35553/1/765F3d01.pdf:jsessionid=19757345B31B63C9239FE01F86DBE748
5. Al-Rifaie W.N., Ahmed W.K. CH30 – Nano cement mortars for construction materials. In *Micro and Nano Technologies, Smart Nanoconcretes and Cement-Based Materials*, Elsevier, 2020, pp. 649–692. 10.1016/B978-0-12-817854-6.00030-1
6. Weng Z., Zhou Y., Lin W., Senthil T., Wu L. Structure-property relationship of nano enhanced stereolithography resin for desktop SLA 3D printer. *Compos Appl Sci Manuf* 2016; 88:234–242.
7. Boparai K.S., Singh R., Fabbrocino F., Fraternali F. Thermal characterization of recycled polymer for additive manufacturing applications. *Compos B Eng* 2016; 106:42–47.
8. Han Y., Wang F., Wang H., Jiao X., Chen D. High-strength boehmite-acrylate composites for 3D Printing: Reinforced filler-matrix interactions. *Compos Sci Technol* 2018; 154:104–109.
9. Zhong J., Zhou G.-X., He P.-G., Yang Z.-H., Jia D.-C. 3D printing strong and conductive geo-polymer nanocomposite structures modified by graphene oxide. *Carbon* 2017; 117:421–426. 10.1016/j.carbon.2017.02.102
10. Wu H., Cheng Y., Liu W., He R., Zhou M., Wu S., Song X., Chen Y. Effect of the particle T.D. Ngo et al. Composites Part B 143 (2018) 172–196 size and the debinding process on the density of alumina ceramics fabricated by 3D Printing based on stereolithography. *Ceram International* 2016; 42(15):17290–17294. https://doi.org/10.1016/j.ceramint.2016.08.024
11. Kazemian A., Yuan X., Cochran E., Khoshnevis B. Cementitious materials for construction-scale 3D printing: Laboratory testing of fresh printing mixture. *Construct Build Mater* 2017; 145:639–647.
12. Koumoulos E.P., Gkartzou E., Charitidis C.A. Additive (nano) manufacturing perspectives: The use of nanofillers and tailored materials. *Manuf Rev* 2017; 4:12. 10.1051/mfreview/2017012.
13. De Ciencias C., Aditiva M., Aplicadas C.D.C., Nacional L., Aditiva D.M. Nanocomposites for additive manufacturing. *Am J Chem Res* 2017, 1(5):1–14. 10.28933/ajcr-2017-04-2701.

14. Kool L., Bunschoten A., Velders A.H., Saggiomo V. Goldnanoparticles embedded in a polymer as a 3D-printable dichroic nanocomposite material. *Beilstein J Nanotechnol* 2019, 10:442–447. 10.3762/bjnano.10.43.

15. Kruger J., Van den Heever M., Cho S., Zeranka S., Van Zijl G. (2019a) High-performance 3D printable concrete enhanced with nanomaterials. In: Proceedings of the International Conference on Sustainable Materials, Systems and Structures (SMSS 2019) 533 New Generation of Construction Materials, 20–22 March 2019, Rovinj, Croatia, pp. 533–540.

16. Cho S., Kruger J., van Rooyen A., Zeranka S., van Zijl G. (2020) Rheology of 3D printable lightweight foam concrete incorporating nano-silica. In: Mechtcherine V., Khayat K., Secrieru E. (eds) *Rheology and Processing of Construction Materials*. Springer International Publishing: New York, pp. 373–381. 10.1007/978-3-030-22566-7_43

17. Panda B., Unluer C., Tan M.J. (2019) Extrusion and rheology characterization of geopolymer nanocomposites used in 3D Printing. *Compos Part B Eng*; 176:1–9.

18. Chougan M., Hamidreza Ghaffar S., Jahanzat M., Albar A., Mujaddedi N., Swash R. The influence of nano-additives in strengthening mechanical performance of 3D printed multi-binder geopolymer composites. *Constr Build Mater* 2020a;250:118928. 10.1016/j.conbuildma t.2020.118928

19. Ramachandran M.G., Rajeswari N. Influence of nano silica on mechanical and tribological properties of additive manufactured pla bio nanocomposite. *Silicon* 2021; 14(20210103): 703–709. 10.1007/s12633-020-00878-4

20. Zhou G.X., Li C., Zhao Z., Qi Y.Z., Yang Z.H., Jia D.C., Zhong J., Zhou Y. 3D printing geopolymer nanocomposites structure: Graphene oxide size effects on a reactive matrix. *Carbon* 2020; 164:215–223. 10.1016/j.carbon.2020.02.021

21. Mendoza Reales O.A., Duda P., Silva E., Paiva M., Filho R. Nanosilica particles as structural buildup agents for 3D Printing with Portland cement pastes. *Constr Build Mater* 2019; 219:91–100. 10.1016/j.conbu ildma t.2019.05.174

22. Kruger J., Zeranka S., van Zijl G. An ab initio approach for thixotropy characterisation of (nanoparticle-infused) 3D printable concrete. *Constr Build Mater* 2019; 224:372–386. 10.1016/j.conbuildmat.2019.07.078

23. Dung H.T., Chinh N.T., Lu L.T., Hoang T. Methylene blue adsorption of chitosan/nano-tio2 films prepared by 3d printing method. *Vietnam Journal of Chemistry* 2021; 59(3):319–325.

24. Konuray O., Altet A., Bonada J., Tercjak A., Fernández-Francos X., Ramis X. Epoxy doped, nano-scale phase-separated poly-acrylates with potential in 3d Printing. *Macromolecular Materials and Engineering* 2021; 306(3):1–10.

25. Aphinyan S., Ang E.Y.M., Yeo J., Ng T.Y., Geethalakshmi K.R. Numerical study of surface agglomeration of ultraviolet-polymeric ink and its control during 3d nano-inkjet printing process. *Journal of Polymer Science Part B: Polymer Physics* 2018; 56(24):1615–1624. 10.1002/polb.24749

26. Wu W. Study on 3d printing technology and mechanical properties of a nano-enhanced composite hydrogel bio-ink. *Micro & Nano Letters* 2020; 15(13):964–968. 10.1049/mnl.2019.0712

27. Ulrich, S., Wang, X., Rottmar, M., Rossi, R.M., Nelson, B.J., Bruns, N., Müller, R., Maniura-Weber, K., Qin, X.-H., Boesel, L.F. Nano-3D-printed photochromic microobjects. *Small* 2021; 17:2101337. 10.1002/smll.202101337

28. Ahn H.-S., Oh A.-S., Kim D.-H., Choi Y., Kim K.-H., Bae H.-C. Fabrication of cu wiring touch sensor via laser sintering of cu nano/microparticle paste on 3d-printed substrate. *Advanced Engineering Materials* 2021; 23(1):1–9. 10.1002/adem.202000688

29. Gulati K., Prideaux M., Kogawa M., Atkins G.J., Findlay D.M., Losic D. Anodized 3d–printed titanium implants with dual micro- and nano-scale topography promote interaction with human osteoblasts and osteocyte-like cells. *Journal of Tissue Engineering and Regenerative Medicine* 2017; 11(12):3313–3325. 10.1002/term.2239

30. Eickelmann S., Tsouka A., Heidepriem J., et al. Microarray synthesizer: A low-cost laser-based nano-3d polymer printer for rapid surface patterning and chemical synthesis of peptide and glycan microarrays (adv. mater. technol. 11/2019). *Advanced Materials Technologies* 2019; 4(11):1–11.

31. Sikora P., Chougan M., Cuevas K., et al. *The Effects of Nano- and Micro-Sized Additives on 3d Printable Cementitious and Alkali-Activated Composites: A Review.* Appl Nanosci, Technische Universität Berlin: Berlin, 2021 July 31, 2021. 12, 805–823(2022). https://doi.org/10.1007/s13204-021-01738-2

32. Thomas D.J. Developing nanocomposite 3d printing filaments for enhanced integrated device fabrication. *The International Journal of Advanced Manufacturing Technology* 2018; 95(9-12):4191–4198. 10.1007/s00170-017-1478-4

33. Gvishi R., Sokolov I. 3D sol–gel printing and sol–gel bonding for fabrication of macro- and micro/nano-structured photonic devices. *Journal of Sol-gel Science and Technology* 2020; 95(3):635–648. 10.1007/s10971-020-05270-7

34. Potnuru A., Tadesse Y., SpringerLink (Online service). *Investigation of Polylactide and Carbon Nanocomposite Filament for 3d Printing* 2018; Day:19. 10.1007/s40964-018-0057-z

35. Silva M., Pinho I.S., Covas José A., Alves Natália M., Paiva M.C. 3D printing of graphene-based polymeric nanocomposites for biomedical applications. *Functional Composite Materials* 2021; 2(1):1–21. 10.1186/s42252-021-00020-6

36. Taylor A.A., Freeman E.L., van der Ploeg M.J.C. Regulatory developments and their impacts to the nano-industry: A case study for nano-additives in 3d Printing. *Ecotoxicology and Environmental Safety* 2021; 207:1–11.

37. Gao F., Zhang Y. Method for preparing micro-nano-particle-reinforced pdms-based fgm using 3d printing single nozzle. *Materials Letters* 2020; 280:1–4. 10.1016/j.matlet.2020.128548

38. Montalbano G., Molino G., Fiorilli S., Vitale-Brovarone C. Synthesis and incorporation of rod-like nano-hydroxyapatite into type i collagen matrix: A hybrid formulation for 3d Printing of bone scaffolds. *Journal of the European Ceramic Society* 2020; 40(11):3689–3697. 10.1016/j.jeurceramsoc.2020.02.018

39. Singh R., Kumar R. Reference module in materials science and materials engineering. In: *Energy Storage Device from Polymeric Waste Based Nano-Composite by 3d Printing.* Elsevier Inc, 2015. 10.1016/B978-0-12-803581-8.11239-1

40. Wei Y., Hu Y., Li M., Li D. Sr-containing micro/nano-hierarchical textured tio2 nanotubes on 3d printing titanium. *Inorganic Chemistry Communications* 2020; 117:1–4. 10.1016/j.inoche.2020.107947

41. Bhattacharyya A., Janarthanan G., Noh I. Nano-biomaterials for designing functional bioinks towards complex tissue and organ regeneration in 3d bioprinting. *Additive Manufacturing* 2021; 37:1–14. 10.1016/j.addma.2020.101639

42. Kumar R., Pandey A.K., Singh R., Kumar V. On nano polypyrrole and carbon nano tube reinforced pvdf for 3d printing applications: Rheological, thermal, electrical, mechanical, morphological characterization. *Journal of Composite Materials* 2020; 54(29):4677–4689.

43. Kumar R., Singh R., Singh M., Kumar P. ZnO nanoparticle-grafted PLA thermoplastic composites for 3D printing applications: Tuning of thermal, mechanical, morphological and shape memory effect. *Journal of Thermoplastic Composite Materials.* May 2020. doi: 10.1177/0892705720925119

44. Domingos M., Gloria A., Coelho J., Bartolo P., Ciurana J. Three-dimensional printed bone scaffolds: The role of nano/micro-hydroxyapatite particles on the adhesion and differentiation of human mesenchymal stem cells. *Proceedings of the Institution of Mechanical Engineers* 2017; 231(6):555–564. 10.1177/0954411916680236

45. Jo W., Yoon B.J., Lee H., Moon M.-W. 3D printed hierarchical gyroid structure with embedded photocatalyst tio$_2$ nanoparticles. *3D Printing and Additive Manufacturing* 2017; 4(4):222–230. 10.1089/3dp.2017.0033

46. Shannon A., O'Connell A., O'Sullivan A., et al. A radiopaque nanoparticle-based ink using polyjet 3d Printing for medical applications. *3D Printing and Additive Manufacturing* 2020; 7(6):259–268. 10.1089/3dp.2019.0160

47. Beloshenko V., Chishko V., Plavan V., et al. Production of filter material from polypropylene/copolyamide blend by material extrusion-based additive manufacturing: Role of production conditions and zro$_2$ nanoparticles. *3D Printing and Additive Manufacturing* 2021; 8(4):253–262. 10.1089/3dp.2020.0195

48. Nadeem I., Memoon S., Khalid R., et al. Fabrication of temperature- and humidity-independent silver nanoparticle's carbon composite-based strain sensor through additive manufacturing process. *3D Printing and Additive Manufacturing* 2021; (20210712). 10.1089/3dp.2021.0032

49. Kim K., Chen S., Sirbuly D.J., et al. 3D optical printing of piezoelectric nanoparticle-polymer composite materials. *Acs Nano* 2014; 8(10):9799–9806. 10.1021/nn503268f

50. Tao J., Zhang J., Du T., et al. Rapid 3d Printing of functional nanoparticle-enhanced conduits for effective nerve repair. *Acta Biomaterialia* 2019; 90:49–59. 10.1016/j.actbio.2019.03.047

51. Thomas N., Kumar M., Palmisano G., et al. Antiscaling 3d printed feed spacers via facile nanoparticle coating for membrane distillation. *Water Research* 2021; 189:1–12. 10.1016/j.watres.2020.116649

52. Xu X., Tao J., Wang S., et al. 3D printing of nerve conduits with nanoparticle-encapsulated rgfp966. *Applied Materials Today* 2019; 16:247–256. 10.1016/j.apmt.2019.05.014

53. Kunchala P., Kappagantula K. 3D printing high density ceramics using binder jetting with nanoparticle densifiers. *Materials & Design.* 2018; 155:443–450. 10.1016/j.matdes.2018.06.009

54. Graf D., Burchard S., Crespo J., et al. Influence of al$_2$o$_3$ nanoparticle addition on a uv cured polyacrylate for 3d inkjet printing. *Polymers* 2019; 11(4):1–13. 10.3390/polym11040633

55. Podstawczyk D., Nizioł M., Szymczyk P., Wiśniewski Piotr, Guiseppi-Elie A. 3D printed stimuli-responsive magnetic nanoparticle embedded alginate-methylcellulose hydrogel actuators. *Additive Manufacturing* 2020; 34:1–12. 10.1016/j.addma.2020.101275

56. Wei Zong, Ningbo Chui, Zhihong Tian, et al. Ultrafine mop nanoparticle splotched nitrogen-doped carbon nanosheets enabling high-performance 3d-printed potassium-ion hybrid capacitors. *Advance Science .* 2021; 8(7): 1–11. 10.1002/advs.202004142

57. Calamak S., Ermis M. In situ silver nanoparticle synthesis on 3D-printed polylactic acid scaffolds for biomedical applications. *Journal of Materials Research* 2020; 36(2020): 166–175. https://doi.org/10.1557/s43578-020-00064-7

58. Dudukovic N.A., Wong L.L., Nguyen D.T., et al. Predicting nanoparticle suspension viscoelasticity for multimaterial 3d Printing of silica–titania glass. *Acs Applied Nano Materials* 2018; 1(8):4038–4044. 10.1021/acsanm.8b00821

59. Nematollahi, B., Xia, M., Vijay, P., Sanjayan, J.G. *Properties of Extrusion-Based 3D Printable Geopolymers for Digital Construction Applications*, Elsevier Inc, 2019. 10.1016/B978-0-12-815481-6.00018-X.

60. Mohajerani A., Burnett L., Smith J.V., et al. Nanoparticles in construction materials and other applications, and implications of nanoparticle use. *Materials* (1996-1944). 2019; 12(19):1–25.

61. Businesswire, https://www.businesswire.com/news/home/20210625005170/en, last accessed June 14, 2021.

62. Gulf News, https://gulfnews.com/business/tourism/uae-to-spend-more-than-300-billion-on-infrastructure-development-by-2030-1.1314180, last accessed June 23, 2021.

63. Expo 2020, https://www.expo2020dubai.com/en/understanding-expo/participants, last accessed July 19, 2021.

64. BBC News, https://www.bbc.com/news/business-56682427, last accessed July 2, 2021.

65. Ahmed, W., Alnajjar, F., Zaneldin, E., Al-Marzouqi, A.H., Gochoo, M., Khalid, S. Implementing FDM 3D printing strategies using natural fibers to produce biomass composite. *Materials* 2020; 13(18):1–23. 10.3390/ma13184065

66. Sikora, P., Chougan, M., Cuevas1, K., Liebscher, M., Mechtcherine, V., Ghaffar, S.H., Liard, M., Lootens, L., Krivenko, P., Sanytsky, M., Stephan, D. The effects of nano- and micro-sized additives on 3D printable cementitious and alkali-activated composites: A review. *Applied Nanoscience* 2021; 12:805–823. 10.1007/s13204-021-01738-2.

67. Prinsloo, J., Sinha, S., von Solms, B. A Review of Industry 4.0 manufacturing process security risks. *Appl Sci* 2019; 9(23):5105. 10.3390/app9235105.

68. Ghaffar, S.H., Corker, J., Fan, M. Additive manufacturing technology and its implementation in construction as an eco-innovative solution. *Automation in Construction* 2018; 93:1–11. 10.1016/j.autcon.2018.05.005.

69. Buswell, R.A., Leal de Silva, W.R., Jones, S.Z., Dirrenberger, J. 3D Printing using concrete extrusion: A roadmap for research. *Cem. Concr. Res* 2018;112:37–49. 10.1 016/j.cemconres.2018.05.006.

70. Hoffmann, M., Skibicki, S., Pankratow, P., Zieliński, A., Pajor, M., Techman, M. Automation in the construction of a 3D-printed concrete wall with the use of a lintel gripper. *Materials* 2020; 13(8):1800. 10.3390/ma13081800.

71. Valente, M., Sibai, A., Sambucci, M. Extrusion-based additive manufacturing of concrete products: Revolutionizing and remodeling the construction industry. *J Compos Sci* 2019; 3:88. 10.3390/jcs3030088.

72. Xia, M., Sanjayan, J. Method of formulating geopolymer for 3D Printing for construction applications. *JMADE* 2016; 110:382–390. 10.1016/j.matdes.2016.07.136.

73. Zareiyan, B. Khoshnevis. Interlayer adhesion and strength of structures in contour crafting - effects of aggregate size, extrusion rate, and layer thickness. *Automation in Construction* 2017; 81:112–121. 10.1016/j.autcon.2017.06.013

74. Xie, J., Kayali, O. Effect of superplasticiser on workability enhancement of Class F and Class C fly ash-based geopolymers. *Constr. Build. Mater* 2016; 122:36–42. 10.1 016/j.conbuildmat.2016.06.067.

75. Zannerni, G.M., Fattah, K.P., Al-Tamimi, A.K. Ambient-cured geopolymer concrete with single alkali activator. *Sustain. Mater. Technol* 2020; 23:1–9. 10.1016/j.susmat.2 019.e00131.

76. Nematollahi, B., Vijay, P., Sanjayan, J., Nazari, A., Xia, M., Nerella, V.N., Mechtcherine, V. Effect of polypropylene fibre addition on properties of geopolymers made by 3D Printing for Digital Construction. *Materials* 2018; 11(12):1–16. 10.3390/ ma11122352.

77. Hambach, M., Volkmer, D. Properties of 3D-printed fiber-reinforced Portland cement paste. *Cem. Concr. Compos* 2017; 79:62–70. 10.1016/j.cemconcomp.2017.02.001.

10 3D Printed Batteries
Architecture, Nanomaterials Processing, Properties, and Performance

Vikas Kumar
Materials Innovation Centre, School of Engineering,
University of Leicester, UK

Shiladitya Paul
Materials Innovation Centre, School of Engineering,
University of Leicester, UK

Materials and Structural Integrity Technology Group, TWI
Ltd, Cambridge, UK

CONTENTS

DOI: 10.1201/9781003189404-10

183

10.1 INTRODUCTION

Electrochemical batteries are a vital source of mobile energy necessary to meet the requirements of modern electronic and electrical appliances, such as electric vehicles, smart phones, and laptops, and serve as an energy storage system for intermittent energy sources (wind, solar etc.) [1]. Some modern electronics like smart cards, sensors, transmitters, medical patches, wireless sensor networks, and other Internet of Things (IoTs) devices required batteries with different shapes and sizes with high-power density, reliability, and longer life. The state-of-the-art commercial battery systems, such as lithium ion batteries (LIBs), are restrictive in terms of their form and are difficult to customize and miniaturize to meet the application criteria of the above-mentioned advanced applications. Three-dimensional (3D) printing has emerged as a novel class of powerful technique that can manufacture these batteries in complex architectures in a layer-by-layer method [2,3]. These 3D-printed systems can achieve high energy density and also can be manufactured as an integrated power source for IoTs, modern personal devices, and remote micro devices. Over the last decade, 3D printing technologies have received increasing academic and industrial attention due to their suitability in manufacturing complex-shaped batteries, as presented in Figure 10.1. Compared to conventional two-dimensional (2D) batteries, low-cost, 3D-printed batteries can achieve higher specific energy and power capability by controlling the component structure (anode, electrolyte, and cathode) for higher weight percentage loading of the active materials within the battery components [4]. Higher loading of active materials in the 3D-printed structures can effectively increase the active surface area by creating a well-controlled porous structure so that more surface is available for reactions. Additionally, a higher loading of active materials will enhance the electrode surface area-to-volume ratio that lowers the ohmic losses due to shorter charge (ions and electrons) transport distances between the electrodes [4]. In the case of lithium-based batteries, the projected output power of a 3D-printed battery is expected to be up to two orders of magnitude higher compared to conventional 2D battery of equal size. This is the result of a complex 3D-printed network of mesoporous structures within the battery electrode that increases the active surface area and provides shorter diffusion pathways for ion transport. In addition to that, 3D printing offers flexibility in: (i) manufacturing complex geometries directly from computer-aided designs (CAD), (ii) single step multi-materials printing, (iii) freeform production of battery electrodes, (iv) precise control over electrode thickness and functionality, (v) lower manufacturing cost due to minimal material wastage, and (vi) a potential to eliminate the expensive packaging via the direct integration of the battery and the external electronics [5,6].

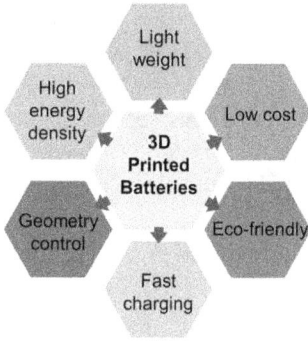

FIGURE 10.1 Schematic representation of the advantages of 3D printing methods for battery applications.

10.2   3D PRINTING TECHNIQUES FOR BATTERY MANUFACTURING

10.2.1   Overview

3D printing was founded in the 1960s to create photopolymers-based 3D objects with the use of laser beams and UV rays to polymerize resin in a layer-by layer way that would later be named as stereolithography (SLA) in 1984 [7]. Later on in the 1990s, several 3D printing techniques were investigated to construct 3D objects either by binding, fusion, and solidification of different class of materials (i.e., resins and powder materials) in a layer-by-layer way using predetermined CAD designs [7]. Currently, ASTM International has classified seven categories of 3D printing techniques including [8]: (1) Vat photopolymerization (e.g., stereolithography [SLA], digital light processing [DLP], and two-photon lithography [TPL]), (2) material extrusion (e.g., direct ink writing [DIW] and fused deposition modeling [FDM]), (3) powder bed fusion (e.g., selective laser sintering [SLS], direct metal laser sintering [DMLS]), (4) material jetting (e.g., inkjet printing [IJP], aerosol jet printing [AJP]), (5) sheet lamination (e.g., laminated object manufacturing [LOM]), (6) binder jetting (BJ) (e.g., 3D printing [3DP]), and (7) directed energy deposition [DED] (e.g., direct metal deposition [DMD], electron beam additive manufacturing (EBAM), as shown in Figure 10.2. Each 3D printing technology has its own application characteristics, and therefore, not every 3D printing technique has the battery manufacturing ability. So far, some advanced 3D printing techniques, such as DIW, IJP, FDM, SLA, BJ,

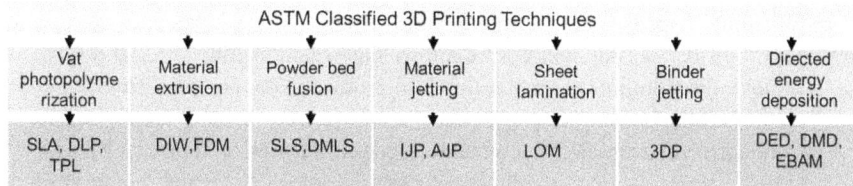

FIGURE 10.2 Schematic illustration of the ASTM classified 3D printing working principles (second row) and the corresponding printing techniques for different class of materials (third row).

SLS, and LOM, have been used to fabricate high-performance batteries with well-controlled geometries. Among them, DIW [9], IJP [10], FDM [11], SLA [12], and SLS [13] have been widely adopted for the manufacturing of 3D-printed batteries for various applications, as summarised in Table 10.1. Therefore, this section mainly emphasizes these 3D printing techniques for the rapid manufacturing of unique 3D material architectures and battery modules with desired set of specifications and performance.

10.2.2 Direct Ink Writing (DIW)

DIW (also known as robocasting) is the most popular, low-cost, easy-to-operate, material extrusion type 3D printing technique that can construct highly accurate and complex 3D architectures from micro-scale to the millimetre-scale (depends on nozzle size) directly from CAD designs. DIW offers a wide choice of liquid ink feedstock materials (or paste) to directly deposit on the printing platform in a layer-by-layer way with the help of a nozzle without any additional masking requirements, as presented in Figure 10.3a. In DIW printing process, feedstock material inks passed through the nozzle that moves horizontally and after printing the CAD-predefined layer, the printing platform moves vertically downward. After that, the extruded ink solidifies through liquid evaporation, gelation, or curing (solvent or thermal). The printing resolution of the battery structure usually depends on nozzle diameter, ink extrusion pressure, and the sheer-thinning properties of the inks. The shear-thinning property of the inks refers to the variation in viscosity under the applied shear rate. The ink viscocity needs to be decreased under high shear rate and flow, like a liquid through the nozzle; however, it behaves like a solid at low shear rate. For complex structures, 3D printing resolution, the ink functionality, and its rheological properties are the key factors that include the yield stress (σ_y), viscoelastic modulus (G'-storage modulus, G''-loss modulus), relaxation time, and viscosity (η). Specific yield stress, suitably high viscosity (10^6–10^8 cP) with a shear thinning behaviour are the key factors that need to be optimised for the battery materials in DIW printing process [14]. DIW 3D printing has been widely used for the manufacture of battery modules because of its advantages, which include:

- The wider materials flexibility is one of the main advantages of DIW 3D printing. This technique has the ability to print different materials, ranging from resins, waxes, metals, colloidal suspensions, ceramics, and novel functional hybrid nanomaterials.
- The high-performing ink formulations have demonstrated the ability to achieve the printing resolution down to 1 μm and the potential to produce integrated battery modules [15].
- Conductive as well as non-conductive ceramic nanomaterials can be used to tune the rheological properties to print complex 3D geometries with an enhanced electrochemical and mechanical properties.
- DIW 3D printing offers a high output, low-cost, easy-to-handle process with an ability to print a multi-materials ink system widely used for battery manufacturing.

TABLE 10.1

Comparative Overview of Different 3D Printing Techniques along with Their Materials Availability, Advantages, and Limitations

3D Printing Technique	Printable Materials	Resolution (µm)	Printing Speed	Cost	Advantages	Limitations
DIW	• Polymers • Waxes • Gels • Ceramics • Metals • Composites	1–250	High	Low	• Low cost operation • Materials flexibility • Accurate printing of complex 3D structures • No mask constraint	• Poor mechanical properties • Additives requirements • High functionality requirement for inks
FDM	• Thermoplastics • Metals • Carbon materials	50–200	High	Low	• Low cost • High speed • Large size printing capability • No post processing requirement	• Poor mechanical Properties • Poor surface quality • Low resolution
IJP	• Conductive Polymers • Metals • Sol-gel • Carbon based materials • Polymers	5–200	Low	Low	• Low cost • High printing resolution • Low material waste • Large area printing • Multimaterials capability • Low temperature process	• Narrow range of printable inks, • Low printing speed • Not suitable for high mass loading • Poor print head • Durability
SLA	Photopolymer materials	0.25–10	Low	High	• High efficiency • High resolution • High surface quality • True arbitrary designs	• Limited to photo-sensitive resins • Requirement of low viscosity • Lower throughput rate

(Continued)

TABLE 10.1 (Continued)

Comparative Overview of Different 3D Printing Techniques along with Their Materials Availability, Advantages, and Limitations

3D Printing Technique	Printable Materials	Resolution (µm)	Printing Speed	Cost	Advantages	Limitations
SLS	• Thermoplastic materials • Metals • Ceramics	80–250	Low	Very high	• Larger range of materials • Easy material Processing • No support structure needed • Large area printing	• Expensive process • Rough surfaces • Poor surface quality • Thermal decay of materials

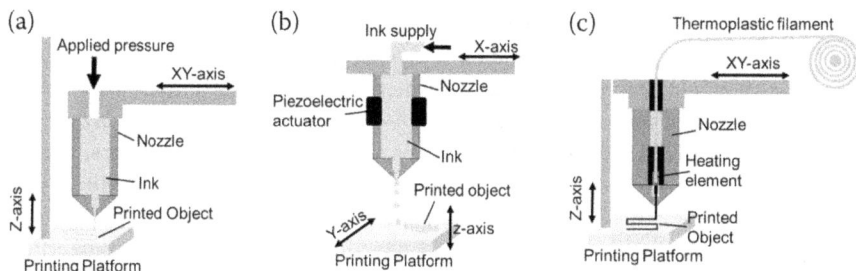

FIGURE 10.3 Systematic overview and features of 3D printing techniques for battery applications; (a) Direct ink writing (DIW) to print 3D freeform structures directly on printing platform, (b) Drop-on-demand (DOD) Inkjet printing (IJP) to print structures on printing substrates, (c) Fused deposition modelling (FDM) to print thermoplastic materials in the form of filaments via heated nozzles onto the printing platform.

10.2.3 INKJET PRINTING (IJP)

IJP is an original and widely studied materials jetting process, where material inks are directly deposited onto the building platform through a nozzle. Similar to DIW, IJP creates objects in a layer-by-layer way by using low viscosity inks (40–100 cP). IJP process has four main steps: (i) ink preparation, (ii) droplet formation and deposition, (iii) wet-film formation, and (iv) curing. IJP mainly use two technologies based on the applications: continuous flow (CIJ) and drop-on-demand (DOD). Droplet-based IJP is mostly used for battery manufacturing and can be further divided into thermal and piezoelectric type DOD inkjet printers, as presented in Figure 10.3b. Thermal inkjet printers deposit the ink droplet onto the printing substrates (paper and plastic) by heating a small portion of the ink that forms a bubble and forces ink droplets to come out through nozzle head. Alternatively, piezoelectric-based DOD inkjet printing uses piezoelectric actuation pressure to push ink droplets to come out of the nozzle. Piezoelectric droplet-based IJP has great industrial potential for battery manufacturing due to its multi-materials capability, wider materials suitability, economical ink formation (viscosity), accurate droplet formation, and high materials utilization with a minimal wastage due to predesigned patterns. IJP has been explored for the different class of materials such as carbon materials, conductive polymer, solid electrolyte materials, metals, and hybrid nanomaterials to print different-sized standalone and integrated batteries with high resolution (5–200 µm) [10]. However, further precision to obtain a nanometre-sized printing structure is achievable by the application of an electric field between the printing head and substrate. Compared to DIW, IJP has relatively lower printing throughput and is not suitable for high mass loading.

10.2.4 FUSED DEPOSITION MODELLING (FDM)

FDM is the most advanced industrial scale 3D printing technique to create complex 3D objects in a layer-by-layer way without any waste. Like DIW, FDM is also an extrusion-based 3D printing technique where solid thermoplastic filaments or metal wires are extruded through the nozzle. Plastic filament heated up to the semiliquid state at the nozzle head is deposited on to the substrate with a computer-controlled design, as shown in Figure 10.3c. The deposited layer solidifies quickly after natural cooling and then is built into a 3D object by the deposition of further layers pre-designed into the CAD software. FDM 3D printing has several advantages, such as cost effectiveness, high-speed printing, and large-size printing capabilities; however, its dependence on thermoplastic filaments (e.g., acrylonitrile butadiene styrene [ABS], thermoplastic urethane [TPU], polylactic acid [PLA], polyamide [PA], polyphenylsulfone [PPSF], and polycarbonate [PC]) limit its application for battery manufacturing due to lower electrical conductivity of plastic filaments [16]. In addition, the low resolution (50–200 μm), poor mechanical properties of the printed structures, and lower flexibility over multi-material application are other issues that deter its use in 3D-printed batteries. However, the recent technological advancement in FDM 3D printing techniques and the development of conductive composite filaments by the addition of carbon-based materials (carbon black, graphene, graphene oxide) into the polymer matrix provides an opportunity for the application of FDM 3D printing for battery manufacturing.

10.2.5 STEREOLITHOGRAPHY (SLA)

SLA 3D printing is gaining tremendous academic and industrial attention for the fabrication of battery modules due to its high longitudinal resolution. The SLA 3D printing process uses a moveable liquid photo-curable resin platform to build 3D structures in a layer-by-layer manner by the application of UV laser, as presented in Figure 10.4a. The movement of the resin platform is divided into two subcategories:

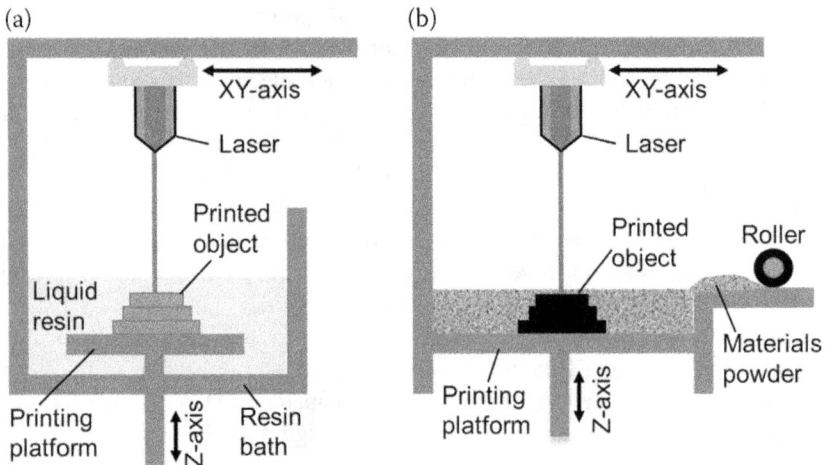

FIGURE 10.4 Systematic overview and features of 3D printing techniques for battery applications; (a) Stereolithography (SLA) to print 3D structures from photopolymer liquid resin through polymerisation via laser source, (b) Selective laser sintering (SLS) to print solid 3D structures onto a printing platform, where rollers spread a uniform powder layer and the laser heat source fuse it to the solid structures.

top-down SLA and bottom-up SLA. For top-down SLA, resin platform moves downward after the UV curing of each new layer, whereas, in bottom-up SLA, the resin platform moves upward. In the SLA printing process, the first step is the preparation of printable resins that contain active materials, photo initiators, and liquid monomers with the desired optical and rheological properties to avoid the printing defects and poor mechanical properties. During the printing process, when the first printed thin layer is photochemically cured, the printing platform relocates to cover the former layer with resin to build up a 3D structure according to the CAD design. Compared to DIW and FDM, the materials-extrusion 3D printing technique, SLA is a nozzle-free 3D printing technique offering the ability to fabricate complex arbitrary designs and battery customization (Figure 10.4a). Since the SLA technique can print customised 3D complex arrangements, it has the potential to fabricate 3D free-standing microporous solid-state electrolytes with unified interfaces and a graphene-based 3D porous electrode for solid-state batteries [17]. Despite these advantages, practical application of SLA printing in battery manufacturing is still limited due to the low throughput rate, high cost, and the requirement of photo curable resins.

10.2.6 SELECTIVE LASER SINTERING (SLS)

SLS is an advanced 3D printing technique where a high-power laser source is used to fuse the powder materials into the selective solid 3D structures. The high-power laser source (i.e., CO_2 or Nd:YAG laser) moves horizontally to fuse the powders into the predefined architectures, whereas the printing platform with the powder bed moves vertically, as presented in Figure 10.4b. SLS 3D printing can fuse a range of powder materials, including metal, polymer, hybrid composite materials, and ceramics [13]. During the printing process, a levelling roller spreads a homogeneous powder layer on the printing platform and then the power source fuses the powder into the solid layer according to the predefined structure. After the solidification of the first layer, the printing platform moves downward and then the roller spreads a new powder layer to fuse the next layer in a layer-by-layer manner to create a 3D pattern. The final predefined 3D structure is obtained by removing the unused power loosely bound to the printed structure. The use of powder materials without any extra treatment as feed stock is the main advantage of SLS 3D printing. This 3D method has the ability to print battery metal current collector, carbon composite electrodes, and the polymeric electrolyte with improved physical, electrical, and mechanical properties. However, the high printing cost, high surface roughness, and the thermal decomposition of heat-sensitive materials could be hurdles for this technique for low-cost 3D-printed batteries.

10.3 3D-PRINTABLE NANOMATERIALS FOR BATTERY COMPONENTS

10.3.1 OVERVIEW

Batteries are an efficient and sustainable energy storage system categorised as primary (non-rechargeable) and secondary (rechargeable) batteries for different applications from

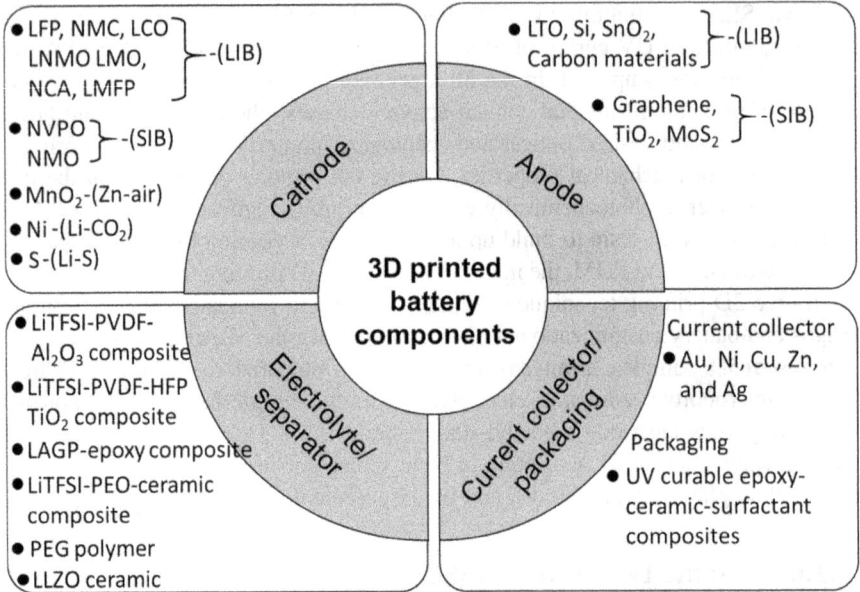

- LFP, NMC, LCO
  LNMO LMO, }-(LIB)
  NCA, LMFP
- NVPO }-(SIB)
  NMO
- MnO$_2$-(Zn-air)
- Ni -(Li-CO$_2$)
- S-(Li-S)

- LTO, Si, SnO$_2$, }-(LIB)
  Carbon materials
- Graphene, }-(SIB)
  TiO$_2$, MoS$_2$

Cathode

Anode

**3D printed
battery
components**

- LiTFSI-PVDF-
  Al$_2$O$_3$ composite
- LiTFSI-PVDF-HFP
  TiO$_2$ composite
- LAGP-epoxy composite
- LiTFSI-PEO-ceramic
  composite
- PEG polymer
- LLZO ceramic

**Electrolyte/
separator**

**Current collector/
packaging**

Current collector
- Au, Ni, Cu, Zn,
  and Ag

Packaging
- UV curable epoxy-
  ceramic-surfactant
  composites

FIGURE 10.5 Overview of the 3D-printed battery components and the list of printable materials for cathode, anode, electrolyte, separator, current collector and packaging materials for different battery systems.

electronics devices to electric vehicles. Rechargeable batteries are the most reliable energy storage system where state-of-the–art LIBs have attracted extensive academic and industrial interest due to their enhanced energy density and long working life. Beyond LIBs, nickel (Ni), sodium (Na), potassium (K), magnesium (mg), and zinc (Zn) based batteries have also been studied. Regardless of their type, size, and shape, all batteries are mainly contained active electrodes, electrolyte socked separator, current collector, and final packaging. Different 3D printing techniques have been extensively used to print the individual components or even the complete battery from start to finish, as summarised in Figure 10.5.

10.3.2 CATHODE MATERIALS

In recent years, various 3D printing techniques have been used to deposit LIB cathode materials. LFP (LiFePO$_4$), NMC (LiNiCoMnO$_2$), LNMO (LiNiMnO$_4$), LCO (LiCoO$_2$), LMO (LiMn$_2$O$_4$), NCA (LiNiCoAlO$_2$), and LMFP (LiMnFePO$_4$) are the typically used cathode materials for LIBs (Figure 10.5). LFP is the commonly used printable cathode material because of its high stability [18]. Owing to the excellent electrochemical properties, carbon-containing materials such as carbon flakes (CFs), carbon nano-tubes (CNTs), graphite, graphene, and reduced graphene oxides (rGOs) have been widely studied to enhance the functionality of the printed cathode materials. So far, DIW is the most advanced and powerful 3D printing technique for the development of various cathode structures; it is followed by FDM and IJP. Beyond LIBs, lithium sulphur (Li-S) batteries are another promising technology with a higher theoretical capacity of 1675 mA h g^{-1} [19]. In case of Li-S batteries, DIW and IJP are the most promising 3D printing techniques that can be used to develop composite cathodes (i.e., carbon and sulphur) with high mass-loading sodium ion battery (SIB) cathode materials, such as Na_3V_2 $(PO_4)_3$ (NVPO) and $NaMnO_2$ (NMO), which are another potential application of DIW and FDM 3D printing techniques. For SIBs, conduction of large sodium ions (diameter: Na^+ ~0.102 nm vs Li^+ ~0.076 nm) is crucial and is responsible for low reversible capacities. Therefore, 3D-printable composite cathode (i.e., NVPO-GO ink) with a high active surface area can effectively shorten the conduction path for sodium ions, as well as electrons, and thus improve the electrochemical performance. In addition to that, various other 3D printable cathode materials include rGO framework embedded with ultrafine nickel nanoparticles for Li-CO_2 batteries, CNT-rGO framework with MnO_2 nanoparticles enabled cathode, and CNT network-enabled MnO_2 cathode for Zn-air batteries.

10.3.3   ANODE MATERIALS

Lithium containing materials are the commonly used anodes in 3D-printed LIBs where lithium titanate ($Li_4Ti_5O_{12}$-[LTO]) is the most adopted due to its minimal volumetric expansion, high electrochemical reversibility, and high stability [20]. In addition, LTO composites (i.e., with CNT, rGO, carbon nanofiber, and graphene), lithiated host matrix, graphite-PLA composite, silicon (Si), tin oxide (SnO_2) composites with acetylene black and graphene are other 3D-printable anode materials for LIBs (see Figure 10.5). In recent years, 3D framework electrodes have gained tremendous attention to enhance the performance of LIBs. Among them, 3D-printed porous frameworks of carbon-hybrid nanomaterials, MXenes, and metals have been studied as a current collector to host Li metal anode [21]. The advantages of 3D-printed framework structures are to accommodate high mass loading of Li metals and simultaneously suppress the Li dendrite growth and volume expansion to enhance electrochemical performance and stability. Beyond LIBs, graphene, titanium dioxide (TiO_2), and molybdenum disulphide (MoS_2)-graphene composites are other 3D-printable anode materials mainly for SIBs.

10.3.4   ELECTROLYTE MATERIALS

Electrolyte is one of the most important components that play a crucial role in determining the battery electrochemical properties, life cycle, and stability. With the continuous ongoing advances in the development of 3D-printed cathode and anode structures, the development of solid electrolytes ise also needed to enhance electrochemical capacity and the overall stability. 3D-printable solid electrolytes have several advantages that include: non-flammability, non-corrosivity, high ionic conductivity, interfacial stability, and the possibility to fabricate complete integrated battery structures. Hybrid composites, polymers, and inorganic ceramics are the main 3D-printable solid electrolytes, whereas hybrid composite electrolytes (summarised in Figure 10.5) have been widely used in various 3D printing techniques. In this regard, hybrid polymer-ceramic composite electrolyte composed of N-methyl-2-pyrrolidone (NMP), poly(vinyldene fluoride) (PVDF), glycerol, and 40–50 nm Al_2O_3 filler has been printed with a controlled porosity via the DIW technique. Another hybrid electrolyte with similar formulation consisting of lithium-bis-trifluoromethanesulfonylimide (LiTFSI), PVDF-hexafluoropropylene (HFP) polymer matrix, and nanosized TiO_2 nanoparticles was also fabricated via the DIW technique to achieve a dense continuous layer with a relatively lower interfacial resistance [22]. $Li_{1.4}Al_{0.4}Ge_{1.6}(PO_4)_3$ (LAGP) and epoxy resin-based hybrid electrolytes with 3D-bicontinuous microchannel structures of various ceramic and epoxy resins also have been developed via the SLA technique [12]. The FDM technique was also employed to print a LiTFSI-based hybrid electrolyte with polyethylene oxide (PEO) matrix where polylactic acid (PLA) and ceramic SiO_2 or Al_2O_3 filler was added to enhance the mechanical and rheological properties. In addition to that, a 3D micro battery has also been assembled by filling the cathode and anode materials into the poly (ethylene glycol) (PEG) based microporous 3D electrolyte developed by the SLA technique. More recently, a garnet-type $Li_7La_3Zr_2O_{12}$ (LLZO) ceramic has attracted attention as a 3D-printable solid electrolyte in a wide variety of 3D-printable structures to enhance the contact with active materials [23].

10.3.5 SEPARATOR, CURRENT COLLECTOR, AND PACKAGING MATERIALS

In addition to the 3D-printable active electrode and electrolyte materials, an appropriate choice of separator, current collector, and final packaging materials is also needed to achieve a fully 3D-printed battery with an improved structural, interfacial, and electrochemical stability. In general, limited studies on 3D printing of these components are reported. However, a modified separator ink, printed into a spiral structure by extrusion-based 3D printing, has been demonstrated for Li-based batteries by mixing boron nitride (BN) nanosheets into PVDF-HFP [24]. The BN-modified separator provides uniform interface thermal distribution and suppresses Li dendrite growth that improves the electrochemical performance further when paired with active materials (LPF cathode and Li anode) to assemble a full battery. Recently, gold (Au) current collector coating has been developed via a combination of different vacuum techniques (such as lithography and electron-beam deposition) followed by the 3D printing of electrodes and electrolyte materials to form a complete battery system [25]. Other 3D-printable materials, such as nickel (Ni), copper (Cu), silver (Ag), and zinc (Zn) based metallic current collectors, have been studied as supercapacitors. 3D-printed complex electrode/electrolyte structures require a flexible packaging material with the printing ability. 3D-printable packaging material inks are a combination of inorganic ceramic, photopolymers, and surfactant (Figure 10.5), which can be printed via extrusion techniques.

10.4 3D CONSTRUCTION STRATEGIES FOR PRINTED BATTERIES

10.4.1 OVERVIEW

The conventional planar electrodes are relatively thick, and therefore, they restrict the conduction pathways of ions as well as electrons. Replacement of the planar thick electrode structure into the 3D-printed structure can significantly shorten the ions and electrons transfer pathways due to their well-defined porosity, conductivity, and the macroscopic diffusivity. Due to the advantage of active materials (cathode, anode, and

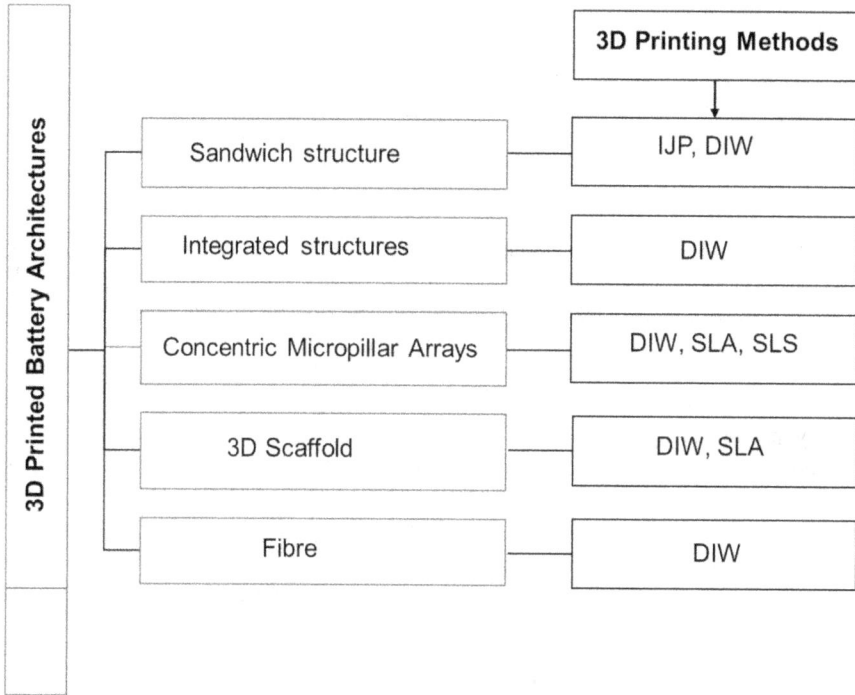

FIGURE 10.6 Overview of 3D-printed battery architectures and the corresponding printing methods.

electrolyte) vertical loading into the 3D matrix, the electrical conductivity of the 3D-printed active electrode materials can be 15 fold greater than the conventional electrodes. In general, 3D-printed batteries can be fabricated with different architectures that significantly affect their electrochemical performance and the applications (see Figure 10.6).

10.4.2 SANDWICH STRUCTURES

The fabrication of sandwich-type batteries using various 3D printing techniques, such as IJP and DIW, is quite common (see Figure 10.7a). In sandwich structure, battery components are current collectors, electrodes, electrolyte, separator, and packaging, which can be printed individually and then stacked together to assemble a complete battery. However, the printing of battery components on top of each other from start to finish is another way that is very cost-effective for mass production. Sandwich-structure batteries can be designed and printed with various shapes (square, circle, etc.) and sizes.

10.4.3 INTEGRATED STRUCTURES

The integrated-structure battery configuration includes two electrodes (cathode and anode) within the same plane, separated by a narrow gap for electrolyte, as shown in

(a)   (b)   (c)   (d)   (e)

Sandwich   Integrated   Concentric   3D Scaffold   Fibre
                                  micropillar arrays

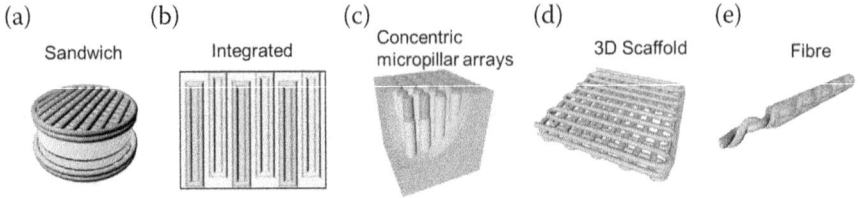

FIGURE 10.7 Schematic illustration of 3D-printed battery structures; (a) Sandwich, (b) Integrated, (c) Concentric micropillar arrays, (d) 3D scaffold, and (e) Fibre. Figures 10.7a, and 10.7c–e reprinted with permission from Reference [26]. Copyright (2020) Elsevier.

Figure 10.7b. For integrated-structure batteries, 3D printing techniques can precisely control the electrode dimensions and their gaps that mainly influence the battery performance. The 3D-integrated electrodes design with a microscale architecture provides large surface area for electrolyte interaction with the electrodes. The large surface area offers a short pathway for ion and electron conduction, resulting in increased performance of such 3D-printed batteries. One of the first 3D-printed micro batteries with an integrated design was demonstrated in 2013, where both the electrodes (LTO anode and LFP cathode) were printed via the DIW technique [25].

10.4.4 CONCENTRIC MICROPILLAR ARRAYS

The concentric 3D micropillar batteries are a unique classic design containing an equally spaced array of electrode materials in the form of micropillars. These micropillars are coated with solid electrolyte materials, and the remaining spaces between the electrolyte-coated pillars are alternately filled with the anode and cathode materials (see Figure 10.7c). The 3D micropillar batteries are getting increased attention because of their miniature design (1–10 mm^3) and their ability to provide high-energy density and power density. Most of the reported 3D micropillar batteries are fabricated with etched and electrodeposited electrodes. There are only few reports on 3D-printed micropillar batteries where the SLA printing technique has been used to fabricate concentric tube-type 3D micropillar batteries on a polymer substrate, and the SLS method has been used to develop an array of micropillar electrodes [27]. A major challenge to develop 3D-printed micropillar batteries is the preparation of conformal electrolyte coatings on electrode materials. These electrolyte coatings should have high ionic conductivity with no or negligible electronic conductivity to form a barrier layer between the electrodes to prevent short circuit.

10.4.5 3D SCAFFOLDS

3D scaffolds contain a sequence of 3D-printed electrode and electrolyte materials with a defined porosity along the vertical direction, which can be fabricated by DIW, FDM, and SLA printing techniques (see Figure 10.7d). Scaffold structures are known to have porosity ranging from nano- to micro-scale that possess advantages for electrolyte infiltration and ion conduction. Some recent studies confirmed that the employment of thicker 3D-printed electrodes (>50 μm) with 3D "scaffold" architecture is a promising approach to improve the electrochemical performance of the batteries with a high mass loading of active electrodes materials. Alternatively, thick electrodes in a traditional electrode structure possess challenges by supressing the electrolyte infiltration and charge conduction that limit the battery cyclability and the power density. Therefore, such structures have attracted attention as a potential solution to address the low recyclability and poor rate-capability issues of Li-O$_2$ batteries by the development of thick electrode (cathode) with a 3D scaffolds geometry using DIW technique [28].

10.4.6 FIBRES

3D-printed batteries with a fibre architecture have received attention for wearable devices due to their structural flexibility, weavability, and stretchability. At present, most of the 3D-printed fibre-based batteries contain electrolyte-coated thread-shaped electrodes that are twisted together to form a battery (see Figure 10.7e). In recent years, "fibre" structure batteries printed through the DIW technique have been extensively researched as a power source for emerging wearable and electronic textiles applications. Recently demonstrated all-fiber quasi-solid-state LIBs were fabricated through the DIW technique, where PVDF (binder) and CNTs (conductive additive) mixed-fibre shaped LFP and LTO worked as a cathode and anode [29]. After that, polymer electrolyte-coated fibre electrodes were weaved together to form a battery for wearable electronics and textiles.

10.5 PERFORMANCE OF 3D-PRINTED BATTERIES

3D-printed batteries can achieve higher energy density and power density by virtue of their complex electrode structures, high mass loading of active electrode materials, and high porosity that increases active surface area for electrochemical reactions. Furthermore, electrode thickness is another factor that affects the energy and power densities of printed batteries. Some studies demonstrate that the thicker electrodes deliver greater areal energy density and also sustain areal power densities equivalent to that of thin electrodes (see Figure 10.8) [30].

Wei and co-workers demonstrated the effect of LTO/LPF electrode thicknesses on areal energy density [30]. The electrodes were printed to a thickness of 0.05 mm

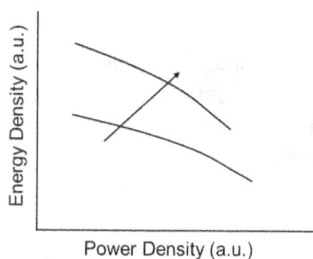

FIGURE 10.8 Ragone plot showing areal energy density versus areal power density for different thicknesses of electrode. Here, a.u. stands for "Arbitrary Unit" and the arrow indicating an increase in electrode thickness.

to 1 mm via DIW technique where a complete 3D-printed LIBs thick, biphasic electrodes (1 mm) delivered an areal energy density of near to 20 mW h cm^{-2}, whereas the thin electrode (0.05 mm) exhibited ~1 mW h cm^{-2}. The complete battery system delivered a high areal capacity and current density of 4.45 mA h cm^{-2} and 0.14 mA cm^{-2}, respectively [30]. In the literature, different 3D-printed batteries' electrochemical performances have been reported using gravimetric, areal, or volumetric parameters. Therefore, it is difficult to make a systematic comparison of different 3D-printed batteries using a single performance parameter. Table 10.2 summarised the progress in 3D-printed batteries fabricated by different methods and materials.

For various 3D printing techniques, DIW is the is the most widely used and popular technique for the fabrication of different electrode structures, followed by IJP and FDM. In a study by Kohlmeyer and co-workers [31], the DIW technique was employed to fabricate structured electrodes for LIBs with high specific capacities of 89 mA h g^{-1} for LTO, 106 mA h g^{-1} for LFP, and 80 mA h g^{-1} for LCO at a rate of 5 C. Hu and co-workers fabricated LFMP-based electrodes for LIBs where the fully assembled 3D-printed battery with a capacity of 108.45 mA h g^{-1} at 100 C rate and reversible capacity of 150.21 mA h g^{-1} at 10 C rate after 1000 cycles [18]. Latter study also revealed that the 3D-printed electrodes' specific capacity was nearly 30% higher compared to the traditional electrodes. Similarly, Sun and co-worker fabricated an integrated structured micro battery with LFP/LTO electrodes via DIW technique with high areal energy of 1.5 mA h cm^{-2} at 1 C and power density of 9.7 J cm^{-2} at 2.7 mW cm^{-2} [20].

In recent study, Bao and co-worker demonstrated a 3D-printed LTO/LPF based patterned electrodes for flexible pouch LIBs and compared the performance with traditional coin cell battery (Figure 10.9) [32]. The 3D-printed pouch and coin battery shows the primary discharge capacities of 120 mA h g^{-1} and 130 mA h g^{-1}, respectively (Figure 10.9a). They also studied the cycle stability where the pouch cell showed superior capacity retention of 83% after 50 cycles, whereas the retention capacity for coin cell was 70%. Further results exhibited that the capacity of the pouch battery decreases nearly 9% with a varied current rate of 0.3 C to 0.6 C and 1.0 C (Figure 10.9b) that demonstrate good rate capability for high-power applications (see Figure 10.9C).

Other battery systems, such as concentric micropillar array and fibre batteries, have also been fabricated by 3D printing techniques. Recently, Hur and co-workers demonstrated 3D-printed micropillar battery with a 3D array of silicon anodes (Figure 10.10a) [33]. These batteries delivered much higher discharge capacities up to 1.80 mAh cm^{-2} (5.2 mWh cm^{-2}) as compared to the corresponding thin-film batteries (Figure 10.10b). Furthermore, Praveen and co-workers fabricated fibre shaped LIBs via the DIW technique [34]. These fibre-shaped batteries demonstrated a discharge capacity of 166 mAh g^{-1} for a rate of 1 C (Figure 10.10c). The areal capacity values were around 2.5 mA h cm^{-2} at 0.1 C rate and 1.9 mA h cm^{-2} at a higher rate of 2 C, as shown in Figure 10.10d. These results therefore established that the 3D printing technique can be a powerful tool for constructing structured batteries with enhanced energy and power densities within the limited space; however, the commercialisation required further optimisation to achieve reliability and reproducibility.

TABLE 10.2

Comparative Overview of Different 3D-Printed Battery Structures

Printing Technique	Cathode	Anode	Electrolyte	Specific Capacity	Ref.
DIW	LFP	LTO	1 M LiClO$_4$ in EC:DMC	116 mAhg^{-1} at 0.5 C	[25]
DIW	LFP	LTO	1 M LiTFSI/PC	133 mAh g^{-1} at 0.2 mA cm^{-2}	[30]
DIW	LFP	LTO	1 M LiPF$_6$, EC/DEC	10 mAh g^{-1} at 50 mA g^{-1}	[29]
DIW	LFP/CNF/PVDF	LTO/CNF/PVDF	Al$_2$O$_3$, PVDF, NMP/ glycerol	154 mAh g^{-1} at 0.2 C	[35]
DIW	LMFP	Li	1 M LiPF$_6$ in EC:DEC	161.36 mAh g^{-1} at 1C	[18]
DIW	LFP/LiCoO$_2$	LTO/CNF/PVDF	1 M LiPF$_6$, EC/PC	150 mAh g^{-1} at 0.2 C	[31]
FDM	Li foil	Graphene/PLA	1 M LiPF$_6$	40 mAh g^{-1} at 120 C	[11]
FDM	Li	Graphene/PLA	1 M LiPF$_6$	215 mAh g^{-1} at 18.6 mA g^{-1}	[36]
FDM	LMO/MWNTs	LTO/graphene nano platelets	1 m LiClO$_4$	7.48 mAh cm^{-3} at 1C	[37]
FDM	LFP with additives	LTO with additives	1 m LiPF$_6$	80 mAh g^{-1} at 2 C	[38]

(Continued)

TABLE 10.2 (Continued)
Comparative Overview of Different 3D-Printed Battery Structures

Printing Technique	Cathode	Anode	Electrolyte	Specific Capacity	Ref.
FDM	NC-Co	Li	0.5 M LiClO$_4$ /DMSO	525 mA h g^{-1} at 0.8 mA cm^{-1}	[39]
IJP	LFP Composite	Li foil	1 M LiPF$_6$, 0.5 M LiTFSI	150 mAh g^{-1} at 1 C	[40]
IJP	LFP	Li foil	1 M LiPF$_6$	134.7 mAh g^{-1} at 0.1 C	[41]
IJP	MWCNT	Li foil	–	1260 mAh g^{-1} at 0.5 C	[42]
IJP	LiCoO$_2$	Li foil	1 M LiPF$_6$	105 mAh g^{-1} at 394 μA cm^{-2}	[43]
SLA	LFP	LTO	1 M LiClO$_4$ in EC:PC	–	[44]
SLA	Li	Li	Li$_{1.4}$Al$_{0.4}$Ge$_{1.6}$(PO$_4$)$_3$	–	[12]

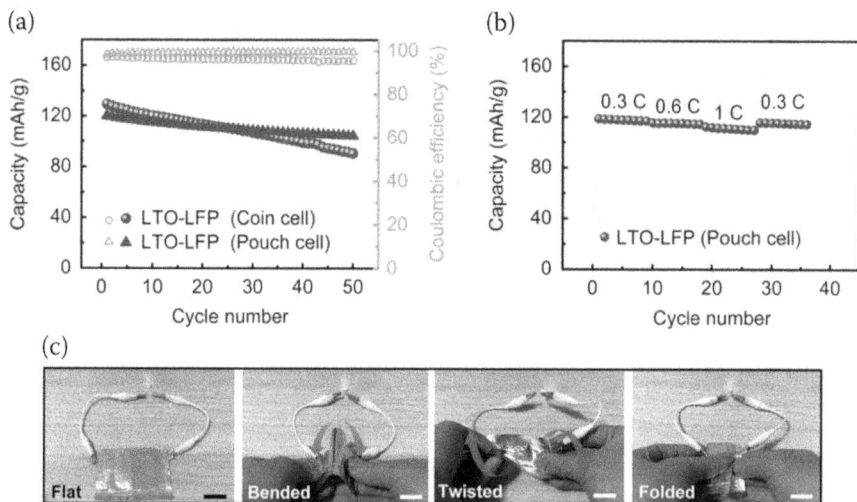

FIGURE 10.9 Electrochemical properties of 3D-printed pouch batteries with patterned LTO/LFP electrodes, (a) Coulombic efficiency and cycling performance of the 3D-printed pouch cell and a comparison with coin cell. (b) C-rate capability of the pouch battery with a varied rate from 0.3 C to 1 C, (c) Mechanical strength demonstration of the pouch cell at flat, bended, twisted, and folded states with LED illumination. The scale bar is 1 cm. Reproduced with permission from reference [32]. Copyright (2020) Elsevier.

10.6 CONCLUSIONS, CHALLENGES, AND FUTURE OUTLOOK

3D printing techniques have drawn great academic and industrial attention for the construction of 3D-structured electrode materials, as well as the full battery from start to finish. The major advantages of the 3D printing process are design flexibility, high controllability, and ability to form complex 3D structures with an enhanced energy and power densities. In addition, 3D-printed structured electrodes promote high charge and discharge rate performances that enable their use for high-power application batteries. Over the last few years, 3D-printed batteries have demonstrated remarkable progress in electrochemical performances to power advanced smart electronics. However, there are still challenges that need to be addressed before 3D printing technologies can be widely employed commercially. At first, the availability of printable active materials (cathode, anode, and electrolyte) that can be used either as ink or filament is a major issue that needs to be tackled at priority. Furthermore, polymeric binder materials are needed to make these inks printable with appropriate rheological properties. These binder materials are mainly insulating in nature and adversely affect the electrochemical properties of the 3D batteries. Therefore, it is essential to develop a new class of active materials and conductive additives to extend the availability of printable materials. 3D-printed battery electrodes are fabricated by layer-by-layer deposition that may cause poor mechanical strength due to weak interlayer bonding. Thus, further research is required to enhance the mechanical robustness, especially for flexible and wearable applications. Lastly, so far, most of the 3D-printed battery research has mainly

(a)   (b)

(c)   (d)

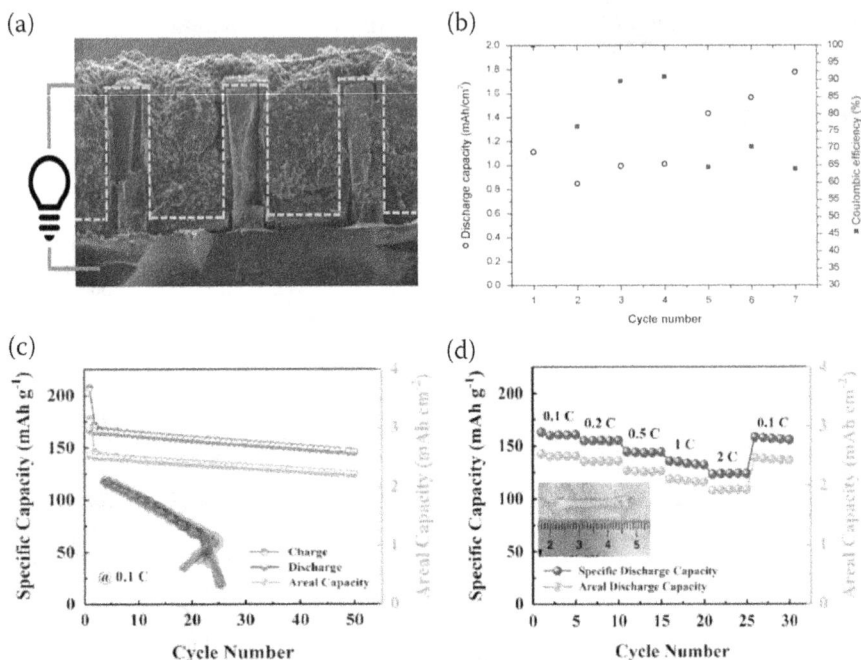

FIGURE 10.10 Electrochemical performance parameters of 3D-printed micropillar arrays and fibre-shaped batteries, (a) Cross-sectional image of concentric micropillar battery showing cathode separated by electrolyte (yellow dashed line), (b) Discharge capacity and the coulombic efficiency of the micropillar battery, (c) Cycling performance of full fibre-shaped battery, (d) C-rate performance of fibre-shaped battery with a varied rate from 0.1 C to 2 C. Figure 10.10a,b reproduced with permission from reference [33]. Copyright (2018) Elsevier. Figure 10.10c,d reproduced with permission from reference [34]. Copyright (2021) Elsevier.

focused on the development and printing of active materials (electrodes and electrolytes). Therefore, further research is required for the development of printable current collectors and packaging materials to make a fully 3D-printed battery for real-world applications.

REFERENCES

1. Dunn, B., Kamath, H., and Tarascon, J.M. (2011). Electrical energy storage for the grid: A battery of choices. *Science* 334, 928–935.
2. Fu, K., Yao, Y., Dai, J., and Hu, L. (2017). Progress in 3D printing of carbon materials for energy-related applications. *Adv. Mater.* 29, 1603486.
3. Chang, P., Mei, H., Zhou, S., Dassios, K.G., and Cheng, L. (2019). 3D printed electrochemical energy storage devices. *J. Mater. Chem. A* 7, 4230–4258.
4. Pang, Y., Cao, Y., Chu, Y., Liu, M., Snyder, K., MacKenzie, D., and Cao, C. (2020). Additive manufacturing of batteries. *Adv. Funct. Mater.* 30, 1906244.
5. Liu, N., Gao, Y. (2017). Recent progress in micro-supercapacitors with in-plane interdigital electrode architecture. *Small* 13, 1701989.

6. Zheng, H., Li, J., Song, X., Liu, G., and Battaglia, V.S. (2012). A comprehensive understanding of electrode thickness effects on the electrochemical performances of Li-ion battery cathodes. *Electrochim. Acta.* 71, 258–265.
7. Wohlers T., and Gornet T. (2012). *History of Additive Manufacturing.* Wohlers Associates, Inc.
8. Active Standard, West Conshohocken: ASTM Int 2012.
9. Gao, X., Sun, Q., Yang, X., Liang, J., Koo, A., Li, W., Liang, J., Wang, J., Li, R., Holness, F.B., et al. (2019). Toward a remarkable Li-S battery via 3D printing. *Nano Energy* 56, 595–603.
10. Milroy, C.A., Jang, S., Fujimori, T., Dodabalapur, A., and Manthiram, A. (2017). Inkjet-printed lithium-sulfur microcathodes for all-printed, integrated nanomanufacturing. *Small* 13, 1603786.
11. Foster, C.W., Down, M.P., Zhang, Y., Ji, X., Rowley-Neale, S.J., Smith, G.C., Kelly, P.J.C.E., and Banks, C.F. (2017). 3D Printed Graphene Based Energy Storage Devices. *Sci. Rep.* 7, 42233.
12. Zekoll, S., Marriner-Edwards, C., Hekselman, A.K.O., Kasemchainan, J., Kuss, C., Armstrong, D.E.J., Cai, D., Wallace, R.J., Richter, F.H., Thijssen, J.H.J., and Bruce, P.G. (2018). Hybrid electrolytes with 3D bicontinuous ordered ceramic and polymer microchannels for all-solid-state batteries. *Energy Environ. Sci.* 11, 185–201.
13. Kim, H., Proell, J., Kohler, Pfleging R.W., and Pique, A. (2012). Laser-printed and processed $LiCoO_2$ cathode thick films for Li-Ion microbatteries. *J. Laser Micro/ Nanoeng.* 7, 320–325.
14. Li, H., and Liang, J. (2020). Recent development of printed micro supercapacitors: printable materials, printing technologies, and perspectives. *Adv. Mater.* 32, e1805864.
15. Gratson, G.M., García-Santamaría, F., Lousse, V., Xu, M., Fan, S., Lewis, J.A., and Braun, P.V. (2006). Direct-write assembly of three dimensional photonic crystals: conversion of polymer scaffolds to silicon hollow-woodpile structures. *Adv. Mater.* 18, 461–465.
16. Wang, X., Jiang, M., Zhou, Z., Gou, J., and Hui, D. (2017). 3D printing of polymer matrix composites: a review and prospective. *Compos. B Eng.* 110, 442–458.
17. Manapat, J.Z., Chen, Q., Ye, P., and Advincula, R.C. (2017). 3D printing of polymer nanocomposites via stereolithography. *Macromol. Mater. Eng.* 302, 1600553.
18. Hu, J., Jiang, Y., Cui, S., Duan, Y., Liu, T., Guo, H., Lin, L., Lin, Y., Zheng, J., Amine, K., et al. (2016). 3D-printed cathodes of $LiMn_{1-x}Fe_xPO_4$ nanocrystals achieve both ultrahigh rate and high capacity for advanced lithium-ion battery. *Adv. Energy Mater.* 6, 1600856.
19. Xie, K., You, y., Yuan, K., Lu, W., Zhang, K., Xu, F. Ye, M., Ke, S., Shen, C., Zeng, X., Fan, X., and Wei, B. (2017). Ferroelectric-enhanced polysulfide trapping for lithium-sulfur battery improvement. *Adv. Mater.* 29, 1604724.
20. Sun, C., Liu, S., Shi, X., Lai, C., Liang, J., and Chen, Y. (2020). 3D printing nano-composite gel-based thick electrode enabling both high areal capacity and rate performance for lithium-ion battery. *Chem. Eng. J.* 381, 122641.
21. Shen, K., Li, B., and Yang, S. (2020). 3D printing dendrite-free lithium anodes based on the nucleated MXene arrays. *Energy Storage Mater.* 24, 670–675.
22. Cheng, M., Jiang, Y., Yao, W., Yuan, Y., Deivanayagam, R., Foroozan, T., Huang, Z., Song, B., Rojaee, R., Shokuhfar, T., et al. (2018). Elevated-temperature 3D printing of hybrid solid-state electrolyte for Li-ion batteries. *Adv. Mater.* 30, e1800615.
23. McOwen, D.W., Xu, S., Gong, Y., Wen, Y., Godbey, G.L., Gritton, J.E., Hamann, T.R., Dai, J., Hitz, G.T., Hu, L., and Wachsman, E.D. (2018). 3D-printing electrolytes for solid-state batteries. *Adv. Mater.* 30, e1707132.

24. Liu, Y., Qiao, Y., Zhang, Y., Yang, Z., Gao, T., Kirsch, D., Liu, B., Song, J., Yang, B., and Hu, L. (2018). 3D printed separator for the thermal management of high-performance Li metal anodes. *Energy Storage Mater.* 12, 197–203.
25. Sun, K., Wei, T.S., Ahn, B.Y., Seo, J.Y., Dillon, S.J., and Lewis, J.A. (2013). 3D printing of interdigitated Li-ion microbattery architectures. *Adv. Mater.* 25, 4539–4543.
26. Lyu, Z., Lim, G.J.H., Koh, J.J., Li, Y., Ma, Y., Ding, Wang, J.J., Hu, Z., Wang, J., Chen, W., and Chen, Y. (2021). Design and Manufacture of 3D-Printed Batteries. *Joule* 5, 89–114.
27. Cohen, E., Menkin, S., Lifshits, M., Kamir, Y., Gladkich, A., Kosa, G., and Golodnitsky, D. (2018). Novel rechargeable 3D-microbatteries on 3D-printed-polymer substrates: feasibility study. *Electrochim. Acta* 265, 690–701.
28. Lacey, S.D., Kirsch, D.J., Li, Y., Morgenstern, J.T., Zarket, B.C., Yao, Y., Dai, J., Garcia, L.Q., Liu, B., Gao, T., et al. (2018). Extrusion-based 3D printing of hierarchically porous advanced battery electrodes. *Adv. Mater.* 30, e1705651.
29. Wang, Y., Chen, C., Xie, H., Gao, T., Yao, Y., Pastel, G., Han, X., Li, Y., Zhao, J., Fu, K.K., and Hu, L. (2017). 3D-printed all-fiber Li-ion battery toward wearable energy storage. *Adv. Funct. Mater.* 27, 1703140.
30. Wei, T.S., Ahn, B.Y., Grotto, J., and Lewis, J.A. (2018). 3D printing of customized Li-ion batteries with thick electrodes. *Adv. Mater.* 30, e1703027.
31. Kohlmeyer, R.R., Blake, A.J., Hardin, J.O., Carmona, E.A., Carpena-Núñez, J., Maruyama, B., Daniel Berrigan, J., Huang, H., and Durstock, M.F. (2016). Composite batteries: a simple yet universal approach to 3D printable lithium-ion battery electrodes. *J. Mater. Chem. A* 4, 16856–16864.
32. Bao, Y., Liu, Y., Kuang, Y., Fang, D., and Li, T. (2020). 3D-printed highly deformable electrodes for flexible lithium ion batteries. *Energy Storage Mater.* 33, 55–61.
33. Hur, J.I., Smith, L.C., and Dunn, B. (2018). High areal energy density 3D lithium-ion microbatteries. *Joule* 2, 1187–1201.
34. Praveen, S., Sang, G., Chang, S., Chang, W.H., and Lee, W. (2021). 3D-printed twisted yarn-type Li-ion battery towards smart fabrics. *Energy Storage Mater.* 41, 748–757.
35. Blake, A.J., Kohlmeyer, R.R., Hardin, J.O., Carmona, E.A., Maruyama, B., Berrigan, J.D., Huang, H., and Durstock, M.F. (2017). 3D printable ceramic-polymer electrolytes for flexible high-performance Li-ion batteries with enhanced thermal stability. *Adv. Energy Mater.* 7, 1602920.
36. Maurel, A., Courty, M., Fleutot, B., Tortajada, H., Prashantha, K., Armand, M., Grugeon, S., Panier, S., and Dupont, L. (2018). Highly loaded graphite-polylactic acid composite-based filaments for lithium-ion battery three-dimensional printing. *Chem. Mater.* 30, 7484–7493.
37. Reyes, C., Somogyi, R., Niu, S., Cruz, M.A., Yang, F., Catenacci, M.J., Rhodes, C.P., and Wiley, B.J. (2018). Three-dimensional printing of a complete lithium ion battery with fused filament fabrication. *ACS Appl. Energy Mater.* 1, 5268–5279.
38. Ragones, H., Menkin, S., Kamir, Y., Gladkikh, A., Mukra, T., Kosa, G. and Golodnitsky, D. (2018). Towards smart free form-factor 3D printable batteries, Sustain. *Energy Fuels* 2, 1542–1549.
39. Lyu, Z., Lim, G.J.H., Guo, R., Kou, Z., Wang, T., Guan, C., Ding, J., Chen, W., and Wang, J. (2019). 3D-printed MOF-derived hierarchically porous frameworks for practical high-energy density Li-O2 batteries. *Adv. Funct. Mater.* 29, 1806658.
40. Delannoy, P.-E., Riou, B., Brousse, T., Le Bideau, J., Guyomard, D., and Lestriez, B. (2015). Ink-jet printed porous composite LiFePO4 electrode from aqueous suspension for microbatteries. *J. Power Sources* 287, 261–268.

41. Gua, Y., Wua, A., Sohnb, H., Nicoletti, C., Iqbal, Z., and Federici, J.F. (2015) Fabrication of rechargeable lithium ion batteries using water-based inkjet printed cathodes. *J. Manuf. Process* 20, 198–205.
42. Milroy, C., and Manthiram, A. (2016). Printed microelectrodes for scalable, high-areal capacity lithium-sulfur batteries. *Chem. Commun.* 52, 4282–4285.
43. Huang, J., Yang, J., Li, W., Cai, W., and Jiang, Z. (2008). Electrochemical properties of LiCoO2 thin film electrode prepared by ink-jet printing technique. *Thin Solid Films* 516, 3314–3319.
44. Chen, Q., Xu, R., He, Z., Zhao, K., and Pan, L. (2017). Printing 3D gel polymer electrolyte in lithium-ion microbattery using stereolithography. *J. Electrochem. Soc.* 164, A1852–A1857.

11 Evaluation of Dimensional Inaccuracy in 3D Printed Products
A Brief Overview

Suman Chatterjee
Department of Mechanical Engineering, National Institute of Technology, Rourkela, India

Ajit Behera
Department of Metallurgical and Materials Engineering, National Institute of Technology, Rourkela, India

Jinyang Xu
State Key Laboratory of Mechanical System and Vibration, School of Mechanical Engineering, Shanghai Jiao Tong University, Shanghai, PR China

CONTENTS

11.1 INTRODUCTION

Additive manufacturing (AM) sector is one of the advanced and rapidly growing sectors recently. The researcher has contributed immensely in this field in terms of both research and its application. Additive shows huge potential in different fields, such as automobile, aviation, space technology, medical science, and artefacts restoration, and with this contribution in so many fields, it attracted recognition and attention from the media. There was a mass survey conducted by various media houses and universities recently; it was stated that AM will have annual turnover up to USD 70 Billion [1]. Survey stated by Buisnesswire has also shown a huge market revenue earned by global AM sector accounts for USD 12 Billion in year 2020 [2]. The report also stated that the market can reach around USD 78 Billion. This huge market potential in AM encourages researchers to contribute, analyze, and research the requirement of improvement in this sector.

Additive manufacturing (AM) is a process of developing (or fabricating) a product by layer-by-layer deposition using CAD 3D models. The AM have incredible potential for fabricating customized, complex, high-value products and parts. The study states that industries in the world started implementing AM technology as a tool to deduct the manufacturing lead time and expand the functionality and quality of the products by reducing the fabricating cost [3]. Adding to this, AM has also contributed to many product developments and contributed to the market; for instance, the food industry, circuit design, biomedical appliances, sportswear, and even in household electrical appliances [3–8].

The recent trends and applicability of AM and AM processes are already discussed in the previous chapters. In addition to this, it is essential to understand the challenges arising during the development of products using AM technology. It is essential to address such challenges before adopting this process. One of the major challenges is dimensional accuracy of the developed products using the AM process. This is one of the concerning factors in AM that needs to be addressed. The dimensional accuracy for biomedical components is one of the important and critical issues where a dimensional and accurate product is required to fit with each person. When manufacturing high value and high-end products like turbine blades and electronic circuit manufacturing, dimensional accuracy is one of the critical factors for avoiding vibration and high efficiency, and circuiting fitting, respectively. Dimensional accuracy in micro and nano product development using AM process is of utmost interest. Even though this is very important, such accuracy issues were not addressed comprehensively. Limited research work has addressed this field. Some additive manufactured parts are illustrated in Figure 11.1, which required highest dimensional accuracy.

Hence, it is essential and desirable to improve and maintain the dimensional accuracy of the products to expand the pertinency of the AM technology, specifically with stereolithography. The present chapter provides the highlights and reviews issues related to the dimensional accuracy of the AM products and processes. The present work is also helpful in identifying the dimensional errors that occur during different AM processes.

11.2   DIMENSIONAL ACCURACY

With the advancement in AM technologies day-by-day, different additive manufacturing techniques have evolved, such as photo polymerization, powder bed fusion, material jetting, sheet metal lamination, fused deposition modeling, and direct energy deposition. There are lot of technical and fabrication issues raised during fabrication and after fabrication of the products, which raises different errors like voids, surface quality, mechanical properties, and dimensional accuracy. The present work mainly focuses on dimensional errors. Figure 11.2 illustrates the dimensional accuracy problem arising during AM. In Figure 11.2, the area under the

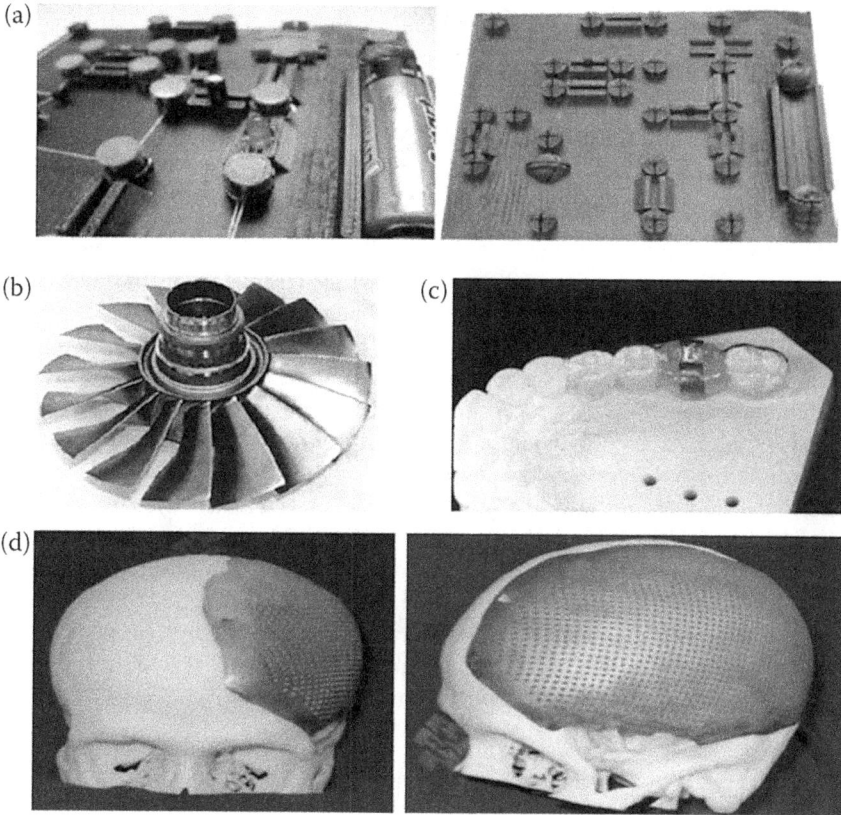

FIGURE 11.1 Additive manufacture products used in different fields (a) electrical and electronics. Reproduced from Lee et al. [3]. Copyright (2014) ASME; (b) turbine blade. Reproduce from Liu and Shin [9]. Copyright (2019) Elsevier; (c) and (d) medical science. Reproduce from Almog et al. [10] and Jardini et al. [11], Copyright (2001) and (2016) Elsevier respectively.

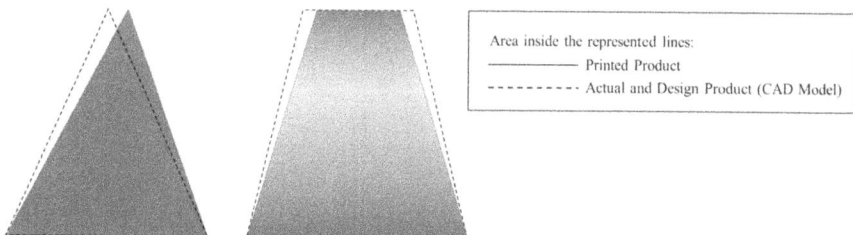

Area inside the represented lines:
————— Printed Product
--------- Actual and Design Product (CAD Model)

FIGURE 11.2 Schematic layout to understand dimensional accuracy of object.

dotted line is the actual figure given by the CAD model, which is asked to fabricate while the shaded area represents the developed area after fabrication of AM product. The difference in the actual expected product and the product after fabrication is clear. This difference is in the form of shape, size, and volume that altogether termed as dimensional error or dimensional inaccuracy. The factors affecting the dimensional error are illustrated in sections below.

11.3 DIMENSIONAL PRECISION IN PHOTO POLYMERIZATION PROCESS

In the photo polymerization process, photo polymers are used as building components for the products. In this technique, visible or ultraviolet rays are used to initiate and start a polymerization reaction and start to harden the resin on the platform layer-by-layer to develop a solid structure. There are different types of photo polymerization techniques, such as liquid crystal display (LCD), polyjet process, digital light processing (DLP), and stereo lithography (SLA) based projections [3].

Stereolithography (SLA) process helps in developing 3D products by focusing ultraviolet laser on a vat of liquid polymer resin to build one layer at a time. The platform shifts and UV laser beam again focuses on the surface of the layer to solidify the polymer resin on the previous layer; thus, another layer has been deposited on the desired 3D part. Hence, the process is repeated until the desired product is developed [12]. Lee et al. [3] in their observation have used the SLA machine, SLA® 5000 by 3D Systems, which has a vertical resolution of 0.2–0.3 mm and position repeatability between ±0.013 mm; it can develop a product having maximum capacity of 508 × 508 × 584 mm in size. Figure 11.3 shows products developed by SLA process and their dimensional errors.

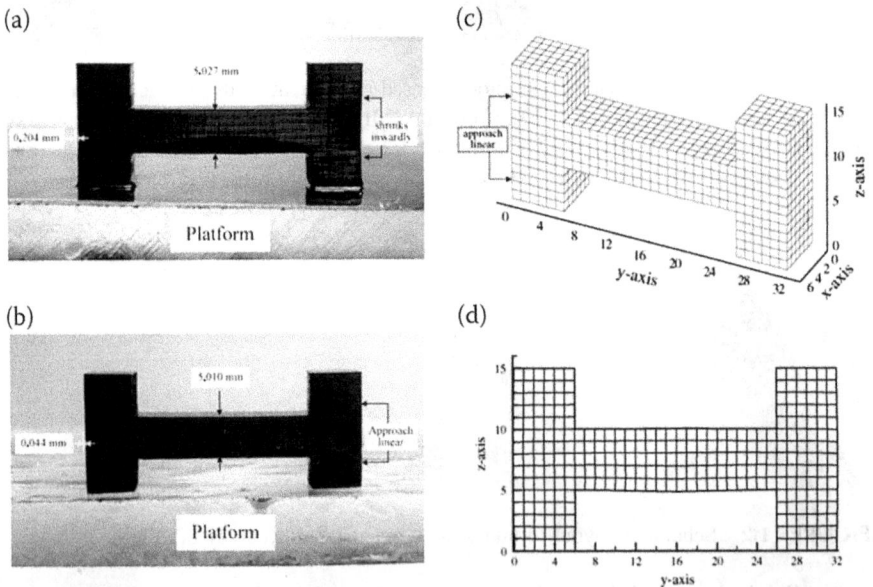

FIGURE 11.3 Actual parts developed by SLA process (representing shrinkage) and CAD model for SLA process (representing 3D and 2D deformation). Reproduce from Huand et al. [12]. Copyright (2005) Elsevier.

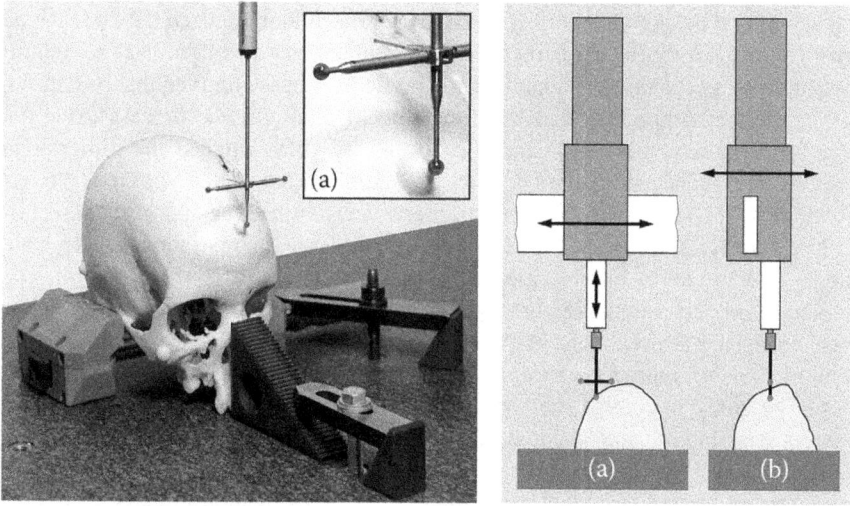

FIGURE 11.4 Dimensional measurement of parts developed by polyjet process. Reproduce from Salmi et al. [13]. Copyright (2013) Elsevier.

The polyjet 3D printing process is similar to inkjet printing of documents. Here, instead of ink, it drops photopolymers getting treated with a UV laser in a desired shape adding layer by layer [13]. The dimensional accuracy of the product developed is about 20–85 μm for features below 50 mm and 200 μm for full size of the developed model [3,13]. Figure 11.4 shows the dimensional measurement of parts developed by the polyjet process.

Similarly, LCD and DLP are also light-based projection methods, but not like SLA; here, the projected image in the vat of resins hardens the resin layer instantly before the final 3D model has been anticipated. KEVVOX, one of the leading manufacturers of DLP 3D printing machines, claims that its SP series DLP machine has 2%~4% dimensional errors [14].

Limited studies have been conducted to improve dimensional accuracy of the product developed by the photo polymerization process, especially the SLA process. Lee et al. [3] have stated the difficulties associated with the development of 3D-printed products. Taft et al. [15] have performed craniofacial reconstruction of neck and head using the SLA process. As there was no standard to ensure the accuracy and acceptability of the craniofacial reconstructed model, Taft et al. [15] have compared the 3D CAD model with the SLA fabricated part using contact probe. The study shows that in the Z-direction (perpendicular to the platform surface) was having much dimensional error as compared to the other printed direction. Huang and Lan [12] have used SLA for fabricating some parts. The study has used CAD/CAM/CAE model to identify the dimensional error of the fabricated parts. The study shows the problems arise due to inward shrinkage and misalignments occurring during fabrication. The authors suggest the development of a reverse compensation 3D model to overcome such errors. Campanelli et al. [16] have used

Taguchi based design of experiment (DOE) approach to minimized the dimensional error of the SLA fabricated parts. The statistical analysis has been used to identify the parameters effecting the quality of the products. It was observed that post curing of the fabricated products can be eliminated to minimize the process cycle and improve the parts accuracy. Chatterjee et al. [17] have studied the dimensional inaccuracy that occurs due to volumetric shrinkage of the 3D-printed product. Chatterjee et al. [17] have used a multicriteria decision-making approach embedded with a statistical approach to minimize the dimensional inaccuracy. The study suggests the effect of raster angle on the product quality.

Khorasani and Baseri [18] have performed a parametric study of the SLA process during the development of H-shape parts. The study uses an artificial neural network (ANN) approach to estimate the factors affecting the dimensional accuracy. The study highlights the shrinkage phenomenon of the 3D products, which reduces the quality and accuracy of the products developed by SLA process. Msallem et al. [19] have used response-surface methodology (RSM) to predict the responses and parametric analysis during the SLA fabrication process. The study shows quite agreeable results for using RSM model.

Other studies on other photo polymerization processes have been conducted to understand the effectiveness of the other processes. Salmi et al. [13] have used polyjet process for craniofacial reconstruction of the skull. The study coordinate-measuring machine (CMM) has been used to measure the coordinates and dimension of the developed skull with the 3D CAD model and the least errors (0.18 ± 0.12% & 0.18 ± 0.13%). The study shows the effectivity of polyjet process, and it can play a critical role in cranio-maxillofacial surgery. It might open new possibilities, but the process is not sufficiently investigated; it may take some time for the implementation. Figure 11.4 illustrates the dimension measurement of the model.

11.4   DIMENSIONAL PRECISION IN OTHER AM PROCESS

A similar trend of irregularities was also observed for other additive manufactured products. Research work is still carried out to address such issues. Researchers have adopted several methodologies to improve the dimensional accuracy of the additive manufactured products.

Turner and Gold [20] have presented a view on irregularities such as surface roughness and dimensional inaccuracy of the AM products. The study states that shrinkage and thermal warping have contributed more toward dimensional inaccuracy of the products. Baturynska and Martinsen [21] have used machine-learning algorithms to predict the dimensional accuracy during fabrication of polyamide 2200 using a powder-based fusion process. The study shows the potential in machine-learning possibilities in AM as thickness-prediction regression model is quite effective. Armillotta et al. [22] have studied the effect of a thermal cycle on the dimensional accuracy of the FDM-fabricated products. A comparative study of experimental and analytical model has been carried out; it states that the thermal cycle created due to the extraction process during FDM leads to wrap (distortion) of materials used for the process.

11.5   CONCLUSION AND FUTURE PROSPECTS

The present work in this chapter highlights the irregularities of additive manufactured products due to dimensional inaccuracy. The study shows the significant increase in curiosity among the researchers in studying dimensional accuracy of the AM product. Considerable research work was done to address the issues related to the dimensional inaccuracy of the AM products.

To achieve good quality and dimensional sound AM products, we still need to improve the developed model by producing a new one. Extensive research work by creating new models and experimentation should be carried out in the near future. It is essential to adapt different process parameters, environmental conditions, different fabricating processes, and varying parts design in AM. Researchers and practitioners should propose recommendations on improving models to minimize the dimension inaccuracy. Machine learning and metaheuristic approaches should be involved more to create a new scope of AM to improve the quality of AM products. The practitioners, authors, and researchers should also put their views on improving the models that can be used in the future.

REFERENCES

1. https://www.metal-am.com/am-market-forecast-to-reach-51-billion-by-2030/
2. https://www.businesswire.com/news/home/20200914005395/en/Global-Additive-Manufacturing-Market-Generated-12-Billion-Revenue-in-2020-and-is-Forecast-to-Reach-78-Billion-by-2028---ResearchAndMarkets.com
3. Lee, P.H., Chung, H., Lee, S.W., Yoo, J. and Ko, J., 2014, June. Dimensional accuracy in additive manufacturing processes. In *International Manufacturing Science and Engineering Conference* (Vol. 45806, p. V001T04A045). American Society of Mechanical Engineers.
4. https://www.3dnatives.com/en/top-10-3d-printing-sport-131120174/
5. Flowers, P.F., Reyes, C., Ye, S., Kim, M.J. and Wiley, B.J., 2017. 3D printing electronic components and circuits with conductive thermoplastic filament. *Additive Manufacturing*, 18, pp. 156–163.
6. Eshraghi, S. and Das, S., 2010. Mechanical and microstructural properties of polycaprolactone scaffolds with one-dimensional, two-dimensional, and three-dimensional orthogonally oriented porous architectures produced by selective laser sintering. *Acta Biomaterialia*, 6(7), pp. 2467–2476.
7. Wrobel, R. and Mecrow, B., 2020. A comprehensive review of additive manufacturing in construction of electrical machines. *IEEE Transactions on Energy Conversion*, 35(2), pp. 1054–1064.
8. Dong, Y., Bao, C. and Kim, W.S., 2018. Sustainable additive manufacturing of printed circuit boards. *Joule*, 2(4), pp. 579–582.
9. Liu, S. and Shin, Y.C., 2019. Additive manufacturing of Ti6Al4V alloy: A review. *Materials & Design*, 164, p. 107552.
10. Almog, D.M., Torrado, E. and Meitner, S.W., 2001. Fabrication of imaging and surgical guides for dental implants. *The Journal of Prosthetic Dentistry*, 85(5), pp. 504–508.
11. Jardini, A.L., Larosa, M.A., Macedo, M.F., Bernardes, L.F., Lambert, C.S., Zavaglia, C.A.C., Maciel Filho, R., Calderoni, D.R., Ghizoni, E. and Kharmandayan, P., 2016. Improvement in cranioplasty: Advanced prosthesis biomanufacturing. *Procedia CIRP*, 49, pp. 203–208.

12. Huang, Y.M. and Lan, H.Y., 2005. CAD/CAE/CAM integration for increasing the accuracy of mask rapid prototyping system. *Computers in Industry*, 56(5), pp. 442–456.
13. Salmi, M., Paloheimo, K.S., Tuomi, J., Wolff, J. and Mäkitie, A., 2013. Accuracy of medical models made by additive manufacturing (rapid manufacturing). *Journal of Cranio-Maxillofacial Surgery*, 41(7), pp. 603–609.
14. KEVVOX, SP Series 3-D Printing. http://kevvox.com
15. Taft, R.M., Kondor, S. and Grant, G.T., 2011. Accuracy of rapid prototype models for head and neck reconstruction. *The Journal of Prosthetic Dentistry*, 106(6), pp. 399–408.
16. Campanelli, S.L., Cardano, G., Giannoccaro, R., Ludovico, A.D. and Bohez, E.L., 2007. Statistical analysis of the stereolithographic process to improve the accuracy. *Computer-Aided Design*, 39(1), pp. 80–86.
17. Mansaram, M.V., Chatterjee, S., Dinbandhu, A.K.S., Abhishek, K. and Mahapatra, S.S., 2021. Analysis of dimensional accuracy of ABS M30 built parts using FDM process. Recent advances in mechanical infrastructure. Proceedings of ICRAM 2020, p. 173.
18. Khorasani, E.R. and Baseri, H., 2013. Determination of optimum SLA process parameters of H-shaped parts. *Journal of Mechanical Science and Technology*, 27(3), pp. 857–863.
19. Msallem, B., Sharma, N., Cao, S., Halbeisen, F.S., Zeilhofer, H.F. and Thieringer, F.M., 2020. Evaluation of the dimensional accuracy of 3D-printed anatomical mandibular models using FFF, SLA, SLS, MJ, and BJ printing technology. *Journal of Clinical Medicine*, 9(3), p. 817.
20. Turner, B.N. and Gold, S.A., 2015. A review of melt extrusion additive manufacturing processes: II. Materials, dimensional accuracy, and surface roughness. *Rapid Prototyping Journal*, 21(3), pp. 250–261. https://doi.org/10.1108/RPJ-02-2013-0017
21. Baturynska, I. and Martinsen, K., 2021. Prediction of geometry deviations in additive manufactured parts: Comparison of linear regression with machine learning algorithms. *Journal of Intelligent Manufacturing*, 32(1), pp. 179–200.
22. Armillotta, A., Bellotti, M. and Cavallaro, M., 2018. Warpage of FDM parts: Experimental tests and analytic model. *Robotics and Computer-Integrated Manufacturing*, 50, pp. 140–152.

12 3D Nanoprinting in Oral Health Care Applications

Gaetano Isola, Alessandro Polizzi, and Simona Santonocito
Department of General Surgery and Surgical-Medical Specialties, School of Dentistry, University of Catania, Catania, Italy

CONTENTS

12.1 INTRODUCTION

Tissue engineering represents a highly innovative search field aimed at recovering an original tissue function by selective replacement of the damaged tissue or organ portion through healthy tissue using specific engineering techniques, through editing or "nano" cell editing. This tissue engineering through nanopoprinting editing technique is obtained by employing three main actors: cell, biomolecule signaling, and a scaffold supporting the process. When the components above are suitably combined through a performing engineering process, a tissue is successfully re-generated. The cells specifically inserted into the scaffold proliferate and differ-entiate and the produced biomolecules quicken the regeneration process; the scaffold provides an artificial structure that must be highly stable to guarantee the creation of the new three-dimensional (3D) tissue. However, to achieve correct and stable tissue regeneration, the scaffolds used during the process must be biocompatible with the cell biotype and have an adequate pore size within the scaffold to allow the cell to grow and interconnect, and, finally, become functional from a biological and biomechanical point of view. In this regard, to obtain an optimal technique, the scaffold used in the fabric should be biodegradable as much as possible, thus avoiding eventually being forced to eliminate the scaffolding after the maturation process has taken place. The engineering-guided tissue regeneration technique has been developed over the past few decades [1–3]. In fact, in the early development period, the scaffolds used were manufactured by methods such as gas foaming, particulate leaching, lyophilization, or other binding method techniques. These did not allow for highly defined long-term catabolic programming and therefore did not allow or control the micrometric diameter of the pores; nor did they control the tissue architecture of the scaffold or the interconnectivity or biodegradability, negatively affecting tissue growth and the final qualitative regeneration of the tissue.

Over the years, various techniques have been used to improve 3D nanoprinting in general medical and scientific fields. Among these, 3D printing technology that uses computer-aided design (CAD) and computer-aided manufacturing (CAM) have definitely represented an important step in optimizing 3D nanoprinting. These methods apply specific computer software to program and manufacture a scaffold. Their inner structure, such as porosity, pore shape and size, and interconnectivity of scaffolds, can be controlled freely and with greater ease [4]. Furthermore, 3D printing assisted by specific software allows to produce highly reproducible products and a standardization of the products and methods to be easily reproduced. This standardisation removes the variability of the inner structure among the scaffolds, thus improving the experiments' validity and repeatability. 3D printing also allows the production of patient-customized scaffolds [5].

Preliminary evidence currently present in the scientific field highlights how this technology has better control on pore size, interconnectivity, and porosity compared to traditional scaffold manufacturing [6]. In this regard, 3D nanoprinting techniques make it possible to manufacture three-dimensional scaffolds identical to the original project with greater accuracy so that they can be used to standardize printing and allow highly predictable reproducibility [7]. Clinical and preclinical studies have extensively validated these technologies to regenerate bone tissues, cartilages, ligaments, muscles, neurons and skin, and organs such as heart, kidney, liver and trachea. Various 3D printing technologies [8,9] have been developed in the biomedical field in recent years, including stereolithography, layering models, inkjet printing, selective laser sintering and electrospinning technology.

Among the various 3D printing methods, two-photon laser-based 3D printing nanofabrication and controlled electrospinning have received attention in various areas of tissue engineering due to their high ability to fabricate structures with a high volume/surface ratio through a porous architecture strongly interconnected at a submicrometric resolution [9,10]. These methods have produced highly precise 3D scaffolds that allow the extracellular matrix to be reproduced at high fidelity and allow a correct cell signaling study in cellular adhesion, proliferation, and differentiation analysis in various preclinical studies. The evidence regarding the 3D nanoprinting method is still limited; however, early results from preclinical and clinical analyses have shown promising results.

This chapter describes the 3D nanoprinting methods for oral health, with particular reference to reconstructive techniques in the restorative, prosthetic, and surgical field. In addition, chapter describes on 3D reconstruction method of a lost or damaged caused by the pathology.

12.2 APPLICATIONS OF 3D NANOPRINTING IN DENTISTRY

During the last few years, there has been a rapid technological growth regarding 3D printing technology, also known as additive manufacturing or solid freeform manufacturing [11,12]. In the dentistry, the 3D printing technology enables the manufacturing of an individualized 3D object with a design and material of your choice. The chance of including living cells has elevated the method of 3D printing and nanoprinting to a superior level and allowed the realization of various tissues. In this regard, new personalized therapeutic opportunities for the patient are emerging. The possibility of producing many types of printed biomaterials allows us to precisely control the inner structure and the external shape of the scaffold. Today's new technologies make it possible to obtain patient reports in 3D. These accurate digital data can be easily transferred to manufacture precise and customizable structures for every dental patient. The expiration of major 3D printing patents has increased further the development of these new technologies at more affordable costs [13].

Following the growing attention toward 3D printing technology over the past decade, its use in regenerative medicine and tissue engineering became the main studied fields of interest. For example, in regenerative medicine, the use of a combination of cells with 3D-printed polymers for drug screening, in vitro disease models, and tissue engineering is emerging. [14]. Artificial tissues for the configuration of 3D in vitro models may be generated though cell ink-based bioprinters or spheroid/microtissue-based systems [15,16]. With this capability, 3D printing allows the production of stem cells in the desired 3D scaffold for regenerative and transplant purposes [17].

Different materials are employed to produce cell-laden 3D-printed scaffolds, such as chitosan [18], calcium silicate complex [19], and controlled-release polymeric materials with bioactive agents [20]. 3D printing has found its application in the generation of optimal volumes of human bone and skin grafts in vitro [21,22]. This technology represents an advantageous possibility to substitute actual self-procurement strategies that are related to donor site morbidity and structure loss [23].

The additive manufacturing process can be broadly subdivided into four stages:

1. The use of CBCT/intraoral scanner/software data for the production of a 3D digital model;
2. The 3D model procession and sectioning into many two-dimensional layers;
3. 3D printing of the final object layer by layer;
4. Post-procession of the new product [24,25].

This process is the base of 3D printing, and it is applicable for different materials, for example ceramics, metals, and polymers. In the next sections, the clinical, experimental, and future didactic dental aspects of this technology will be analyzed.

12.3 3D NANOPRINTING AND DIGITAL DENTAL WORKFLOW: FROM DIGITAL MODEL TO 3D PRINTING

During the last ten years, thanks to progress in 3D imaging and modelling technologies such as cone-beam computed tomography and intraoral scanning, and with the refinement of CAD-CAM technologies in dentistry, 3D printing technology has had a major resonance in all branches of dentistry [13]. 3D printers must be supported by computer-aided design (CAD) software that enables medical devices to get designed in a virtual setting. Therefore, CAD software is nowadays commonly found in the dental laboratory and also in many dental practices. Along with CAD software, significant importance is also attached to computed tomography (CT), cone-beam computed tomography (CBCT), and intraoral or laboratory optical scan data, which provide access to volumetric data necessary for 3D design and printing [13].

In detail, the workflow for making a 3D print involves the acquisition of a 3D model of the patient, which also can be quickly taken utilising an intra-oral scanner or a CBCT to generate a three-dimensional image of the dental arches. It may also be obtained in the traditional way using impression material from which the plaster model is obtained, which is then digitised in 3D using a scanner. From the dental impression taken, the model/appliance/prosthesis designed in CAD software is processed. The virtual models designed with appropriate software are loaded into 3D printers that may be applied the dental laboratory or, directly, in the doctor's office. These machines allow the physical production of a large number of artefacts, such as dental models, surgical guides, splints, wax-ups, stereolithographic models, and more [16] (Figure 12.1). Of course, it must be taken into account that the accuracy of 3D prints can be limited in the presence of artefacts in the instrumental image or in the digital impression, due to metal components such as teeth, restorations or implants [26–28].

12.3.1 3D PRINTING TECHNOLOGIES

There are currently several distinct printing technologies on the market, each with advantages and disadvantages (Table 12.1).

- Stereolithography (SLA, SL) is a manufacturing method that uses photo-induced polymerisation to build layered structures, utilising high cross-linked polymers, in a pool of photopolymer resin [29]. Every layer is traced by laser on the surface of the liquid resin and cured, at which point a "build platform" drops down a gap equivalent to the depth of one layer and builds the next layer until the 3D object is printed [30]. A 3D-digitised model, used as a template for the manufacturing process, drives the SLA

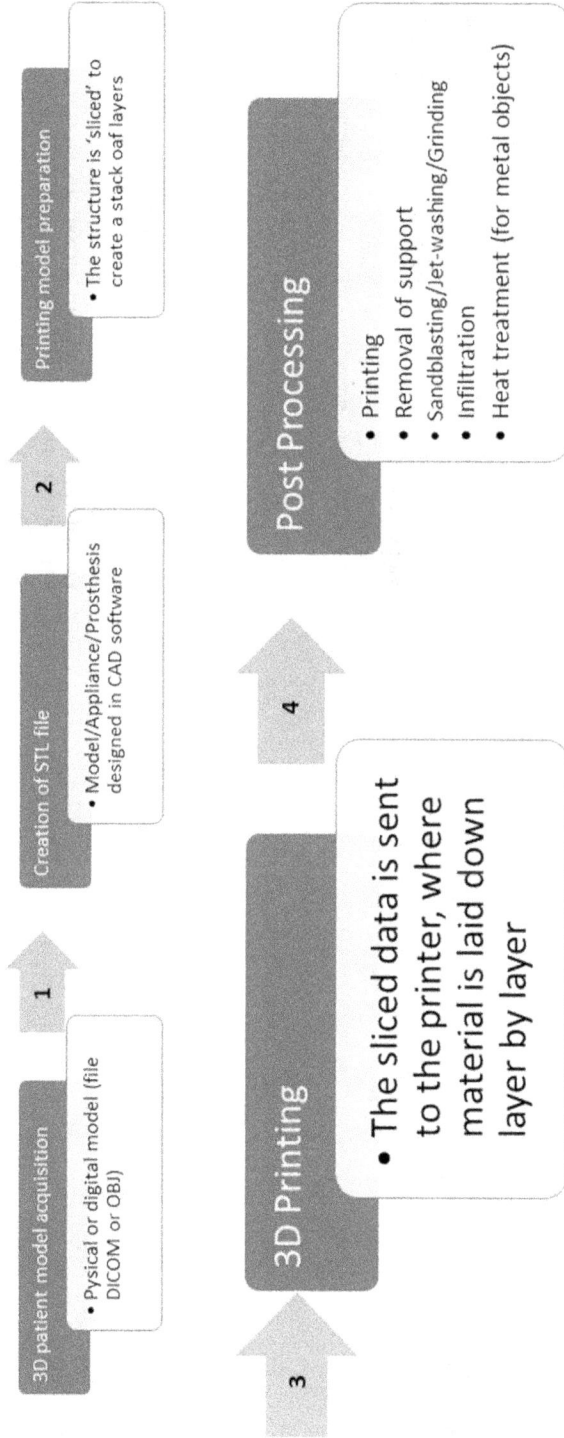

3D patient model acquisition
- Pysical or digital model (file DICOM or OBJ)

Creation of STL file
- Model/Appliance/Prosthesis designed in CAD software

Printing model preparation
- The structure is 'sliced' to create a stack oaf layers

3D Printing
- The sliced data is sent to the printer, where material is laid down layer by layer

Post Processing
- Printing
- Removal of support
- Sandblasting/Jet-washing/Grinding
- Infiltration
- Heat treatment (for metal objects)

FIGURE 12.1 3D printing process.

TABLE 12.1

Characteristics and Features of 3D and 4D Printing Methods in Dentistry

3D/4D Printing System	Material Characteristics	Materials	Advantages	Disadvantages
Stereolithography (SLA)	Light curable resin	Epoxy and methacrylate monomers	• Product resolution • Efficiency • Short working time	• Over-curing • Lack of surface smoothness • Limited mechanical strength • Irritant
Selective laser sintering (SLS) Powder	• Polymers • Ceramics	• Metals	• Structures are fully self-supporting • Protective gas in not needed • Vast variety of materials can be selected • Little to no thermal stresses are accumulated on the component • Components exhibit excellent mechanical properties • Relatively fast method	• Sample surfaces are porous and rough • Harmful gases release during fabrication • Materials waste is relatively high • Raw powders are expensive • Post-processing is often expensive and long
Fused deposition modeling (FDM)Thermoplastic polymer and composites, low melting temperature metal alloys	• Paste	• Wire	• Filaments are cheap and arrive in various colors • Easy to change materials • Cost-effective maintenance • Capable of fast production of shelled structures • Fundamental for thinner layers up to 0.1 mm thick • Released fumes are not toxic	• The seam between layers is visible • Discontinuous extrusion results in formation of defects • Support structure is required in some cases • Delamination between layers may occur due to low extrusion temperature • Printed component may curl off the build platform because of induced thermal stresses

Technique	Material	Material types	Advantages	Disadvantages
Photopolymer jetting	Light curable resin	• Biocompatible (MED610) VeroDentPlus (MED690) and VeroDent (MED670) (all are natural looking medically –approved photopolymers)	• High resolution due to thin layer printing (~16 microns per layer) • Short working time • Excellent surface features • No need for post-modification • Supporting wide range of materials	• Irritant • High cost
Powder binders	Materials which are available in powder	• Metal • Ceramic • Plastics	• Safe material • Short working time • Suitable mechanical performance • Low cost	• Low resolution • Low strength • Cannot be soaked/heat sterilized
Digital light projection	Light curable resin	• Resin	• High complexity and excellent surface finish • Short timeframe • Good accuracy • Smooth surfaces	• Limited material selection • Photocurable resin can cause skin sensitization, and maybe irritant by contact
Computed axial lithography	Light curable resin	• Resin containing dissolved oxygen	• It can be used in specific conditions where existing methods fall short, such as: Printing soft materials that cannot maintain the forces applied during layerwise printing, • Creating lenses with smooth curved surfaces, • Encapsulating other objects in three dimensions	• Limited material selection

machinery and is generated by CAD software. The digitised model generates the substrates, which, once printed, must withstand gravity rubbing
and must then be stripped from the final product. Post-processing includes
removing of surplus resin and a curing process in a UV oven [30,31]. At
present, SLA is used in the production of temporary and permanent crowns
and bridges, temporary restorations, surgical guides, models, and replicas
of dental models [32]. SLA has been shown to have greater clinical precision than other digital/analogue techniques in the manufacture of plaster
casts, proviting an accepted alternative for the diagnosis, treatment, and
production of prosthetic devices [33]. But devices printed with ALS appear
to have poor mechanical characteristics, determined by the restricted selection of resins that can be light-cured [30]. To overcome this limitation,
new resins with nanoparticles embedded in the polymer matrix are being
created in recent years to improve their mechanical properties [34].
Ceramic fillers also provide protection of the moulded structure from
fracture by improving stress distribution. In conclusion, ALS is a rapid,
cost-effective, and multifunctional technology in dental 3D printing, allowing the production of accurate and reliable devices for long-term use,
thanks to high-precision measures recoved from the scan data. The long-
evity of printed artefacts can be affected by the leakage of nonreactive
resin monomers present [35].

- Photopolymer jetting (PPJ) employs light-cured resin materials and dynamic
  print heads. The light-sensitive polymer is sprayed into a build platform by
  an inkjet print head and cured layer by layer on the downward platform, and
  an ultraviolet light source is used to cure each layer of resin or wax that has
  been sprayed [35]. Also in this technique, a support structure is printed,
  which can eventually be removed. This technology is utilised to print a
  broad range of resins and waxes for casting highly detailed crowns and
  anatomical study models with a resolution of ~16 microns [13,36]. The
  benefits of this technology are fast production, smoothness of the surface,
  and cost-effectiveness, while the disadvantages include the impossibility of
  heat sterilisation and the high cost of the material [37].

- Selective laser sintering (SLS) utilises a high-energy laser beam to fuse
  material in the form of a fine powder and create a solid layer from the dust.
  The platform is moved down each time to allow room for the laser to sinter
  the next layer of dust. This structure creation technique does not need additional material backing during printing because the backing is given by the
  surrounding powder [38]. Using SLS, it is possible to create multi-use studio
  models, punch and cut guides, and metal structures. The advantages of this
  technique include the use of autoclavable substrates, increased mechanical
  strength of the moulded object, and reduced production costs. Disadvantages
  comprise the high initial cost of installation and the necessary for additional
  consumable, like air pressure for proper SLS operation [39,40].

- Powder binder printers (PBP) utilise droplets of liquid adhesive inside a
  modified ink-jet printhead. The inkjet head delivers these droplets to

infiltrate a layer of powder beneath. This phase is continued by re-integrating a new layer of powder until the the end product is produced. The primary use of printing powder binders in dentistry is in the printing of models or study models. However, printed objects are brittle and lack precision. As it is a low-cost process, the technique is used in non-sterilisation applications, like printing study models. Nevertheless, the production procedure is untidy due to the use of powder [41,42].

- Fused deposition modelling (FDM) is one of the first 3D printing technologies consisting essentially of a robotic adhesive gun, characterised by an extruder passing through a fixed platform or a platform moving under the extruder. Objects are "sliced" into layers by software and the coordinates are passed to the printer [13]. Thermoplastic polymers and their compounds (e.g., acrylonitrile-butadiene-styrene (ABS), polycarbonates and polysulphones) together with lower melten metal alloys (e.g., bronze metal filaments) are used in the production of thermoplastics, bronze metal filaments) are the most commonly used FDM filaments [43]. The construction of complex geometry typically requires the laying of supporting structures that can be made from the same material, or from a second material laid by a second extruder. The precision will dependent on the travel speed of the extruder, as well as the material stream and the size of each "step" [13].

- Computerized axial lithography has some similarities with the DLP approach as both use the projector as a light source to light-cure the resin. The difference between computer axial lithography and other traditional printing methods is that light curing is sprayed at different angles to the substrate, whereas other AM printing methods use layer-by-layer curing. In other words, computed axial lithography can output entire target at once (not layer by layer) [37].

- Digital light projection (DLP) is similar to SLA printers. The substantial distinction between the two technologies is the time required to produce the object since in DLP, the resin layer, hit by the light beam from the projector, cures simultaneously in all the points belonging to the same layer, while in printers using the laser (SLA), the layer is composed more slowly since the light beam generated by the beam must "draw/pass" the entire layer [2,44].

12.4 3D NANOPRINTING IN DENTAL MATERIALS AND PROSTHESIS

12.4.1 MATERIALS USED FOR 3D/4D PRINTING IN DENTISTRY

The most widely adopted materials in dental 3D printing are polymers, polyesters, metals, and ceramics [31]. Among the most commonly used polymers for dental 3D printing are vinyl polymers, requiring sintering (e.g., SLS) or photopolymerisation (e.g., SLA). Poly(methyl methacrylate) (PMMA) is the most widely used vinyl polymer in dental 3D printing to fabricate prosthetic bases. The merits of PMMA are its easy hanging, lower cost, light weight, stability in the oral cavity setting, and aesthetic properties. Nevertheles, PMMA has low surface characteristics and poor mechanical characteristics. Therefore, in recent years, additives such as poly-etheretherketone, SiO_2, and Al_2O_3 have been used to overcome the mechanical restrictions of the material [13,42,45]. Polyesters, in particular polycarbonate (PC), polycaprolactone (PCL), and polylactic acid (PLA), are extensively employed in the field of 3D printing. PCL, because of its low fusion point ($\sim63°C$), is widely used in 3D printing using a printing technique such as FDM. 3D printed PCL is utilized in bone tissue regeneration like alveolar bone augmentation [46]. On the other hand, PLA possesses excellent processability that allows its use in different 3D printing techniques and for different appliances, such as FDM printing of drill guides for surgical insertion of dental implants [47] and temporary restores to protect teeth after crown preparation [48].

The metal alloys primarily used in 3D printing techniques are CoCr, Ti, and tantalum powders [49]. The 3D printing techniques used are SLS and SLM, which exploit an initial blend of metal alloy dusts consisting of spherical particles and have a rigorous distribution of particle dimensions to achieve smooth packing behaviour. With the use of metal-based additive modelling techniques, fully functional components can be fabricated using laser sintering in a layer-by-layer procedure. Several studies observed that CoCrMo alloys fabricated with SLS had better mechanical characteristics and fewer metal ion dissolutions than cast alloys [3]. Metallic frameworks that can be 3D printed create removable partial dentures, overdentures and fixed prosthetic structures, manual and rotary endodontic files, and metal implants [50]. The most widely applied ceramics in 3D printing are zirconia powder [51] and alumina [52]. Yttria-stabilised zirconia powder (3Y-TZP) is mixed with acrylates and methacrylats, photoinitiator and dispersing agent to make a homogeneous hybrid sol, which can be light-cured by a 3D printer with light-curing capabilities. SL has been used for printing dental bridges [51].

12.4.2 3D PRINTING APPLICATIONS IN PROSTHETICS

CAD/Computer-aided manufacturing (CAD/CAM) systems have totally revolutionised the way dentures are made. In the fabrication of total dentures, the workflow involves: taking virtual impressions using intra-oral and extra-oral scanners and recording the patient's occlusion; the design of the denture using computer software; and the fabrication of the denture using additive or subtractive manufacturing methods [53,54]. With the additive manufacturing technique, both the denture bases and the denture teeth are now printed with 3D printing technology using methacrylate-based light-curing resins [55]. The prosthetic and artificial teeth are printed independently and then fixed in place using a light-curing adhesive [55,56]. Recently, 3D printing technology has been used to fabricate crowns and bridges. In addition, lower cost 3D printers have been utilised to manufacture accurate temporary crown and bridge restorations using SLA printing technology [57], which provides high efficiency and a high level of precision [58]. Using the SLA technique, it is possible to obtain a printed object with a resolution of up to 0.05 mm, resulting in a superior technician in terms of precision, even compared to some newer techniques like digital light projection (DLP) [59]. However, the photopolymer-curing procedure in the DLP printer is quicker than SLA, as SLA employs laser light photopolymer arrays working layer by layer [60]. The workflow involves use of intraoral optical scanners or laboratory scanners that allow the development of accurate virtual models of the preparatin tooth [61,62], implant position [63], and dental arch [64]. This allows prosthetic or implant-prosthetic treatment to be planned using CAD software. This scan information and CAD design can be utilised to cut or print crown or bridge caps, implant abutments, bridge frames, and prostheses. 3D printing can also be used to metal structures either indirectly by printing in burn-out resins or wax for a lost wax procedure, or directly in metals or metallic alloys [39,65]. Direct metal printing involves the need for more expensive technology and requires a large amount of post-processing before the parts are finished [66], compared to printing in resin/wax and later using a traditional casting approach [67]. When printing intricate implant bridge structures, 3D printing may be used in combination with milling/machining techniques to create a very precise mechanical attachment to the implant. [13].

12.5 3D NANOPRINTING IN ORAL AND MAXILLOFACIAL SURGERY, BONE REGENERATION, AND TISSUE ENGINEERING

The development of technology led to the spreading of innovative digital workflows in dentistry, all characterized by three phases:

1. Data digitalization by scanning;
2. Data processing by computer-aided design (CAD) software;
3. Fabrication through computer-aided-manufacturing (CAM) [68].

The technology initially involved in realising the artefacts consisted of subtractive manufacturing or milling that allowed to create accurate products but with some limitations: time consumption, material waste, and the impossibility of reproducing complex anatomies [69]. These limitations can be overcome through 3D printing, an additive manufacturing characterized by objects' realization layer by layer [70]. This promising technology can be used in various fields of dentistry, not only in prosthodontics and restorative dentistry, but also in orthodontics, endodontics, periodontics, implantology, and oral and maxillofacial surgery (Figure 12.2).

12.5.1 ORAL AND MAXILLOFACIAL SURGERY

CBCT and 3D printing enabled the realization of guided surgical procedures, based on a greatly accurate diagnosis and personalized treatment planning, thus reducing intra- and post-operative risks. The treatment of maxillofacial defects, orthognathic surgery, sinus lift, and implant placement and fabrication can be improved through 3D devices [72].

Maxillofacial surgery. Maxillofacial defects may be caused by tumor resections, infections, trauma, and degenerative diseases. These defects, more or less extensive and severe, deeply affect patients' quality of life and may benefit from

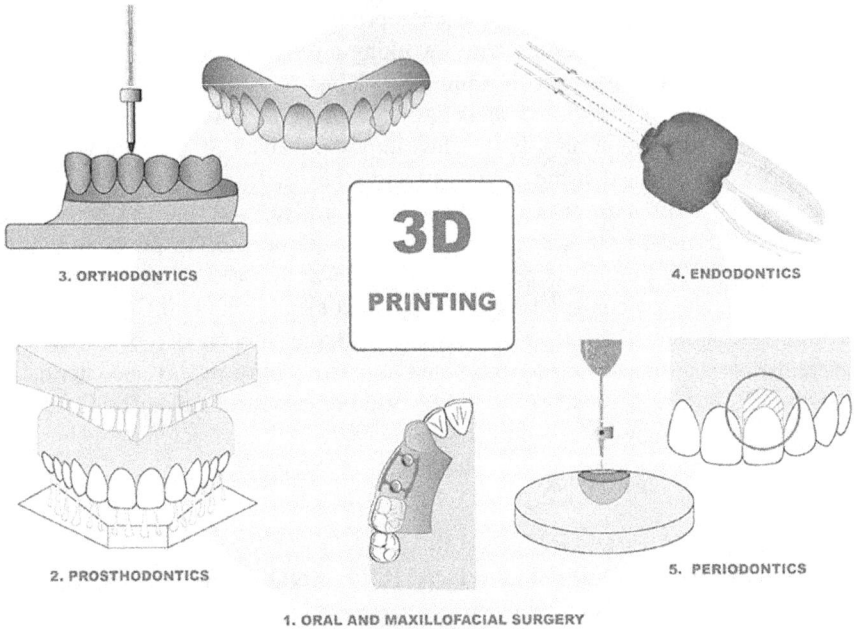

FIGURE 12.2 3D printing possible applications in dentistry. (1) 3D-printed guides and scaffolds for bone defects; (2) 3D-printed dental prosthesis; (3) clear aligners and dental models printed for orthodontic patients; (4) CBCT-derived guides for endodontic treatments; (5) 3D-printed periodontal scaffolds. From Oberoi et al. [71] under the agreement of the terms of the Creative Commons Attribution License (CC BY).

maxillofacial prostheses. Thanks to the use of scanners, maxillofacial prostheses can be realized more comfortably and precisely. It has been shown that maxillofacial prostheses made by additive manufacturing are characterized by an improvement in precision and adaptation on the defect area [70,71]. Furthermore, by scanning the surgical site, 3D-printed models can be realized to better pre-plan complex cases and test the fabricated prostheses before the surgical procedure. These innovative protocols allow a medium time saving of 30–90 minutes in operating procedures [73].

Orthognathic surgery. Adult patients with severe maxillary-mandibular discrepancies, malocclusions, or esthetic deformities require orthognathic surgery. 3D printing may be useful to improve surgical accuracy and to reduce temporomandibular joint post-operative instability [71,72]. For example, 3D-printed osteotomy surgical guides may reduce operative timing and the risk of nerve injuries compared to traditional operations [74]. Therefore, since software pre-determines the position of the bone segments and drill holes for screws, the customized 3D-printed titanium plates can be placed when the bone elements are precisely in the predetermined position [75,76].

Sinus lift. Sinus position and maxillary tooth extraction may be a cause of vertical bone height reduction, which can complicate prosthetic implant treatment [77]. Sinus lift surgery can be improved through the use of grafting materials that are important to increase implant survival rates [78]. Several types of grafting materials have been used for sinus lift surgery; for example, allografts, autogenous bone, mixtures of various biomaterials, synthetics, and xenografts [79]. 3D printing and additive manufacturing make it possible to replicate the bony architecture and the graft macroporosity with minimal wastage of material. Furthermore, ethical concerns, risk of infection, and availability limits are surpassed [80]. However, the use of 3D scaffolds for sinus lift is only documented in case reports and in vivo studies, but randomized controlled clinical trials on the subject are lacking [77,81].

Implant placement and customized implants. The anomalous implants positioning causes unpredictability and instability of the prosthesis [82]. 3D printing technology allows the creation of precise guides for the pre-planned surgical implant insertion [83]. 3D-printed surgical guides have been shown to improve hard and soft tissue healing around the screw and increase implant placement accuracy. To have stability during implant placement, 3D-printed surgical guides, basing on the remaining teeth, tissues support, and location of the surgical site, may be distinguished in tooth/mucosa/bone/pin-supported guides [84]. Furthermore, thanks to 3D printing technology, customized implants with complex geometry and porous surfaces can be produced [70]. The use of printed titanium and zirconium implants have been investigated in vitro and in vivo [85–88], but their employment in the clinic cannot be justified due to laking of long-term studies [71].

12.5.2   BONE REGENERATION AND TISSUE ENGINEERING

3D printing is also emerging in periodontology through the realization of 3D CBCT-based printed guides for soft tissue esthetic improvement and the use of tissue scaffolding in regenerative periodontology [70,71]. These scaffolds can consist of the various elements of periodontum: periodontal ligament, gingiva, cementum, or bone [89].

Bone regeneration. The need for bone defects therapy for implant-supported prosthesis can be managed employing autologous bone (osteoconductive, osteoinductive properties, no immune reactions, increased surgical invasiveness and limited availability based on the sampling site) or other materials such as allografts, xenografts, and alloplastics [90,91]. 3D-printed scaffolds may be produced with a precise 3D architecture and spacing of pores [92,93] and can be loaded with stem cells, allowing a better vascularization and osteogenesis [70,72].

Soft tissue engineering. Regarding oral soft tissue regeneration, the elective material for obtaining the best clinical results is the autologous graft, which has several disadvantages: different dimensional collection limits for each patient, complex collection technique, and prolonged post-operative discomfort [94]. In order to reduce these limitations, an ideal non-autologous soft tissue graft should be biocompatible, stable, biodegradable, and low cost. Soft tissue engineering allowed the realization of 3D-printed soft tissue grafts that may be used in the management of large defects, avoiding post-operation discomfort of the donor site with a reduction of intervention times and costs. However, these scaffolds must be made to measure for the single defect; they cannot reproduce the inner architecture of oral soft tissues and their surgical management remains challenging [90].

A great challange in the research is the realization of scaffolds able to reproduce the dentin/cementum-periodontal ligament-alveolar bone interface through the realization of 3D-printed multi-phasic scaffolds able to use different types of periodontal cells during the healing process. Different strategies and materials have been proposed to guide the multiphasic periodontal regeneration through delivering various differentiated cells and signaling proteins [71,90].

Although hard and soft tissue engineering strategies are very promising, much more long-term RCT and clinical studies must be conducted to establish the use of 3D-printed scaffolds in clinical practice [95].

12.6 3D NANOPRINTING IN EDNODONTICS, ORTHODONTICS, AND ORAL MEDICINE

3D printing technology is gradually involving all branches of dentistry, including endodontics, orthodontics, and oral medicine.

12.6.1 3D PRINTING IN ENDODONTICS

3D printing in endodontics involves different procedures, such as regenerative endodontics, surgical and non-surgical endodontic treatment, and educative approaches.

Regenerative endodontics. 3D printing in regenerative endodontics is a recent research field. It found different applications, such as stem cell delivery, injectable calcium phosphates cements, pulp scaffolds, gene therapy, and growth factors [96]. For example, different 3D-printed calcium phosphates cement have been realized as porous scaffolds to induce pulp regeneration [97]. 3D-printed polyepsilon-caprolactone and hydroxyapatite scaffolds have allowed the generation of anatomically shaped tooth-like tissue [98]. Furthermore, other proposed regenerative protocols involved using scaffold-free techniques consisting of dental pulp cell-derived spheroids [99–103].

Micro-guided endodontics. 3D printing found clinical application also in nonsurgical and surgical root canal treatment. In particular, through CBCT data, it is possible to design custom-made guides to improve the accuracy of access into the dental canals, especially in case of root canal anatomy anomalies, which would greatly increase the difficulties during shaping, disinfection, and obturation phases. In these cases, it is possible to print 3D tooth models with the reproduction of the internal root canal anatomy useful to manufacture the guide [104]. This protocol that uses 3D CBCT data may also be useful in molars with complex root canal anatomy that cannot be observed through 2D radiography [105]. Furthermore, 3D-printed guides are very useful in calcified canal cases and in surgical endodontics of posterior teeth [72,106].

Educative approaches. Dental students have been often trained through the use of typodonts that have idealistic root canal anatomies. Additive manufacturing plays an important role in endodontic training [71]. Transparent 3D resin teeth can be printed using CBCT data of extracted teeth. The 3D-printed resin teeth may be useful for the educative purpose of students and to enhance the skills of non-specialists. These approaches are very useful because it is possible to observe the working length, the anatomy of the apical delta, the gauging of preparation, and canal transportation [72]. Furthermore, 3D-printed models may be useful to simulate critical elements like blood vessels and nerves near the roots or thick cortical bone-covering root apieces. These approaches may help the surgeon to prepare for challenging situations [107].

12.6.2   3D Printing in Orthodontics and Oral Medicine

Orthodontics is another field in which 3D printing is flourishing. In particular, additive manufacturing finds application in the realization of clear aligners, splints, and 3D-printed brackets and appliances.

Clear aligners and splints. 3D printing finds clinical application in the production of orthodontic aligners used to correct the position of the teeth and for interceptive treatment [71,108]. Clear aligners are considered innovative alternatives for fixed brackets because they can be easily inserted and removed for eating, drinking, and tooth brushing and because they does not affect patients' esthetic [72,109]. Physicochemical and mechanical properties of thermoformed aligners have been studied and described [109,110]. It has been shown that resin-printed clear aligners are geometrically more accurate, mechanically stronger, and more elastic than conventional thermoformed aligners [111]. 3D-printed aligners are produced through intraoral scanning, the virtual placing of teeth in the desired position, and additive manufacturing technology [71,72]. This workflow is characterized by a reduction of production times because of fewer stages in manufacturing steps (Figure 12.3). Similarly, the creation of 3D-printed splints for the management of patients affected by temporomandibular joint dysfunction allowed less time-consuming protocols and increased precision [112].

3D-printed brackets and appliances. A new objective in orthodontic treatment is to reduce the number of appointments and to reduce dental complications related to orthodontic treatment, such as enamel demineralization and root resorption [113]. In this regard, it is possible to produce 3D-printed brackets through a digital tooth movement planning to obtain during brackets manufacturing adjustments personalized for the patient in terms of material selection, angulation, and bending [114,115]. Some authors showed an excellent intraoral fit of 3D-printed appliances, such as sleep apnea appliances, Andresen and Herbst [116,117]. However, further investigations must be conducted to support the clinical use of 3D-printed brackets and appliances [71].

3D printing in oral medicine. In oral medicine, the use of 3D-printed individual splints allow the local delivery of preloaded drugs with exceptional precision [118,119]. For example, chlorhexidine-coated mouthguards have been used to suppress oral bacteria [120]. 3D-printed individual splints allow to overcome the main drawback of conventional topical therapy: the inability to provide continuous contact between the agent and the treated areas. Therefore, the use of this technology has the potential to personalize drug delivery and improve the efficacy of local therapy in oral medicine.

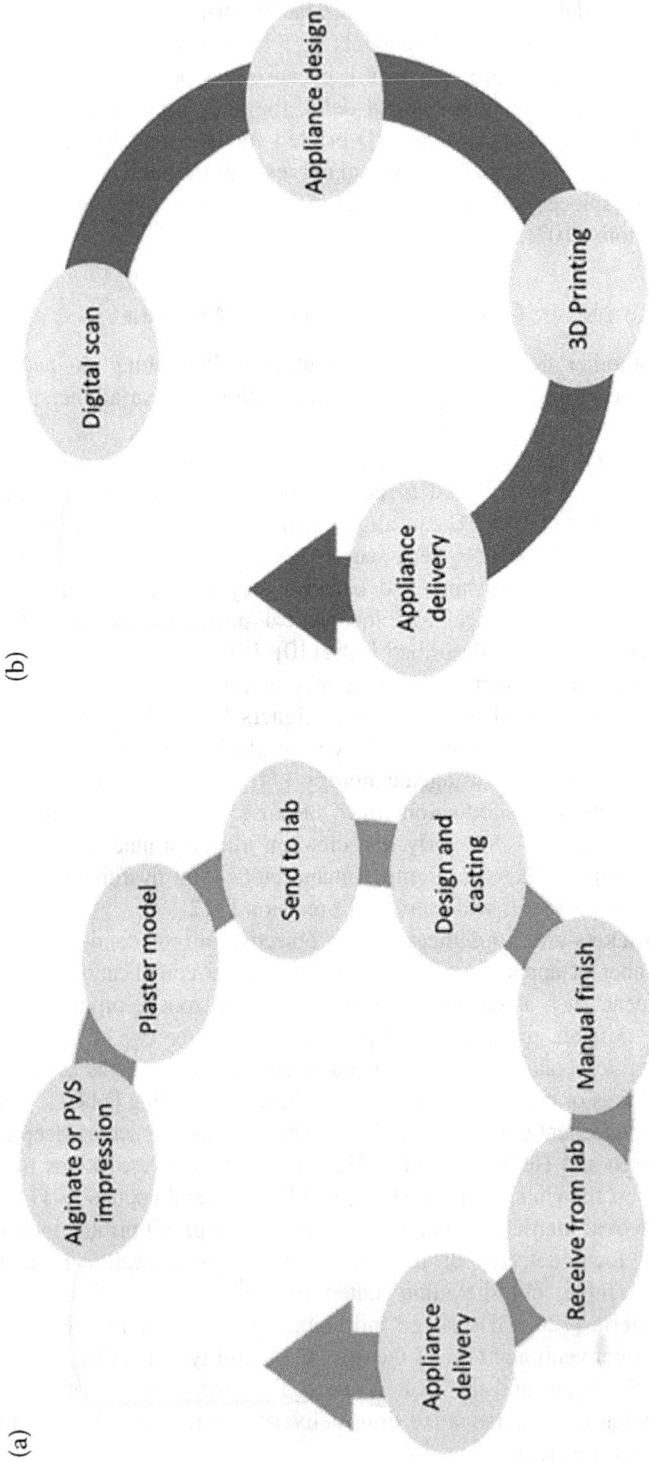

FIGURE 12.3 (a) Phases for conventional aligners processing; (b) workflow in 3D printing. Reprinted with permission from Khorsandi, D. et al., 3D and 4D printing in dentistry and maxillofacial surgery: Printing techniques, materials, and applications. Acta biomaterialia, 2020. [72] Copyright Elsevier (2020).

12.7 FUTURE DIRECTIONS

Three-dimensional printing was used for rapid prototyping of various elements and equipment. In the past decade, this technology has rapidly evolved in several medical and dental industries. New materials and methods are being studied to decrease time consumption, accelerate manufacturing, and improve technique performances. Qualitative characteristics such as particle shape, size, polymer density and viscosity, and powder distribution of the feed material determines the mechanical efficiency of the printed object. A challange in this recent technology is due to the viscosity-dependent flowability of polymers and metallic materials. Patient-specific accurate manufacts are needed in dentistry; they can greatly benefit from this recent technology, as already explained. However three-dimensional printing needs improvement regarding surface quality, geometric accuracy, and mechanical properties. The improvement of technical processes, layers' richness, and cure depth represents the key point for the production of objects with superior quality. Their mechanical properties are directly affected by surface quality and porosity (related to processing temperature and participle size). Material infiltration can be a solution to exceed this difficulty. The contraction of materials during production can lead to an important decrease of mechanical properties. Different printing techniques determine alterations in geometric accuracy, implying that dental artifacts may be affected when used for patients. Future developments in dentistry will need to focus on improving geometric accuracy, property gradients within materials, and process reliability, possibly employing low-cost materials and requiring short working times in the next years.

REFERENCES

1. Chen, W., Y. Tabata, and Y.W. Tong, *Fabricating tissue engineering scaffolds for simultaneous cell growth and drug delivery. Curr Pharm Des*, 2010. **16**(21): p. 2388–2394.
2. Pan, Y., C. Zhou, and Y. Chen, *A fast mask projection stereolithography process for fabricating digital models in minutes.* 2012.
3. Xin, X., et al., *In vitro biocompatibility of Co–Cr alloy fabricated by selective laser melting or traditional casting techniques. Materials Letters*, 2012. **88**: p. 101–103.
4. Zein, I., et al., *Fused deposition modeling of novel scaffold architectures for tissue engineering applications. Biomaterials*, 2002. **23**(4): p. 1169–1185.
5. Harris, G.M., et al., *Strategies to direct angiogenesis within scaffolds for bone tissue engineering. Curr Pharm Des*, 2013. **19**(19): p. 3456–3465.
6. Dutta Roy, T., et al., *Performance of hydroxyapatite bone repair scaffolds created via three-dimensional fabrication techniques. J Biomed Mater Res A*, 2003. **67**(4): p. 1228–1237.
7. Pati, F., et al., *Printing three-dimensional tissue analogues with decellularized extracellular matrix bioink. Nat Commun*, 2014. **5**: p. 3935.
8. Cooke, M.N., et al., *Use of stereolithography to manufacture critical-sized 3D biodegradable scaffolds for bone ingrowth. J Biomed Mater Res B Appl Biomater*, 2003. **64**(2): p. 65–69.
9. Lee, J.W., et al., *Fabrication and characteristic analysis of a poly(propylene fumarate) scaffold using micro-stereolithography technology. J Biomed Mater Res B Appl Biomater*, 2008. **87**(1): p. 1–9.

10. Lan, P.X., et al., *Development of 3D PPF/DEF scaffolds using micro-stereolithography and surface modification. J Mater Sci Mater Med*, 2009. **20**(1): p. 271–279.
11. Sun, W., et al., *The bioprinting roadmap. Biofabrication*, 2020. **12**(2): p. 022002.
12. Matai, I., et al., *Progress in 3D bioprinting technology for tissue/organ regenerative engineering. Biomaterials*, 2020. **226**: p. 119536.
13. Dawood, A., et al., *3D printing in dentistry. British dental journal*, 2015. **219**(11): p. 521–529.
14. Zhang, Y.S., et al., *3D Bioprinting for tissue and organ fabrication. Ann Biomed Eng*, 2017. **45**(1): p. 148–163.
15. Athirasala, A., et al., *A dentin-derived hydrogel bioink for 3D bioprinting of cell laden scaffolds for regenerative dentistry. Biofabrication*, 2018. **10**(2): p. 024101.
16. Ip, B.C., et al., *The bio-gripper: A fluid-driven micro-manipulator of living tissue constructs for additive bio-manufacturing. Biofabrication*, 2016. **8**(2): p. 025015.
17. Murphy, S.V. and A. Atala, *3D bioprinting of tissues and organs. Nat Biotechnol*, 2014. **32**(8): p. 773–785.
18. Intini, C., et al., *3D-printed chitosan-based scaffolds: An in vitro study of human skin cell growth and an in-vivo wound healing evaluation in experimental diabetes in rats. Carbohydr Polym*, 2018. **199**: p. 593–602.
19. Chen, Y.W., et al., *Osteogenic and angiogenic potentials of the cell-laden hydrogel/ mussel-inspired calcium silicate complex hierarchical porous scaffold fabricated by 3D bioprinting. Mater Sci Eng C Mater Biol Appl*, 2018. **91**: p. 679–687.
20. Rahman, S.U., et al., *Nanoscale and macroscale scaffolds with controlled-release polymeric systems for dental craniomaxillofacial tissue engineering. Materials (Basel)*, 2018. **11**(8).
21. Lee, V., et al., *Design and fabrication of human skin by three-dimensional bio-printing. Tissue Eng Part C Methods*, 2014. **20**(6): p. 473–484.
22. Almela, T., et al., *3D printed tissue engineered model for bone invasion of oral cancer. Tissue Cell*, 2018. **52**: p. 71–77.
23. Chiarello, E., et al., *Autograft, allograft and bone substitutes in reconstructive or- thopedic surgery. Aging Clin Exp Res*, 2013. **25 Suppl 1**: p. S101–S103.
24. Chia, H.N. and B.M. Wu, *Recent advances in 3D printing of biomaterials. J Biol Eng*, 2015. **9**: p. 4.
25. Ligon, S.C., et al., *Polymers for 3D printing and customized additive manufacturing. Chem Rev*, 2017. **117**(15): p. 10212–10290.
26. Salmi, M., et al., *Accuracy of medical models made by additive manufacturing (rapid manufacturing). Journal of Cranio-Maxillofacial Surgery*, 2013. **41**(7): p. 603–609.
27. Liang, X., et al., *A comparative evaluation of cone beam computed tomography (CBCT) and multi-slice CT (MSCT): Part I. On subjective image quality. European Journal of Radiology*, 2010. **75**(2): p. 265–269.
28. Pinto, J.M., et al., *Sensitivity analysis of geometric errors in additive manufacturing medical models. Medical Engineering & Physics*, 2015. **37**(3): p. 328–334.
29. Fan, D., et al., *Progressive 3D printing technology and its application in medical materials. Frontiers in Pharmacology*, 2020. **11**: p. 122.
30. Sakly, A., et al., *A novel quasicrystal-resin composite for stereolithography. Materials & Design (1980–2015)*, 2014. **56**: p. 280–285.
31. Khorsandi, D., et al., *3D and 4D printing in dentistry and maxillofacial surgery: Printing techniques, materials, and applications. Acta Biomaterialia*, 2021. **122**: p. 26–49.
32. Hazeveld, A., J.J.H. Slater, and Y. Ren, *Accuracy and reproducibility of dental replica models reconstructed by different rapid prototyping techniques. American Journal of Orthodontics and Dentofacial Orthopedics*, 2014. **145**(1): p. 108–115.

33. Aly, P. and C. Mohsen, *Comparison of the accuracy of three-dimensional printed casts, digital, and conventional casts: An in vitro study. European Journal of Dentistry*, 2020. **14**(02): p. 189–193.

34. Bustillos, J., et al., *Stereolithography-based 3D printed photosensitive polymer/ boron nitride nanoplatelets composites. Polymer Composites*, 2019. **40**(1): p. 379–388.

35. Noorani, R., *3D Printing: Technology, Applications, and Selection.* 2017. CRC Press.

36. Lipson, H. and M. Kurman, *Fabricated: The New World of 3D Printing.* 2013. John Wiley & Sons.

37. Kessler, A., R. Hickel, and M. Reymus, *3D printing in dentistry—State of the art. Operative Dentistry*, 2020. **45**(1): p. 30–40.

38. Olakanmi, E.O., R. Cochrane, and K. Dalgarno, *A review on selective laser sintering/melting (SLS/SLM) of aluminium alloy powders: Processing, microstructure, and properties. Progress in Materials Science*, 2015. **74**: p. 401–477.

39. Kruth, J.-P., et al., *Rapid manufacturing of dental prostheses by means of selective laser sintering/melting. Proceedings of the AFPR S*, 2005. **4**: p. 176–186.

40. Liu, J., Zhang, B., Yan, C. and Shi, Y., *The effect of processing parameters on characteristics of selective laser sintering dental glass-ceramic powder. Rapid Prototyping Journal*, 2010, 16(1): p. 138–145. https://doi.org/10.1108/13552541011025861

41. Mostafaei, A., et al., *Binder jet printing of partial denture metal framework from metal powder. Mater. Sci. Technol*, 2017: p. 1–3.

42. Stansbury, J.W. and M.J. Idacavage, *3D printing with polymers: Challenges among expanding options and opportunities. Dental Materials*, 2016. **32**(1): p. 54–64.

43. Cuan-Urquizo, E., et al., *Characterization of the mechanical properties of FFF structures and materials: A review on the experimental, computational and theoretical approaches. Materials*, 2019. **12**(6): p. 895.

44. Sun, C., et al., *Projection micro-stereolithography using digital micro-mirror dynamic mask. Sensors and Actuators A: Physical*, 2005. **121**(1): p. 113–120.

45. Jockusch, J. and M. Özcan, *Additive manufacturing of dental polymers: An overview on processes, materials and applications. Dental Materials Journal*, 2020: p. 2019–2123.

46. Park, S.A., et al., *In vivo evaluation of 3D-printed polycaprolactone scaffold implantation combined with β-TCP powder for alveolar bone augmentation in a beagle defect model. Materials*, 2018. **11**(2): p. 238.

47. David, O.T., et al., *Polylactic acid 3D printed drill guide for dental implants using CBCT. Rev. Chim.-Bucharest*, 2017. **68**: p. 341–342.

48. Molinero-Mourelle, P., et al., *Polylactic acid as a material for three-dimensional printing of provisional restorations. International Journal of Prosthodontics*, 2018. **31**(4): p. 349–350.

49. Wauthle, R., et al., *Additively manufactured porous tantalum implants. Acta Biomaterialia*, 2015. **14**: p. 217–225.

50. Baciu, A., et al. *Influence of process parameters for selective laser melting on the roughness of 3D printed surfaces in Co-Cr dental alloy powder.* In *IOP Conference Series: Materials Science and Engineering.* 2019. IOP Publishing.

51. Li, H., et al., *Dental ceramic prostheses by stereolithography-based additive manufacturing: Potentials and challenges. Advances in Applied Ceramics*, 2019. **118**(1–2): p. 30–36.

52. Khabas, T., L. Maletina, and K.S. Kamyshnaya. *Influence of nanopowders and pore–forming additives on sintering of alumma–zircorna ceramics.* In *IOP Conference Series: Materials Science and Engineering.* 2014. IOP Publishing.

53. Zimmermann, M., et al., *Fracture load of three-unit full-contour fixed dental prostheses fabricated with subtractive and additive CAD/CAM technology. Clinical Oral Investigations*, 2020. **24**(2): p. 1035–1042.

54. Nishiyama, H., et al., *Novel fully digital workflow for removable partial denture fabrication. Journal of Prosthodontic Research*, 2020. **64**(1): p. 98–103.

55. Hussein, M.O. and L.A. Hussein, *Novel 3D modeling technique of removable partial denture framework manufactured by 3D printing technology. Int J Adv Res*, 2014. **9**: p. 686–694.

56. inaya Pereyra, N.M., et al., *Comparison of patient satisfaction in the fabrication of conventional dentures vs. DENTCA (CAD/CAM) dentures: A case report. Journal of the New Jersey Dental Association*, 2015. **86**(2): p. 26–33.

57. Tahayeri, A., et al., *3D printed versus conventionally cured provisional crown and bridge dental materials. Dental Materials*, 2018. **34**(2): p. 192–200.

58. Etemad-Shahidi, Y., et al., *Accuracy of 3-Dimensionally printed full-arch dental models: A systematic review. Journal of Clinical Medicine*, 2020. **9**(10): p. 3357.

59. Hada, T., et al., *Effect of printing direction on the accuracy of 3D-printed dentures using stereolithography technology. Materials*, 2020. **13**(15): p. 3405.

60. Seprianto, D., R. Sugiantoro, and M. Erwin. *The effect of rectangular parallel key manufacturing process parameters made with stereolithography DLP 3D printer technology against impact strength.* In *Journal of Physics: Conference Series.* 2020. IOP Publishing.

61. Logozzo, S., et al., *Recent advances in dental optics–Part I: 3D intraoral scanners for restorative dentistry. Optics and Lasers in Engineering*, 2014. **54**: p. 203–221.

62. Akyalcin, S., et al., *Diagnostic accuracy of impression-free digital models. American Journal of Orthodontics and Dentofacial Orthopedics*, 2013. **144**(6): p. 916–922.

63. Lin, W.-S., et al., *Use of intraoral digital scanning for a CAD/CAM-fabricated milled bar and superstructure framework for an implant-supported, removable complete dental prosthesis. The Journal of Prosthetic Dentistry*, 2015. **113**(6): p. 509–515.

64. Ender, A. and A. Mehl, *Accuracy of complete-arch dental impressions: A new method of measuring trueness and precision. The Journal of Prosthetic Dentistry*, 2013. **109**(2): p. 121–128.

65. Venkatesan, G., A. Uppoor, and D.G. Naik, *Redefining the role of dendritic cells in periodontics. Journal of Indian Society of Periodontology*, 2013. **17**(6): p. 700.

66. Kasparova, M., et al., *Possibility of reconstruction of dental plaster cast from 3D digital study models. Biomedical Engineering Online*, 2013. **12**(1): p. 1–11.

67. Örtorp, A., et al., *The fit of cobalt–chromium three-unit fixed dental prostheses fabricated with four different techniques: A comparative in vitro study. Dental Materials*, 2011. **27**(4): p. 356–363.

68. Katkar, R.A., R.M. Taft, and G.T. Grant, *3D Volume rendering and 3D printing (additive manufacturing). Dent Clin North Am*, 2018. **62**(3): p. 393–402.

69. Petzold, R., H.F. Zeilhofer, and W.A. Kalender, *Rapid protyping technology in medicine–Basics and applications. Comput Med Imaging Graph*, 1999. **23**(5): p. 277–284.

70. Barazanchi, A., et al., *Additive technology: Update on current materials and applications in dentistry. J Prosthodont*, 2017. **26**(2): p. 156–163.

71. Oberoi, G., et al., *3D printing-encompassing the facets of dentistry. Front Bioeng Biotechnol*, 2018. **6**: p. 172.

72. Khorsandi, D., et al., *3D and 4D printing in dentistry and maxillofacial surgery: Printing techniques, materials, and applications. Acta Biomaterialia*, 2020.

73. Wilde, F., Plail, M., Riese, C., Schramm, A. & Winter, K., *Mandible reconstruction with patient-specific pre-bent reconstruction plates: comparison of a transfer key method to the standard method—Results of an in vitro study. International Journal of Computer Assisted Radiology and Surgery*, 2012. Jan; **7**(1): p. 57–63. doi: 10.1 007/s11548-011-0599-8. Epub 2011 May 19. PMID: 21594568.

74. Lin, H.-H., D. Lonic, and L.-J. Lo, *3D printing in orthognathic surgery – A literature review. Journal of the Formosan Medical Association*, 2018. **117**(7): p. 547–558.

75. Li, B., et al., *A new approach of splint-less orthognathic surgery using a personalized orthognathic surgical guide system: A preliminary study. International Journal of Oral and Maxillofacial Surgery*, 2017. **46**(10): p. 1298–1305.

76. Philippe, B., *Custom-made prefabricated titanium miniplates in Le Fort I osteotomies: principles, procedure and clinical insights. International Journal of Oral and Maxillofacial Surgery*, 2013. **42**(8): p. 1001–1006.

77. Tamimi, F., et al., *Craniofacial vertical bone augmentation: A comparison between 3D printed monolithic monetite blocks and autologous onlay grafts in the rabbit. Biomaterials*, 2009. **30**(31): p. 6318–6326.

78. Del Fabbro, M., et al., *Systematic review of survival rates for implants placed in the grafted maxillary sinus. International Journal of Periodontics & Restorative Dentistry*, 2004. **24**(6): p. 565–577.

79. Grasso, G., et al., *Histological and histomorphometric evaluation of new bone formation after maxillary sinus augmentation with two different osteoconductive materials: A randomized, parallel, double-blind clinical trial. Materials (Basel)*, 2020. **13**(23): p. 97–103.

80. Yen, H.H. and P.G. Stathopoulou, *CAD/CAM and 3D-printing applications for alveolar ridge augmentation. Curr Oral Health Rep*, 2018. **5**(2): p. 127–132.

81. Mangano, C., et al., *In vivo behavior of a custom-made 3d synthetic bone substitute in sinus augmentation procedures in sheep. J Oral Implantol*, 2015. **41**(3): p. 240–250.

82. Ruppin, J., et al., *Evaluation of the accuracy of three different computer-aided surgery systems in dental implantology: Optical tracking vs. stereolithographic splint systems. Clinical Oral Implants Research*, 2008. **19**(7): p. 709–716.

83. Papaspyridakos, P. and K. Lal, *Complete arch implant rehabilitation using subtractive rapid prototyping and porcelain fused to zirconia prosthesis: A clinical report. The Journal of Prosthetic Dentistry*, 2008. **100**(3): p. 165–172.

84. D'haese, J., et al., *Current state of the art of computer-guided implant surgery. Periodontology 2000*, 2017. **73**(1): p. 121–133.

85. Figliuzzi, M., F. Mangano, and C. Mangano, *A novel root analogue dental implant using CT scan and CAD/CAM: Selective laser melting technology. International Journal of Oral and Maxillofacial Surgery*, 2012. **41**(7): p. 858–862.

86. Mangano, F.G., et al., *Immediate, non-submerged, root-analogue direct laser metal sintering (DLMS) implants: A 1-year prospective study on 15 patients. Lasers in Medical Science*, 2014. **29**(4): p. 1321–1328.

87. Mangano, C., et al., *The osteoblastic differentiation of dental pulp stem cells and bone formation on different titanium surface textures. Biomaterials*, 2010. **31**(13): p. 3543–3551.

88. Moin, D.A., et al., *A novel approach for computer-assisted template-guided autotransplantation of teeth with custom 3D designed/printed surgical tooling. An Ex Vivo Proof of Concept. Journal of Oral and Maxillofacial Surgery*, 2016. **74**(5): p. 895–902.

89. Castillo-Dalí, G., et al., *Importance of poly(lactic-co-glycolic acid) in scaffolds for guided bone regeneration: A focused review. J Oral Implantol*, 2015. **41**(4): p. e152–e157.

90. Nesic, D., et al., *3D Printing approach in dentistry: The future for personalized oral soft tissue regeneration. Journal of Clinical Medicine*, 2020. **9**(7): p. 2238.

91. Mancini, L., et al., *Biomaterials for Periodontal and Peri-Implant Regeneration. Materials (Basel)*, 2021. **14**(12): p. 1–15.

92. Ge, Z., Z. Jin, and T. Cao, *Manufacture of degradable polymeric scaffolds for bone regeneration. Biomedical Materials*, 2008. **3**(2): p. 022001.

93. Obregon, F., et al., *Three-dimensional bioprinting for regenerative dentistry and craniofacial tissue engineering. J Dent Res*, 2015. **94**(9 Suppl): p. 143s–152ss.

94. Pilipchuk, S.P., et al., *Integration of 3D printed and micropatterned polycaprolactone scaffolds for guidance of oriented collagenous tissue formation in vivo. Advanced Healthcare Materials*, 2016. **5**(6): p. 676–687.

95. Gul, M., A. Arif, and R. Ghafoor, *Role of three-dimensional printing in periodontal regeneration and repair: Literature review. J Indian Soc Periodontol*, 2019. **23**(6): p. 504–510.

96. Murray, P.E., F. Garcia-Godoy, and K.M. Hargreaves, *Regenerative endodontics: A review of current status and a call for action. Journal of Endodontics*, 2007. **33**(4): p. 377–390.

97. Xu, H.H., et al., *Calcium phosphate cements for bone engineering and their biological properties. Bone Research*, 2017. **5**(1): p. 1–19.

98. Kim, K., et al., *Anatomically shaped tooth and periodontal regeneration by cell homing. Journal of Dental Research*, 2010. **89**(8): p. 842–847.

99. Xiao, L. and T. Tsutsui, *Characterization of human dental pulp cells-derived spheroids in serum-free medium: Stem cells in the core. Journal of Cellular Biochemistry*, 2013. **114**(11): p. 2624–2636.

100. Dissanayaka, W., et al., *Scaffold-free prevascularized microtissue spheroids for pulp regeneration. Journal of Dental Research*, 2014. **93**(12): p. 1296–1303.

101. Dissanayaka, W.L., et al., *In vitro analysis of scaffold-free prevascularized microtissue spheroids containing human dental pulp cells and endothelial cells. Journal of Endodontics*, 2015. **41**(5): p. 663–670.

102. Neunzehn, J., et al., *Dentin-like tissue formation and biomineralization by multicellular human pulp cell spheres in vitro. Head & Face Medicine*, 2014. **10**(1): p. 1–11.

103. Janjić, K., et al., *Formation of spheroids by dental pulp cells in the presence of hypoxia and hypoxia mimetic agents. International Endodontic Journal*, 2018. **51**: p. e146–e156.

104. Byun, C., et al., *Endodontic treatment of an anomalous anterior tooth with the aid of a 3-dimensional printed physical tooth model. Journal of Endodontics*, 2015. **41**(6): p. 961–965.

105. Rodrigues, C.T., et al., *Prevalence and morphometric analysis of three-rooted mandibular first molars in a Brazilian subpopulation. Journal of Applied Oral Science*, 2016. **24**: p. 535–542.

106. Connert, T., et al., *Microguided Endodontics: A method to achieve minimally invasive access cavity preparation and root canal location in mandibular incisors using a novel computer-guided technique. International Endodontic Journal*, 2018. **51**(2): p. 247–255.

107. Shah, P. and B. Chong, *3D imaging, 3D printing and 3D virtual planning in endodontics. Clinical Oral Investigations*, 2018. **22**(2): p. 641–654.

108. Ravera, S., et al., *Short term dentoskeletal effects of mandibular advancement clear aligners in Class II growing patients. A prospective controlled study according to STROBE Guidelines. Eur J Paediatr Dent*, 2021. **22**(2): p. 119–124.

109. Daniele, V., et al., *Thermoplastic disks used for commercial orthodontic aligners: complete physicochemical and mechanical characterization. Materials (Basel)*, 2020. **13**(10): p. 1–18.

110. Memè, L., et al., *ATR-FTIR analysis of orthodontic invisalign(®) aligners subjected to various in vitro aging treatments. Materials (Basel)*, 2021. **14**(4): p. 1–10.
111. Jindal, P., et al., *Mechanical and geometric properties of thermoformed and 3D printed clear dental aligners. American Journal of Orthodontics and Dentofacial Orthopedics*, 2019. **156**(5): p. 694–701.
112. Salmi, M., et al., *A digital process for additive manufacturing of occlusal splints: a clinical pilot study. Journal of the Royal Society Interface*, 2013. **10**(84): p. 20130203.
113. Abella, F., et al., *Outcome of autotransplantation of mature third molars using 3-dimensional–printed guiding templates and donor tooth replicas. Journal of Endodontics*, 2018. **44**(10): p. 1567–1574.
114. Normando, D., *3D orthodontics-from verne to shaw. Dental Press Journal of Orthodontics*, 2014. **19**: p. 12–13.
115. Kreya, K., et al., *3D-printed orthodontic brackets–Proof of concept Dreidimensional gedruckte kieferorthopädische Brackets–eine Machbarkeitsstudie. International Journal of Computerized Dentistry*, 2016. **19**(4): p. 351–362.
116. Farronato, G., et al., *The digital-titanium Herbst. Journal of Clinical Orthodontics: JCO*, 2011. **45**(5): p. 263–267.
117. Al Mortadi, N., et al., *CAD/CAM/AM applications in the manufacture of dental appliances. American Journal of Orthodontics and Dentofacial Orthopedics*, 2012. **142**(5): p. 727–733.
118. Miri, A.K., et al., *Bioprinters for organs-on-chips. Biofabrication*, 2019. **11**(4): p. 042002.
119. Liang, K., et al., *3D printing of a wearable personalized oral delivery device: A first-in-human study. Science Advances*, 2018. **4**(5): p. eaat2544.
120. D'Ercole, S., et al., *The use of chlorhexidine in mouthguards. Journal of Biological Regulators and Homeostatic Agents*, 2017. **31**(2): p. 487–493.

13 3D Printing of Smart Materials

A Path toward Evolution of 4D Printing

Manila Mallik
Veer Surendra Sai University of Technology, Sambalpur,
Odisha, India

CONTENTS

13.1 INTRODUCTION

The evolution of additive manufacturing (AM) is a boon for the traditional manufacturing process. AM is commonly dubbed as 3D printing or rapid prototyping

DOI: 10.1201/9781003189404-13

239

(Choi et al. 2015) (Gao et al. 2015). The advancement in additive manufacturing is still progressing since 1980 (O'Donnell et al. 2014). Chuck Hull, the father of 3D printing, invented the first 3D-printed part in 1983. It is an emerging inter-disciplinary topic that fascinates both academicians and industrialists. It consumes less time, less energy, and is associated with a low-waste generation that are the basic manufacturing gaps faced in the conventional manufacturing process. The advancement in 3D printing has a high potential, which escalates the manufacturing process to a great height. The most appreciated advancement to 3D printing is the evolution of 4D printing. The theme of 4D printing was first announced during a TED talk by Skylar Tibbits (Choi et al. 2015) (Kuang et al. 2019). 4D printing is a new perspective to 3D printing that is accomplished with the addition of a fourth dimension, i.e., time. It is simply morphing the function of 3D-printed objects with the influence of external stimuli like heat, light, water, pH, pressure, electricity, and magnetic fields. The static 3D-printed objects behave dynamically under the in-fluence of these predetermined stimuli. The obligatory printing material is a smart material that should be receptive to external stimuli. The smart materials exhibit many smart behaviors like self-healing, self-diagnosing, self-actuating, self-sensing, and self-assembling with the privilege of dimensional shape change. The wide application of 4D printing includes areas like aerospace, robotic, and biomedical sectors. The comparison between 3D and 4D printing is shown in Figure 13.1.

The transformation of a 3D-printed structure to a 4D-printed structure is attainable because of the shape renovation capability of smart material. The catalog of different shape-morphing behavior is revealed in Figure 13.2. This shape-changing action is characterized into three sections. One is rudimentary shape change, the second is complex or intricate shape change, and the third is a combination of all basic shape change (Nam and Pei 2019). When the transformation is completed through a single-stage activity like rolling, bending, twisting, buckling, curving, expansion, and contraction, then it is known

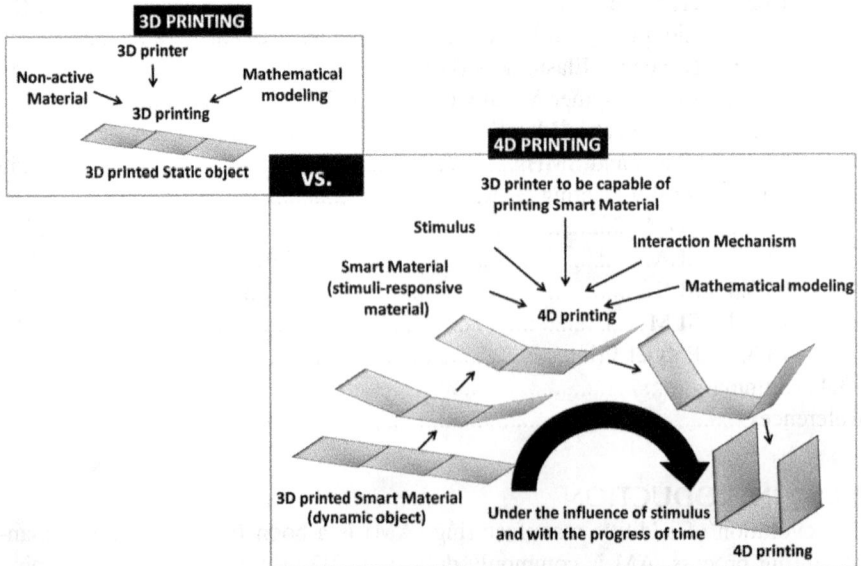

FIGURE 13.1 Schematic diagram shows the comparison between 3D and 4D printing.

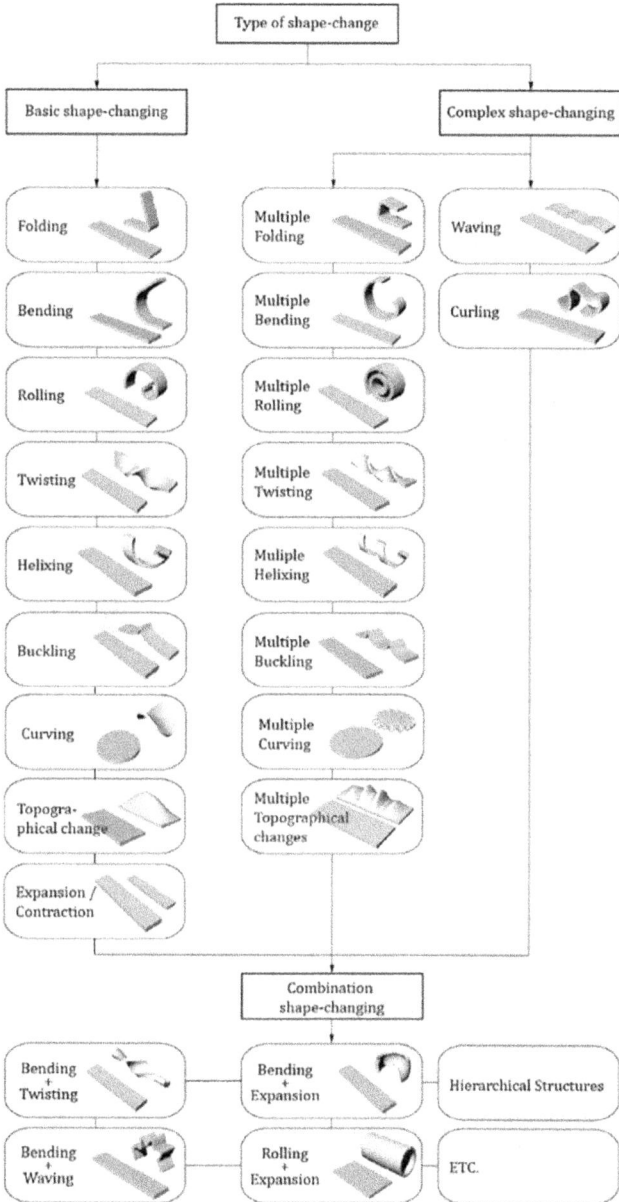

FIGURE 13.2 Catalogue of different shape-changing behavior. Reproduced with permission from Nam, Seokwoo, and Eujin Pei. 2019. "A Taxonomy of Shape-Changing Behavior for 4D Printed Parts Using Shape-Memory Polymers." *Progress in Additive Manufacturing* 4 (2): 167–184.

as an elementary shape change. However, recurring action of basic shape change develops intricate shape alteration, such as multiple rolling, multiple bending, and multiple twisting, etc. On the contrary, consolidation of one or more basic shape-changing behaviors evolves the third category of shape-changing behavior.

The shape-changing behavior terminates with 1D (one dimension) to 1D (Raviv et al. 2014), 1D to 2D (two dimension) (Tibbits et al. 2014) (Goo, Hong, and Park 2020), 1D to 3D (three dimension) (Ding et al. 2018), 2D to 2D (Villar, Graham, and Bayley 2013) (Q. Zhang et al. 2015), 2D to 3D (Sydney Gladman et al. 2016), (Q. Zhang, Zhang, and Hu 2016), and 3D to 3D (Bakarich et al. 2015) (Kokkinis, Schaffner, and Studart 2015) transformation through different deformation processes like bending, twisting, stretching, expansion, contraction, folding, and curling (Momeni et al. 2017). Eventually, the dimensional change of 3D-printed smart material inculcates the evolution of 4D printing. The dimensional shape-changing transformation behavior that is experimented through different researchers is presented in Table 13.1.

13.2 CLASSIFICATION OF SMART MATERIALS

The materials that perform smartly are known as smart materials. 4D printing technology is a synopsis of intelligent material, 3D printers, and smart design (Shin, Kim, and Kim 2017; Khoo et al. 2015; Suriano et al. 2019; J.-J. Wu et al. 2018; Momeni et al. 2017). One of the major components in 4D printing is the use of smart materials that are quite different from the material used for 3D printing. The selection criteria of 4D printing material are of utmost importance. The commercial 3D-printed objects are produced from metal, polymer, and ceramic. The passiveness of these conventional materials toward the external stimuli is the foremost cause for the use of smart material in 4D printing. The materials used for 4D printing behave smartly under the impact of external stimuli (Lee, An, and Chua 2017). They display shape-morphing phenomena when exposed to external stimuli like light, heat, electric fields, load, magnetic fields, pH, and water or moisture. The materials that derive their shape based on temperature are known as thermo-responsive materials; similarly, materials responsive to light, electric fields, magnetic fields, load, pH, and water or moisture are acknowledged as photo-responsive, electro-responsive, magneto-responsive, mechanically-responsive, pH-responsive, and moisture-responsive material, respectively. When the classification of the smart materials is built on the conception of peripheral stimuli, they are broadly classified as physical and chemical stimuli-responsive material, as presented in Figure 13.3(a). On the other hand, based on shape-morphing behavior, the broad division is shape memory and other smart material. Under the shape-memory effect, the nominations are shape memory alloys (SMAs), shape memory gel (SMG), shape memory ceramics, shape memory polymers (SMPs), and shape memory hybrids (SMHs), shape memory composites (SMCs), and other smart polymers. The categorization of smart materials according to the shape memory effect is shown in Figure 13.3(b).

13.2.1 SHAPE MEMORY ALLOYS (SMAs)

The shape memory feature in SMAs is classified into three categories and is offered as one-way shape memory effect, two-way shape memory effect, and pseudoelasticity

TABLE 13.1

Different Types of Dimensional Shape-Changing Transformation

Shape Shifting Dimension	Smart Material	Stimulus	Schematic Diagram for Shape Transformation
1D to 1D	Hydrogel and rigid disk	Water	
1D to 2D bending	ABS	Heat	

(Continued)

TABLE 13.1 (Continued)
Different Types of Dimensional Shape-Changing Transformation

Shape Shifting Dimension	Smart Material	Stimulus	Schematic Diagram for Shape Transformation
1D to 3D	Elastomer and glass polymer composite	Heat	
2D to 2D	PLA	Heat	

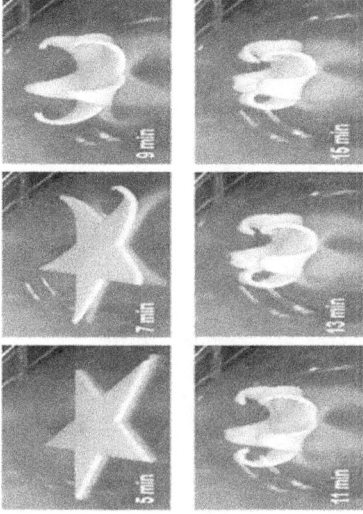 

(Continued)

2D to 3D curling and
bending

ABS

Heat

2D to 3D folding

Composite sheet of PLA strip printed on paper

Heat

TABLE 13.1 (Continued)
Different Types of Dimensional Shape-Changing Transformation

Shape Shifting Dimension	Smart Material	Stimulus	Schematic Diagram for Shape Transformation
2D to 3D bending	Composite sheet of PLA strip printed on paper	Heat	
2D to 3D twisting	Composite sheet of PLA strip printed on paper	Heat	

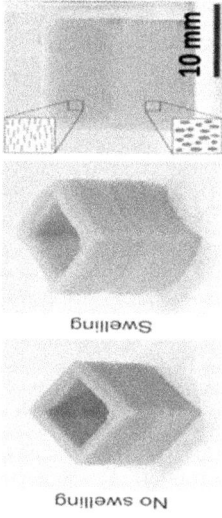

3D to 3D bending

Cross linking polymer with magnetized platelet aligned in different orientation

Chemical (ethyl acetate)

No swelling

Swelling

10 mm

(a)

(b)

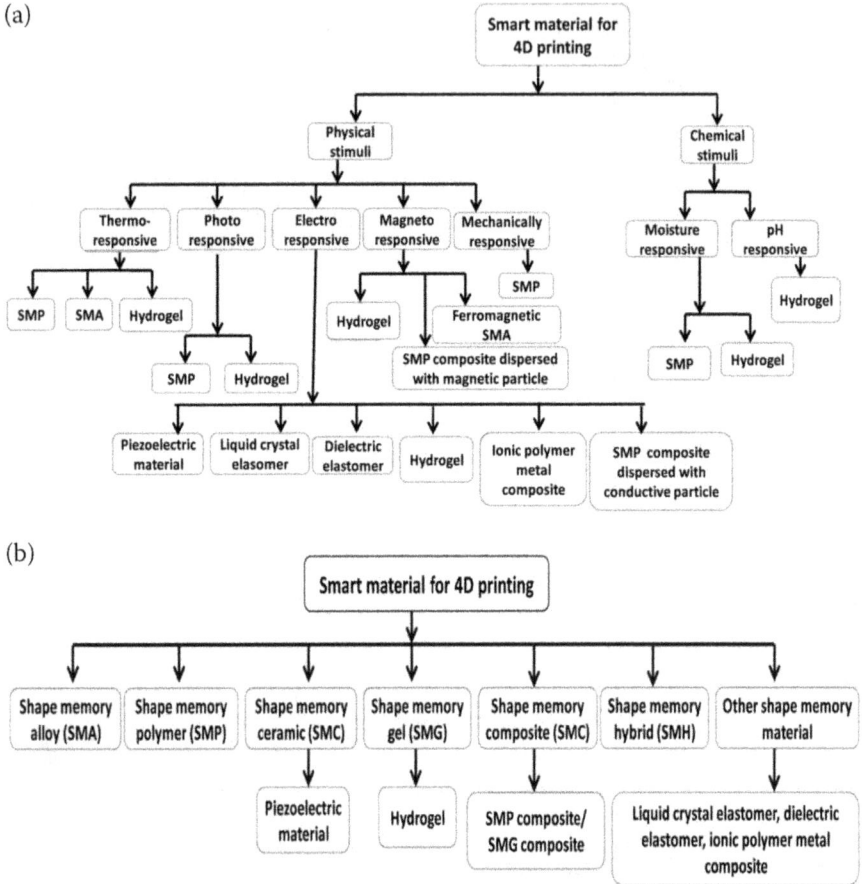

FIGURE 13.3 Classification of (a) stimuli responsive and (b) shape memory based smart material.

(PE) (Lee, An, and Chua 2017). The smartness lies in the memory, which remembers its high-temperature shape or both low and high-temperature shape, depending on whether it belongs to one-way or two-way shape memory alloy, respectively. Essentially, the shape memory effect is a consequence of a stimulus, i.e., temperature. It is the transformation of the low-temperature phase to the high-temperature phase, i.e., from martensite to austenite and the reverse phenomenon happens when the temperature is lowered. With the effect of loading, it transforms from twinned martensite to detwinned martensite, and detwinned martensite gets transformed into austenite with the effect of heat. The reversible transformation from the high-temperature phase, i.e., austenite to twinned low-temperature phase martensite, happens with the consequence of cooling. However, the contribution from PE toward the shape memory effect is minimal.

Some SMAs are responsive to magnetic fields; those are known as a magnetic shape memory alloy or ferromagnetic SMA. These SMAs are otherwise recognized as magneto-responsive shape memory alloy (Zafar and Zhao 2020) (Lee, An, and Chua 2017). This was originally discovered by Ullakko at MIT in 1996 (Ullakko 1996). A magnetic field is responsible for the change in orientation of the structure i.e., the movement in the twin boundary in the martensitic structure. The shape transformation is due to magnetization, which is an aftermath of the magnetically induced reorientation (MIR) (Pons et al. 2008). 3D printing of SMA plots a route for the production of multifarious geometry that is not viable through the other manufacturing method. Nitinol (50 wt% Ni–50 wt% Ti) is one of the promising SMA that is vastly used in various fields, such as the aerospace, automotive, and biomedical industries. Especially its biocompatibility makes it suitable for making appropriate devices for diverse medical sectors, such as orthopedics, cardiology, odontology, and neurology. Many researchers have used 3D printing techniques like selective laser melting (SLM) (Bormann et al. 2012; Dadbakhsh et al. 2014; Habijan et al. 2013; Meier, Haberland, and Frenzel 2011), electron beam melting (EBM) (Elahinia et al. 2012; Zhou et al. 2018), and laser metal deposition (Khademzadeh, Parvin, and Bariani 2015) to produce near-net-shape intricate structures by using nitinol. Other groups like Cu-Al-Ni-Mn (Gargarella et al. 2015) (Mazzer et al. 2014), Fe (Niendorf et al. 2016), and Cu-based SMAs (Gustmann et al. 2017) have also been printed through SLM 3D printing technique.

13.2.2 SHAPE MEMORY POLYMERS (SMPs)

SMP can recoup the original or programmed state from its temporary deformed shape when exposed to external stimuli. SMPs are active polymers that are responsive to various stimuli like light, heat, water, electric, and magnetic fields.

13.2.2.1 Thermo-Responsive SMP

Thermo-responsive SMP utilizes the molecular switch i.e., "T_{trans}, the transition temperature" for a transition from the deformed state to the original state. It is deformed at a temperature $T > T_{trans}$ to get a temporary shape and cooled below T_{trans} to retain the temporary shape. The original shape reverts when it is heated above T_{trans} (Zarek et al. 2016). T_{trans} varies according to the network structure. It is "glass transition temperature, T_g" for thermoplastic SMPs (TP SMPs) and "melting temperature, T_m" for thermosetting SMPs (TS SMPs) (Mora et al. 2019). Therefore, SMP has two divisions: one is TP SMPs that are physically crosslinked and the other is TS SMPs that are chemically or covalently crosslinked. Both programming and recovery constitute the cycle for the shape memory effect. Based upon these two steps, the mechanisms are categorized as dual-state mechanism and dual-composite mechanism. *Dual-state mechanism (DSM)*: This is dependent on T_g. Generally, elastomers have T_g that lies below room temperature. Therefore, the basic behavior that comes from the elastomers is due to their rubbery state at room temperature and a reverse case, i.e., brittle nature is tracked below T_g. With the applied constrained, these polymers deform above T_g and still, that deformed shape remains though it is unburdened below T_g because that temperature freezes the micro Brownian motion. The deformed state reverts to its original figure only when it is heated above T_g where micro Brownian motion becomes dynamic (Xie 2011)

(Lee, An, and Chua 2017). *Dual-component mechanism (DCM)*: This is applicable in the case of dual-component structure where one phase is elastic (hard segment) as compared to other segments whose stiffness is temperature-dependent. Therefore, the overall deformed structure doesn't come to the original structure even if it is unloaded. Nevertheless, the deformed structure of the dual-component system is regained when it is heated. This is attributed to the shape regained by the temperature-dependent soft segment whose elastic energy is triggered by the rise in temperature. *Partial-transition mechanism*: In this mechanism, unlike DSM and DCM, the polymer is exposed to a temperature that lies within the transition zone. Under this category, the elastic component is the unsoftened segment, whereas the soft part acts as the transition segment.

The idea of SMP started in Japan. Originally in the year 1984, Nippon Zeon introduced a polynorbornene-based SMP in Japan. Later, Kuraray and the Asahi company developed Mitsubishi Heavy Industries (MHI) polyurethane SMP supporting a wide range of T_g (Joshi et al. 2020). This SMP has distinctive features. Its application includes a Braille pen, which rectifies the mistake with the help of a pointer heater. Owing to the thermo-responsive nature, the SMP is setback to its original condition (Joshi et al. 2020). Monzon et al. have used an FDM 3D printer with thermoplastic polyurethane for the printing of a mechanical actuator (Monzón et al. 2017). The printed object showed a good shape memory performance viable with an 80% recovery ratio. Polyurethane SMP has also been used to build a 4D scaffold to be used in medical applications (Hendrikson et al. 2017). Not only polyurethane, but also various types of other SMPs like polycaprolactone (PCL) (Zarek et al. 2016) (Zarek et al. 2017), polylactic acid (PLA) (Senatov et al. 2017) (Yang Liu et al. 2018), epoxy (J. Wang et al. 2018) (K. Yu et al. 2015), and acrylate based resins (R. Yu et al. 2017) (Ge et al. 2016) (H. Wu et al. 2019) are also used to formulate smart structures for various applications.

13.2.2.2 Moisture-Responsive SMP

The journey of 4D printing started with moisture-responsive SMP, and the research was initiated by Tibbit (Tibbits 2014). The smart material was expanding with the exposure to water. Originally, he started with a 1D strand, and it self-assembled into the 2D word "MIT" when it was immersed in water.

13.2.2.3 Chemo-Responsive SMP

Here the stimulus is a chemical where the polymer is engrossed in a chemical. The mechanisms that are responsible for the shape memory effect are softening, swelling, and dissolving.

13.2.2.4 Photo-Responsive SMP

The photo-responsive SMP contains a photosensitive group that acts as a switch. Under loading conditions, the photosensitive group forms a crosslink when exposed to a UV light with a wavelength greater than a fixed wavelength, and the new shape is retained even after the load is removed. The original state recoups when it is exposed to a UV light of a higher wavelength (Lee, An, and Chua 2017) because the cross-link breaks.

13.2.3 HYDROGEL

Hydrogel is another active material that is used for 3D printing. Hydrogels come under the category of hydrophilic polymeric material (Joshi et al. 2020). Usually, it is an amalgamation of poly 2-hydroxymethyl methacrylate (PHEMA) or poly-isopropyl acrylamide (PNIPAAm) with poloxamers, or it contains biological polymer only (Shafranek et al. 2019). Alginate, gelatine, and hyaluronic are a class of biological polymers. These hydrogels change their volume when exposed to external stimuli. One problem that is faced with hydrogel is that they are relatively weak, and this weakness can be beaten with the incorporation of another polymeric network. Hydrogel with interpenetrating network structures is used to print hinges optimized with bending ability. The amalgamation of these hydrogels with other polymers creates a vast space in biomedical sciences. Compounds of hydrogel are used for kidney cells (Hong et al. 2015) and vascular networks (Miller et al. 2012). A combination of pHEMA (poly 2-hydroxymethyl methacrylate) and HEMA hydrogel are used as bio-ink that is recommended for the printing of 3D scaffold. Hydrogels react to various stimuli like temperature, light, pH, electricity, and magnetic fields (S. Wang, Lee, and Yeong 2015). Recently, pH and thermo-responsive block copolymer hydrogels have been explored to prevent clogging during injection by developing complex ions with anionic or cationic biomolecules (Singh and Lee 2014). Photo-responsive PVA can form a crosslink structure under the influence of UV light, which is pertinent in tissue engineering as a scaffold (Schmedlen, Masters, and West 2002). Pluronic/methacrylic acid sodium salt (PLMANa) electro-responsive hydrogel has been proved as an efficient candidate for biomedical application (Jackson and Stam 2015). Toth et al. have developed magnetic hydrogel by a co-precipitation method, where they have inserted magnetic nanoparticles in hyaluronate (HyA) gel (Tóth et al. 2015). These are the smart hydrogels for 3D printing (S. Wang, Lee, and Yeong 2015). Projection micro stereolithography 3D-printed electroactive hydrogel has been used as actuating material for soft robots (Han et al. 2018).

13.2.4 LIQUID CRYSTAL ELASTOMERS (LCES)

LCEs are a very good stimuli-responsive material. The shape-changing behaviour of the soft polymeric LCEs correspondence to expansion in one direction and contraction in the other direction when exposed to external stimuli (Kotikian et al. 2018) (Ambulo et al. 2017) (Ma et al. 2020) (Chen et al. 2019) (Hao Zeng et al. 2014) (Zhibin, Yang, and Raquez 2020). The contraction is in their direction along which liquid crystal molecules are oriented or mesogens are oriented, i.e., nematic director, on the other hand, expansion happens in the perpendicular direction. There is a transition from nematic to isotropic phase. The transition happens when it is heated above the nematic isotropic transformation temperature (T_{ni}). Irradiation to light also results in huge macroscopic shape change, and that is for the instigation of trans-cis photoisomerization mechanism in case of the light-responsive mesogens. The shape-morphing behaviour makes LCE for the development of sensors, robots, actuators, and artificial muscles. Kotikian et al. have reported the 3D printing of LCE actuators. They have employed LCE ink with high operating temperature direct-ink writing (HOT-DIW) to orient the mesogens along the direction of the printing path. Under the effect of the thermal stimulus, the printed object was able to morph its structure from planer to 3D and 3D to 3D′ (Kotikian et al. 2018). Ambulo et al. have also printed thermo-responsive LCE through DIW where a reversible contraction of 40% was achieved along the print direction (Ambulo et al. 2017). Zeng et al. have worked on 3D printing of submicron structures like rings and wood piles (Hao Zeng et al. 2014). A lot of researchers have also printed light-responsive LCEs. A bionic walker has been developed by 3D printing from light-responsive LCE where the walker's body was printed with light-sensitive azo-dye (H Zeng et al. 2015).

13.2.5 Dielectric Elastomers (DEs)

DE can deform their configuration due to their stimuli-responsive behaviour. As these are electroactive polymers, therefore, large strain is generated under an electric field actuation. Its lightweight, large strain-carrying capability, and high energy density enable it to mimic human muscle (Cai 2016). It resembles a capacitor because a sheet of elastomer is sandwiched between two compliant electrodes (Ma et al. 2020) (Rossiter, Walters, and Stoimenov 2009). With the applied voltage, the elastomer undergoes in-plane stretching with an increase in area and a decrease in thickness. Many researchers have worked on the 3D printing of DE. 3D printing of DE actuator has been reported by Rossiter et al. (Rossiter, Walters, and Stoimenov 2009). McCoul et al. have used an inkjet printer to produce DE actuators (McCoul et al. 2017).

13.2.6 Ionic Polymer Metal Composites (IMPCs)

IMPC is considered under the class of electroactive polymer. Not only its bendable structure but also the insertion of low driving voltage and the competence to actuate under an aqueous environment build a space for biomimetic application (Stalbaum et al. 2017). IMPCs are ionic polymers, typically nafion, consisting of an ion-exchange membrane sandwiched between two noble metals, i.e., Pt or Au electrode. Generally, the membrane is neutralized with Li or Na cations. When it is hydrated with water and a potential difference is applied across the electrodes, the cations along with solvent molecules start migrating toward the cathode. Swelling ensues as a result of the accumulation of charge at the negative side. Force generated due to swelling causes bending in IMPC. This type of electrochemical behaviour allows it to deploy as a sensor or actuator. 3D printing creates a platform for manufacturing actuators for soft robotics by using IMPC. Many researchers have used particularly FDM 3D printers for the manufacturing process (Carrico and Leang 2017) (Carrico et al. 2015) (Yin et al. 2021). Even researchers are trying to explore a new ionomer that can replace nafion. In this regard, Aquivion is found to be the potential candidate (Trabia, Olsen, and Kim 2017).

13.2.7 Piezoelectric Materials

These materials come under the division of electroactive material. They can convert mechanical to electrical energy with the application of stress and vice versa

(Grinberg et al. 2018). It has huge applications in the medical sector, particularly for implant technology. Different types of piezoelectric material such as barium titanate ($BaTiO_3$), lead zirconate titanate ($Pb[Zr_xTi_{1-x}]O_3, 0<x<1$), lead titanate ($PbTiO_3$), lithium tantalite ($LiTaO_3$), and lithium niobate ($LiNbO_3$) are man-made ceramic with a perovskite crystal structure. $BaTiO_3$ (BTO) is widely used because of its high dielectric constant. 3D printing of BTO has been carried out by Cheng et al. for ultrasonic transducer application (Cheng et al. 2019). Piezoelectric composite has been developed by inserting nanoparticles of BTO (Grinberg et al. 2018).

13.2.8 SMCs AND SMHs

One of the constituents in SMCs should be shape memory material like SMP or SMG. It is true to say that 3D printing of smart material is a way forward to 4D printing. However, non-active materials with suitable compositions play a vital role in shape-changing behaviour. The incorporation of nanomaterial into these smart materials also helps in evolving the shape-changing behaviour. Carbon nanotube (CNT) (Yang Liu et al. 2019) (Wan et al. 2019) (Dong et al. 2021), graphene (Garces and Ayranci 2018) (Jakus et al. 2015) as reinforcement increases the strength of the composite. Other non-active reinforcement like carbon fiber (Shen et al. 2019) (An and Yu 2019) (Dickson, Ross, and Dowling 2018) (C. Zeng et al. 2020), carbon black (CB) particle (Rosales et al. 2018) (Yang et al. 2017), Fe_3O_4 particle (F. Zhang et al. 2019) (Wei et al. 2017) (Zhao et al. 2019) (Yanju Liu et al. 2017) (De Santis, D'Amora, et al. 2015), carbonyl iron particle (CIP) (Hassan, Jo, and Seok 2018), SiC and graphite particle (W. Liu, Wu, and Pochiraju 2018) have been added to achieve large shape-changing behaviour in smart materials. Nowadays, many researchers have used hydroxyapatite (De Santis, Russo, et al. 2015) (Sui et al. 2018) as reinforcement particularly applicable for biomedical applications. Not only SMP composites are candidates for self-morphing structures, but also composites of hydrogel are (Demirtaş, Irmak, and Gümüşderelioğlu 2017) (Castro, O'Brien, and Zhang 2015) (Sayyar et al. 2017) and SMH (Sun et al. 2019) have also been developed in this regard. In addition to this, smart materials like SMA and SMP together contribute their shape memory effect in the generation of shape memory composites (Sun et al. 2019). A list of stimuli-responsive SMCs and SMHs has been given in Figure 13.2.

13.3 PRINTERS FOR 4D PRINTING

The smart materials used for 4D printing are printed with the help of different varieties of 3D printers. The printers are categorized according to the form of starting material, such as solid, liquid, and powder. Fusion-deposition modeling (FDM) is a solid based technique, whereas selective laser sintering (SLS), selective laser melting (SLM), and electron beam melting (EBM) comes under powder-based technique. However, stereolithography (SLA) and direct ink writing (DIW), and direct light processing (DLP) states liquid-based methods. 4D printing uses FDM, SLA, SLS, and inkjet printers in recent times (Quanjin et al. 2020) (Ali, Abilgaziyev, and Adair 2019).

13.3.1 FDM

FDM is based on an extrusion mechanism where a motor-driven extruded filament is deposited on the build plate after hardening (Ali, Abilgaziyev, and Adair 2019) (Sadasivuni, Deshmukh, and Almaadeed 2020). The nozzle moves in a raster manner, i.e., in the X and Y printing axes for printing one layer on the build plate. With subsequent printing of one layer on the plate, the nozzle moves upward, i.e., Z direction to print the next layer. In a nutshell, the movement of the nozzle in the X, Y, and Z axes constitutes a 3D printing object. The process continues until the printing ends for the whole object. However, it is associated with discrepancies related to the anisotropic behavior of mechanical properties (X. Wang et al. 2017). The mechanical property of the printed object varies according to the printing orientation of the filament. Nozzle blockage and structural defects like voids and surface irregularities are also detected in the printed objects. Voids are generated in the printed object due to the poor adhesion of interlayers. This process is relevant to thermoplastic polymer only and also with the desired viscosity that will provide easy flow from the nozzle. The loopholes can be excluded by the optimization of different parameters like printing speed, the temperature of the print plate, and filament. Nevertheless, this process is frequently used because of its cost-effectiveness and high speed. With the aid of a multi-nozzle, it is also applicable for multi-materials like composite fabrication. It is applicable for printing SMPs (Bajpai et al. 2020) and composites of SMP, like acrylonitrile butadiene styrene (ABS) (Goo, Hong, and Park 2020), PLA (F. Zhang et al. 2019) (Zhao et al. 2019), and polyamide (Dickson, Ross, and Dowling 2018).

13.3.2 SLA

This technique, developed in 1986, uses a liquid resin or photocurable polymer, which is sensitive to light (Falahati et al. 2020). Therefore, it transforms into solid after exposure to UV light. The viscosity of the resin changes during curing, and the curing action is headed with the transformation from a gel state to a solid cross-linked polymer. Henceforth, the solid structure of the printed objects ends with cross-linking. A fresh uncured layer of resin is applied over the curable 2D polymer layer. Nozzle clogging is not an issue when compared to FDM. Intricate parts with an improved surface finish can easily be printed through this process. The quality of the printing depends on the intensity of the laser, exposure, and curing time. Photo activators are added for effective polymerization. The high cost of this process limits its application in the industrial sector. It is well used for the preparation of SMP and its nanocomposite (Hassan, Jo, and Seok 2018) (Bajpai et al. 2020). Zhao et al. have used photopolymers to manufacture shape memory polymer by SLA techniques. Photopolymer was synthesized by adding a radical photoinitiator to polyurethane acrylate which was also blended with epoxy acrylate and isobornyl acrylate. Shape memory polymer with a high degree of performance was manufactured through SLA (SLA2) by creating a network of tBA-co-DEGDA where tBA (tert-Butyl acrylate) and DEGDA (diethylene glycol diacrylate) were acting as the soft and hard segment, respectively.

13.3.3   SLS

This is another 3D printing technique that supports powder material as its precursor. This printer can work with different types of material without any discrepancy regarding its colour and hardness. Sintering of one layer is feasible with the help of laser light. Sintering is done layer by layer, and it continues to get the final 3D object. The materials that are suggested for SLM are SMP and SMCs (Bajpai et al. 2020). The efficiency of this method depends on the size distribution and the packing of powder.

13.3.4   SLM

This is another category of powder bed diffusion process. Unlike SLS, a laser source helps in the melting of the powder. Different types of laser sources like CO2 and Nd:YAG, and fiber lasers are used in this method. Laser intensity at beam speed and hatch pattern are the operating parameter. The main advantage of SLM over SLS is that it does not need any post-processing treatment because direct melting helps in creating a strong bond between the layers. SLM is applicable for the printing of SMA alloys, which include NiTi alloys, Cu-based, and Fe-based alloys (Ngo et al. 2018).

13.3.5   Polyjet Printer

This is based on the photo-jetting principle. It carries a print head with multiple nozzles from which a jet of resin is deposited onto the platform, and it is cured by UV light before the second layer is jetted onto it. It makes printing easy for multi-material; conversely, nozzle clogging is a problem, and the viscosity of the polymer should be low. This facilitates the technology to expand its applicability for multi-material. It is applicable for SMPs and SMCs (Khoo et al. 2015) (Bajpai et al. 2020). (Table 13.2).

13.4   SUMMARY

4D printing amplifies the avenue of advancement for the smart devices that are engrossed in multifarious areas like aerospace, engineering, and biomedical. This is doable with smart design, smart material, and smart printers. Custom-designed parts with complex architecture are accessible with 3D printing of smart materials. Though the field of 4D printing offers a flight of steps to many diverse fields, many challenges remain to surmount. Still, it is far away from many demands and hunger of the commercial as well as the industrial user. Inaccuracy is observed during the conversion of the CAD model into real 3D-printed parts. Defects like voids are detected between the layers of feed material, which in turn put impacts the interfacial strength. It is also considered that the meticulous research and advancement in smart technology, and novel materials would surely resolve the issues concerned so far.

TABLE 13.2

List of Stimuli Responsive SMCs and SMHs

Stimuli Responsive Smart Material Composites	Matrix	Reinforcement	Printing Method	Application	References
Electroactive or electro-responsive SMP composites	PLA filament	CNT	FDM	Preform	(Yang Liu et al. 2019)
	Poly(D,L-lactide-co-trimethylene carbonante ((PLMC)	Multi-walled CNT	Direct ink writing (DIW)	Sensor	(Wan et al. 2019)
	PLA + TPU	CNT	FDM	Trusses, adaptive energy absorption devices and orthopaedic	(Dong et al. 2021)
	PLA	Graphene	Extrusion	Biomedical stent, sport equipment, unmanned air vehicles (UAVs)	(Garces and Ayranci 2018)
	Polyester polylactide-co-glycolide (PLG)	Graphene flakes	Extrusion	Scaffold for electronic and biomedical application	(Jakus et al. 2015)
	SMP with glass transition temp. of 76°	Continuous carbon fiber (CCF)	FDM	Aerospace	(Shen et al. 2019)
	Thermoset (epoxy) & thermoplastic (PLA + PU)	Carbon fiber	FDM	Fabrication for complex structure	(An and Yu 2019)
	Polymide 66	Carbon fiber	FDM	Aerospace application	(Dickson, Ross, and Dowling 2018)

SMP category	Matrix	Filler	3D printing technique	Application	Reference
	PLA filament	Carbon fiber	FDM	Building block for electrically activated and deployable structure	(C. Zeng et al. 2020)
	Commercial SMP	CB nanoparticle	Material extrusion (ME)	Complex functional devices	(Rosales et al. 2018)
	PLA	SiC and C (carbon)	Screw extruder for producing filament for FDM	Structures with shape memory response at various rate	(W. Liu, Wu, and Pochiraju 2018)
Photo-responsive SMP composite	PU	CB	FDM	Biomimetic smart devices and soft robotics	(Yang et al. 2017)
Magnetic- responsive SMP composites	PLA	Fe_3O_4 particle	FDM	Biomedical	(F. Zhang et al. 2019)
	PLA	Fe_3O_4 particle	DW	Intracascular stent	(Wei et al. 2017)
	PLA	Fe_3O_4 particle	FDM		(Zhao et al. 2019)
	Poly(ε-caprolactone) (PCL)	Magnetic nanoparticle (MNP) i.e., iron oxide (Fe_3O_4) particle	Direct write (DW) assembly	Scaffold	(Yanju Liu et al. 2017)
			FDM and stereolithography	Scaffold	(De Santis, D'Amora, et al. 2015)
	Poly(ethylene glycol) dimethacrylate (PEGDMA)	CIP	Stereolithography	Medical and soft robotics	(Hassan, Jo, and Seok 2018)
	PCL	Iron-doped hydroxyapatite (FeHA)nanoparticle	FDM	Scaffold for bone regeneration	(De Santis, Russo, et al. 2015)
Thermo-mechanical responsive SMP composite	PLA	HAp	FFF	Bone replacement and other biomedical engineering	(Sui et al. 2018)

(Continued)

TABLE 13.2 (Continued)
List of Stimuli Responsive SMCs and SMHs

Stimuli Responsive Smart Material Composites	Matrix	Reinforcement	Printing Method	Application	References
Hydrogel composites	Alginate, chitosan	HA	Extruder based bioprinter	Bone tissue engineering	(Demirtaş, Irmak, and Gümüşderelioğlu 2017)
	Polyethylene glycol diacrylate (PEG)	Nanocrystalline hydroxyapatie(nHA) and coreshell PLG acid nanosphere encapsulated with transforming growth factor β1 (TGF-β1)	Stereolithography	Tissue and organ regeneration	(Castro, O'Brien, and Zhang 2015)
	Metha crylated chitosan (ChiMA)	Graphene oxide (GO)	Extrusion	Tissue engineering	(Sayyar et al. 2017)
Thermo-responsive Hybrid	PLA + polyethylene glycol (PEG)		FDM	Functionally graded structure	(Sun et al. 2019)
Thermo-responsive shape memory composite	SMP filament (Nylon 12) + SMA (Nitinol wire)		FDM	Stent and valve controllers	(Kang et al. 2018)

REFERENCES

Ali, Md Hazrat, Anuar Abilgaziyev, and Desmond Adair. 2019. "4D Printing: A Critical Review of Current Developments, and Future Prospects." *International Journal of Advanced Manufacturing Technology* 105 (1–4): 701–717. doi:10.1007/s00170-019-04258-0

Ambulo, Cedric P, Julia J Burroughs, Jennifer M Boothby, Hyun Kim, M Ravi Shankar, and Taylor H Ware. 2017. "Four-Dimensional Printing of Liquid Crystal Elastomers." *ACS Applied Materials & Interfaces* 9 (42). American Chemical Society: 37332–37339. doi:10.1021/acsami.7b11851

An, Yongsan, and Woon-Ryeol Yu. 2019. *Three-Dimensional Printing of Continuous Carbon Fiber-Reinforced Shape Memory Polymer Composites. AIP Conference Proceedings.* Vol. 2113. doi:10.1063/1.5112513

Bajpai, Ankur, Anna Baigent, Sakshika Raghav, Conchúr Ó Brádaigh, Vasileios Koutsos, and Norbert Radacsi. 2020. "4D Printing: Materials, Technologies, and Future Applications in the Biomedical Field." *Sustainability* 12 (24): 1–32. doi:10.3390/su122410628

Bakarich, Shannon E, Robert Gorkin, Marc In Het Panhuis, and Geoffrey M Spinks. 2015. "4D Printing with Mechanically Robust, Thermally Actuating Hydrogels." *Macromolecular Rapid Communications* 36 (12): 1211–1217. doi:10.1002/marc.201500079

Bormann, Therese, Ralf Schumacher, Bert Müller, Matthias Mertmann, and Michael de Wild. 2012. "Tailoring Selective Laser Melting Process Parameters for NiTi Implants." *Journal of Materials Engineering and Performance* 21 (12): 2519–2524. doi:10.1007/s11665-012-0318-9

Cai, Jiyu. 2016. "4D Printing Dielectric Elastomer Actuator Based Soft Robots," 88. http://scholarworks.uark.edu/etd/1680/.

Carrico, James D, and Kam K Leang. 2017. "Fused Filament 3D Printing of Ionic Polymer-Metal Composites for Soft Robotics." In *Electroactive Polymer Actuators and Devices (EAPAD) 2017*, edited by Yoseph, Bar-Cohen, 10163:70–82. SPIE. 10.1117/12.2259782

Carrico, James D, Nicklaus W Traeden, Matteo Aureli, and Kam K Leang. 2015. "Fused Filament 3D Printing of Ionic Polymer-Metal Composites (IPMCs)." *Smart Materials and Structures* 24 (12). IOP Publishing: 125021. doi:10.1088/0964-1726/24/12/125021

Castro, Nathan J, Joseph O'Brien, and Lijie Grace Zhang. 2015. "Integrating Biologically Inspired Nanomaterials and Table-Top Stereolithography for 3D Printed Biomimetic Osteochondral Scaffolds." *Nanoscale* 7 (33). The Royal Society of Chemistry: 14010–14022. doi:10.1039/C5NR03425F

Chen, Ling, Yuqing Dong, Chak-Yin Tang, Lei Zhong, Wing-Cheung Law, Gary C P Tsui, Yingkui Yang, and Xiaolin Xie. 2019. "Development of Direct-Laser-Printable Light-Powered Nanocomposites." *ACS Applied Materials & Interfaces* 11 (21). American Chemical Society: 19541–19553. doi:10.1021/acsami.9b05871

Cheng, Jian, Yan Chen, Jun Wei Wu, Xuan Rong Ji, and Shang Hua Wu. 2019. "3d Printing of BaTiO3 Piezoelectric Ceramics for a Focused Ultrasonic Array." *Sensors (Switzerland)* 19 (19): 1–8. doi:10.3390/s19194078

Choi, Jin, O Chang Kwon, Wonjin Jo, Heon Ju Lee, and Myoung Woon Moon. 2015. "4D Printing Technology: A Review." *3D Printing and Additive Manufacturing* 2 (4): 159–167. doi:10.1089/3dp.2015.0039

Dadbakhsh, Sasan, Mathew Speirs, Jean-Pierre Kruth, Jan Schrooten, Jan Luyten, and Jan Humbeeck. 2014. "Effect of SLM Parameters on Transformation Temperatures of Shape Memory Nickel Titanium Parts." *Advanced Engineering Materials* 16 (September): 1140–1146. doi:10.1002/adem.201300558

Demirtaş, Tuğrul Tolga, Gülseren Irmak, and Menemşe Gümüşderelioğlu. 2017. "A Bioprintable Form of Chitosan Hydrogel for Bone Tissue Engineering." *Biofabrication* 9 (3). IOP Publishing: 35003. doi:10.1088/1758-5090/aa7b1d

Dickson, Andrew N, Keri-Ann Ross, and Denis P Dowling. 2018. "Additive Manufacturing of Woven Carbon Fibre Polymer Composites." *Composite Structures* 206: 637–643. doi:10.1016/j.compstruct.2018.08.091

Ding, Zhen, Oliver Weeger, H Jerry Qi, and Martin L Dunn. 2018. "4D Rods: 3D Structures via Programmable 1D Composite Rods." *Materials & Design* 137: 256–265. doi:10.1016/j.matdes.2017.10.004

Dong, Ke, Mahyar Panahi-Sarmad, Ziying Cui, Xiayan Huang, and Xueliang Xiao. 2021. "Electro-Induced Shape Memory Effect of 4D Printed Auxetic Composite Using PLA/TPU/CNT Filament Embedded Synergistically with Continuous Carbon Fiber: A Theoretical & Experimental Analysis." *Composites Part B: Engineering* 220: 108994. doi:10.1016/j.compositesb.2021.108994

Elahinia, M, Mahdi Hashemi, Majid Tabesh, and S Bhaduri. 2012. "Manufacturing and Processing of NiTi Implants: A Review." *Progress in Materials Science* 57: 911–946.

Falahati, Mojtaba, Parvaneh Ahmadvand, Shahriar Safaee, Yu-Chung Chang, Zhaoyuan Lyu, Roland Chen, Lei Li, and Yuehe Lin. 2020. "Smart Polymers and Nanocomposites for 3D and 4D Printing." *Materials Today* 40: 215–245. doi:10.1016/j.mattod.2020.06.001

Gao, Wei, Yunbo Zhang, Devarajan Ramanujan, Karthik Ramani, Yong Chen, Christopher B Williams, Charlie C L Wang, Yung C Shin, Song Zhang, and Pablo D Zavattieri. 2015. "The Status, Challenges, and Future of Additive Manufacturing in Engineering." *Computer-Aided Design* 69: 65–89. 10.1016/j.cad.2015.04.001

Garces, I T, and C Ayranci. 2018. "A View into Additive Manufactured Electro-Active Reinforced Smart Composite Structures." *Manufacturing Letters* 16: 1–5. doi:10.1016/j.mfglet.2018.02.008

Gargarella, Piter, Cláudio Kiminami, Eric Marchezini Mazzer, Regis Cava, Leonardo Basilio, C Bolfarini, Walter Botta, Jürgen Eckert, Tobias Gustmann, and Simon Pauly. 2015. "Phase Formation, Thermal Stability and Mechanical Properties of a Cu-Al-Ni-Mn Shape Memory Alloy Prepared by Selective Laser Melting." *Materials Research* 18 (December): 35–38. doi:10.1590/1516-1439.338914

Ge, Qi, Amir Hosein Sakhaei, Howon Lee, Conner K Dunn, Nicholas X Fang, and Martin L Dunn. 2016. "Multimaterial 4D Printing with Tailorable Shape Memory Polymers." *Scientific Reports* 6 (1): 31110. doi:10.1038/srep31110

Goo, Bona, Chae-Hui Hong, and Keun Park. 2020. "4D Printing Using Anisotropic Thermal Deformation of 3D-Printed Thermoplastic Parts." *Materials & Design* 188: 108485. doi:10.1016/j.matdes.2020.108485

Grinberg, Daniel, Sabrina Siddique, Minh-Quyen Le, Richard Liang, Jean-Fabien Capsal, and Pierre-Jean Cottinet. 2018. "4D Printing Based Piezoelectric Composite for Medical Applications." *Journal of Polymer Science Part B: Polymer Physics* 57 (November): 109–115. doi:10.1002/polb.24763

Gustmann, T, J M dos Santos, P Gargarella, U Kühn, J Van Humbeeck, and S Pauly. 2017. "Properties of Cu-Based Shape-Memory Alloys Prepared by Selective Laser Melting." *Shape Memory and Superelasticity* 3 (1): 24–36. doi:10.1007/s40830-016-0088-6

Habijan, T, C Haberland, H Meier, J Frenzel, J Wittsiepe, C Wuwer, C Greulich, T A Schildhauer, and M Köller. 2013. "The Biocompatibility of Dense and Porous Nickel–Titanium Produced by Selective Laser Melting." *Materials Science and Engineering: C* 33 (1): 419–426. 10.1016/j.msec.2012.09.008

Han, Daehoon, Cindy Farino, Chen Yang, Tracy Scott, Daniel Browe, Wonjoon Choi, Joseph W Freeman, and Howon Lee. 2018. "Soft Robotic Manipulation and Locomotion with a 3D Printed Electroactive Hydrogel." *ACS Applied Materials & Interfaces* 10 (21). American Chemical Society: 17512–17518. doi:10.1021/acsami.8b04250

Hassan, Rizwan Ul, Soohwan Jo, and Jongwon Seok. 2018. "Fabrication of a Functionally Graded and Magnetically Responsive Shape Memory Polymer Using a 3D Printing Technique and Its Characterization." *Journal of Applied Polymer Science* 135 (11): 45997. doi:10.1002/app.45997

Hendrikson, Wilhelmus J, Jeroen Rouwkema, Federico Clementi, Clemens A van Blitterswijk, Silvia Farè, and Lorenzo Moroni. 2017. "Towards 4D Printed Scaffolds for Tissue Engineering: Exploiting 3D Shape Memory Polymers to Deliver Time-Controlled Stimulus on Cultured Cells." *Biofabrication* 9 (3). IOP Publishing: 31001. doi:10.1088/1758-5090/aa8114

Hong, Sungmin, Dalton Sycks, Hon Fai Chan, Shaoting Lin, Gabriel P Lopez, Farshid Guilak, Kam W Leong, and Xuanhe Zhao. 2015. "3D Printing of Highly Stretchable and Tough Hydrogels into Complex, Cellularized Structures." *Advanced Materials* 27 (27): 4035–4040. doi:10.1002/adma.201501099

Jackson, Nathan, and Frank Stam. 2015. "Optimization of Electrical Stimulation Parameters for Electro-Responsive Hydrogels for Biomedical Applications." *Journal of Applied Polymer Science* 132 (12): 1–8. doi:10.1002/app.41687

Jakus, Adam E, Ethan B Secor, Alexandra L Rutz, Sumanas W Jordan, Mark C Hersam, and Ramille N Shah. 2015. "Three-Dimensional Printing of High-Content Graphene Scaffolds for Electronic and Biomedical Applications." *ACS Nano* 9 (4). American Chemical Society: 4636–4648. doi:10.1021/acsnano.5b01179

Joshi, Siddharth, Krishna Rawat, Karunakaran C., Vasudevan Rajamohan, Arun Tom Mathew, Krzysztof Koziol, Vijay Kumar Thakur, and Balan A.S.S. 2020. "4D Printing of Materials for the Future: Opportunities and Challenges." *Applied Materials Today* 18. Elsevier Ltd.: 100490. doi:10.1016/j.apmt.2019.100490

Kang, Minkyu, Youngjun Pyo, Joon young Jang, Yunchan Park, Yeon-Ho Son, MyungChan Choi, Joo wan Ha, Young-Wook Chang, and Caroline Sunyong Lee. 2018. "Design of a Shape Memory Composite(SMC) Using 4D Printing Technology." *Sensors and Actuators A: Physical* 283: 187–195. doi:10.1016/j.sna.2018.08.049

Khademzadeh, Saeed, Nader Parvin, and Paolo F Bariani. 2015. "Production of NiTi Alloy by Direct Metal Deposition of Mechanically Alloyed Powder Mixtures." *International Journal of Precision Engineering and Manufacturing* 16 (11): 2333–2338. doi:10.1007/s12541-015-0300-1

Khoo, Zhong Xun, Joanne Ee Mei Teoh, Yong Liu, Chee Kai Chua, Shoufeng Yang, Jia An, Kah Fai Leong, and Wai Yee Yeong. 2015. "3D Printing of Smart Materials: A Review on Recent Progresses in 4D Printing." *Virtual and Physical Prototyping* 10 (3). Taylor & Francis: 103–122. doi:10.1080/17452759.2015.1097054

Kokkinis, Dimitri, Manuel Schaffner, and André R Studart. 2015. "Multimaterial Magnetically Assisted 3D Printing of Composite Materials." *Nature Communications* 6: 1–10. doi:10.1038/ncomms9643

Kotikian, Arda, Ryan L Truby, John William Boley, Timothy J White, and Jennifer A Lewis. 2018. "3D Printing of Liquid Crystal Elastomeric Actuators with Spatially Programed Nematic Order." *Advanced Materials* 30 (10): 1706164. doi:10.1002/adma.201706164

Kuang, Xiao, Devin J. Roach, Jiangtao Wu, Craig M. Hamel, Zhen Ding, Tiejun Wang, Martin L Dunn, and Hang Jerry Qi. 2019. "Advances in 4D Printing: Materials and Applications." *Advanced Functional Materials* 29 (2): 1–23. doi:10.1002/adfm.201805290

Lee, Amelia Yilin, Jia An, and Chee Kai Chua. 2017. "Two-Way 4D Printing: A Review on the Reversibility of 3D-Printed Shape Memory Materials." *Engineering* 3 (5): 663–674. doi:10.1016/J.ENG.2017.05.014

Liu, Wenbo, Nan Wu, and Kishore Pochiraju. 2018. "Shape Recovery Characteristics of SiC/C/PLA Composite Filaments and 3D Printed Parts." *Composites Part A: Applied Science and Manufacturing* 108: 1–11. doi:10.1016/j.compositesa.2018.02.017

Liu, Yang, Wei Zhang, Fenghua Zhang, Xin Lan, Jinsong Leng, Shuang Liu, Xinqiao Jia, et al. 2018. "Shape Memory Behavior and Recovery Force of 4D Printed Laminated Miura-Origami Structures Subjected to Compressive Loading." *Composites Part B: Engineering* 153: 233–242. doi:10.1016/j.compositesb.2018.07.053

Liu, Yang, Wei Zhang, Fenghua Zhang, Jinsong Leng, Shaopeng Pei, Liyun Wang, Xinqiao Jia, Chase Cotton, Baozhong Sun, and Tsu-Wei Chou. 2019. "Microstructural Design for Enhanced Shape Memory Behavior of 4D Printed Composites Based on Carbon Nanotube/Polylactic Acid Filament." *Composites Science and Technology* 181: 107692. doi:10.1016/j.compscitech.2019.107692

Liu, Yanju, Hongqiu Wei, Qiwei Zhang, Yongtao Yao, Liwu Liu, and Jinsong Leng. 2017. "4D Printing of Poly (Lactic Acid)-Based Shape Memory Polymers and Shape Memory Nanocomposites," *21st International Conference on Composite Materials Xi'an, 20–25th August 2017.* no. August: 21–23.

Ma, Su Qian, Yun Peng Zhang, Meng Wang, Yun Hong Liang, Lei Ren, and Lu Quan Ren. 2020. "Recent Progress in 4D Printing of Stimuli-Responsive Polymeric Materials." *Science China Technological Sciences* 63 (4): 532–544. doi:10.1007/s11431-019-1443-1

Mazzer, E M, Claudio Shyinti Kiminami, P Gargarella, R D Cava, L A Basilio, C Bolfarini, W J Botta, J Eckert, T Gustmann, and S Pauly. 2014. "Atomization and Selective Laser Melting of a Cu-Al-Ni-Mn Shape Memory Alloy." In Filho, Francisco Ambrozio and Klein, Aloisio Nelmo (Eds.), *Advanced Powder Technology IX*, 802: 343–348. Materials Science Forum. Trans Tech Publications Ltd. doi:10.4028/www.scientific.net/MSF.802.343

McCoul, David, S Rosset, Samuel Schlatter, and H Shea. 2017. "Inkjet 3D Printing of UV and Thermal Cure Silicone Elastomers for Dielectric Elastomer Actuators." *Smart Materials and Structures* 26: 125022.

Meier, Horst, Christoph Haberland, and Jan Frenzel. 2011. *Structural and Functional Properties of NiTi Shape Memory Alloys Produced by Selective Laser Melting.* doi:10.1201/b11341-47

Miller, Jordan S, Kelly R Stevens, Michael T Yang, Brendon M Baker, Duc-Huy T Nguyen, Daniel M Cohen, Esteban Toro, et al. 2012. "Rapid Casting of Patterned Vascular Networks for Perfusable Engineered Three-Dimensional Tissues." *Nature Materials* 11 (9): 768–774. doi:10.1038/nmat3357

Momeni, Farhang, Seyed M, Mehdi Hassani N, Xun Liu, and Jun Ni. 2017. "A Review of 4D Printing." *Materials & Design* 122: 42–79. 10.1016/j.matdes.2017.02.068

Monzón, M, R Paz, E Pei, F Ortega, L Suárez, Z Ortega, M E Alemán, T Plucinski, and N Clow. 2017. "4D Printing: Processability and Measurement of Recovery Force in Shape Memory Polymers." *The International Journal of Advanced Manufacturing Technology* 89: 1827–1836.

Mora, Phattarin, Hannes Schäfer, Chanchira Jubsilp, Sarawut Rimdusit, and Katharina Koschek. 2019. "Thermosetting Shape Memory Polymers and Composites Based on Polybenzoxazine Blends, Alloys and Copolymers." *Chemistry – An Asian Journal* 14 (November): 4129–4139. doi:10.1002/asia.201900969

Nam, Seokwoo, and Eujin Pei. 2019. "A Taxonomy of Shape-Changing Behavior for 4D Printed Parts Using Shape-Memory Polymers." *Progress in Additive Manufacturing* 4 (2): 167–184. doi:10.1007/s40964-019-00079-5

Ngo, Tuan D., Alireza Kashani, Gabriele Imbalzano, Kate T Q Nguyen, and David Hui. 2018. "Additive Manufacturing (3D Printing): A Review of Materials, Methods, Applications and Challenges." *Composites Part B: Engineering* 143 (February). Elsevier: 172–196. doi:10.1016/j.compositesb.2018.02.012

Niendorf, Thomas, Florian Brenne, Philipp Krooß, Malte Vollmer, Johannes Günther, Dieter Schwarze, and Horst Biermann. 2016. "Microstructural Evolution and Functional Properties of Fe-Mn-Al-Ni Shape Memory Alloy Processed by Selective Laser Melting." *Metallurgical and Materials Transactions A* 47 (6): 2569–2573. doi:10.1007/s11661-016-3412-z

O'Donnell, John, Farzad Ahmadkhanlou, Hwan-Sik Yoon, and Gregory Washington. 2014. "All-Printed Smart Structures: A Viable Option?" *Active and Passive Smart Structures and Integrated Systems 2014* 9057: 905729. doi:10.1117/12.2045284

Pons, J, E Cesari, C Seguí, F Masdeu, and R Santamarta. 2008. "Ferromagnetic Shape Memory Alloys: Alternatives to Ni–Mn–Ga." *Materials Science and Engineering: A* 481–482: 57–65. doi:10.1016/j.msea.2007.02.152

Quanjin, Ma, M R M Rejab, M S Idris, Nallapaneni Manoj Kumar, M H Abdullah, and Guduru Ramakrishna Reddy. 2020. "Recent 3D and 4D Intelligent Printing Technologies: A Comparative Review and Future Perspective." *Procedia Computer Science* 167. Elsevier B.V.: 1210–1219. doi:10.1016/j.procs.2020.03.434

Raviv, D, Wei Zhao, Carrie Mcknelly, A Papadopoulou, A Kadambi, Boxin Shi, Shai Hirsch, et al. 2014. "Active Printed Materials for Complex Self-Evolving Deformations." *Scientific Reports* 4: 1–8.

Rosales, Carlos A Garcia, Mario F Garcia Duarte, Hoejin Kim, Luis Chavez, Deidra Hodges, Paras Mandal, Yirong Lin, and Tzu-Liang (Bill) Tseng. 2018. "3D Printing of Shape Memory Polymer (SMP)/Carbon Black (CB) Nanocomposites with Electro-Responsive Toughness Enhancement." *Materials Research Express* 5 (6). IOP Publishing: 65704. doi:10.1088/2053-1591/aacd53

Rossiter, Jonathan, Peter Walters, and Boyko Stoimenov. 2009. Printing 3D Dielectric Elastomer Actuators for Soft Robotics, *Proceedings Volume 7287, Electroactive Polymer Actuators and Devices (EAPAD) 2009; 72870H.* https://doi.org/10.1117/12.815746

Sadasivuni, Kishor Kumar, Kalim Deshmukh, and Mariam Alali Almaadeed, eds. 2020. "Front Matter." In *3D and 4D Printing of Polymer Nanocomposite Materials*, i–ii. Elsevier. doi:10.1016/B978-0-12-816805-9.09990-7

Santis, Roberto De, Ugo D'Amora, Teresa Russo, Alfredo Ronca, Antonio Gloria, and Luigi Ambrosio. 2015. "3D Fibre Deposition and Stereolithography Techniques for the Design of Multifunctional Nanocomposite Magnetic Scaffolds." *Journal of Materials Science: Materials in Medicine* 26 (10): 250. doi:10.1007/s10856-015-5582-4

Santis, Roberto De, Alessandro Russo, Antonio Gloria, Ugo D'Amora, Teresa Russo, Silvia Panseri, Monica Sandri, et al. 2015. "Towards the Design of 3D Fiber-Deposited Poly (-Caprolactone)/Iron-Doped Hydroxyapatite Nanocomposite Magnetic Scaffolds for Bone Regeneration." *Journal of Biomedical Nanotechnology* 11 (July): 1236–1246. doi:10.1166/jbn.2015.2065

Sayyar, Sepidar, Sanjeev Gambhir, Johnson Chung, David L Officer, and Gordon G Wallace. 2017. "3D Printable Conducting Hydrogels Containing Chemically Converted Graphene." *Nanoscale* 9 (5). The Royal Society of Chemistry: 2038–2050. doi:10.1039/C6NR07516A

Schmedlen, Rachael H, Kristyn S Masters, and Jennifer L West. 2002. "Photocrosslinkable Polyvinyl Alcohol Hydrogels That Can Be Modified with Cell Adhesion Peptides for Use in Tissue Engineering." *Biomaterials* 23 (22): 4325–4332. doi:10.1016/S0142-9612(02)00177-1

Senatov, F S, M Y u Zadorozhnyy, K V Niaza, V V Medvedev, S D Kaloshkin, N Y u Anisimova, M V Kiselevskiy, and Kai-Chiang Yang. 2017. "Shape Memory Effect in 3D-Printed Scaffolds for Self-Fitting Implants." *European Polymer Journal* 93: 222–231. 10.1016/j.eurpolymj.2017.06.011

Shafranek, Ryan, Siyami Millik, Patrick Smith, Chang-Uk Lee, Andrew Boydston, and Alshakim Nelson. 2019. "Stimuli-Responsive Materials in Additive Manufacturing." *Progress in Polymer Science* 93 (June): 36–67. doi:10.1016/j.progpolymsci.2019.03.002

Shen, Xinxin, Baoxian Jia, Hanxing Zhao, Xing Yang, and Zhengxian Liu. 2019. "Study on 3D Printing Process of Continuous Carbon Fiber Reinforced Shape Memory Polymer Composites." *IOP Conference Series: Materials Science and Engineering* 563 (August): 22029. doi:10.1088/1757-899X/563/2/022029

Shin, Dong-Gap, Tae-Hyeong Kim, and Dae-Eun Kim. 2017. "Review of 4D Printing Materials and Their Properties." *International Journal of Precision Engineering and Manufacturing-Green Technology* 4 (3): 349–357. doi:10.1007/s40684-017-0040-z

Singh, Narendra K, and Doo Sung Lee. 2014. "In Situ Gelling PH- and Temperature-Sensitive Biodegradable Block Copolymer Hydrogels for Drug Delivery." *Journal of Controlled Release* 193: 214–227. 10.1016/j.jconrel.2014.04.056

Stalbaum, Tyler, Taeseon Hwang, Sarah Trabia, Qi Shen, Robert Hunt, Zakai Olsen, and Kwang Kim. 2017. "Bioinspired Travelling Wave Generation in Soft-Robotics Using Ionic Polymer-Metal Composites." *International Journal of Intelligent Robotics and Applications* 1 (June): 167–179. doi:10.1007/s41315-017-0015-9

Sui, Tan, Enrico Salvati, H. Zhang, Kirill Niaza, Fedor Senatov, Alexey Salimon, and Alexander Korsunsky. 2018. "Probing the Complex Thermo-Mechanical Properties of a 3D-Printed Polylactide-Hydroxyapatite Composite Using in Situ Synchrotron X-Ray Scattering." *Journal of Advanced Research* 16 (November): 113–122. doi:10.1016/j.jare.2018.11.002

Sun, Yu-Chen, Yimei Wan, Ryan Nam, Marco Chu, and Hani E Naguib. 2019. "4D-Printed Hybrids with Localized Shape Memory Behaviour: Implementation in a Functionally Graded Structure." *Scientific Reports* 9 (1): 18754. doi:10.1038/s41598-019-55298-1

Suriano, Raffaella, Roberto Bernasconi, Luca Magagnin, and Marinella Levi. 2019. "4D Printing of Smart Stimuli-Responsive Polymers." *Journal of The Electrochemical Society* 166 (9): B3274–B3281. doi:10.1149/2.0411909jes

Sydney Gladman, A, Elisabetta A Matsumoto, Ralph G Nuzzo, L Mahadevan, and Jennifer A Lewis. 2016. "Biomimetic 4D Printing." *Nature Materials* 15 (4): 413–418. doi:10.1038/nmat4544

Tibbits, Skylar. 2014. "4D Printing: Multi-Material Shape Change." *Architectural Design* 84 (1): 116–121. doi:10.1002/ad.1710

Tibbits, Skylar, Carrie McKnelly, Carlos Olguin, Daniel Dikovsky, and Shai Hirsch. 2014. "4D Printing and Universal Transformation." *ACADIA 2014 – Design Agency: Proceedings of the 34th Annual Conference of the Association for Computer Aided Design in Architecture* 2014 (October): 539–548.

Tóth, Ildikó Y, Gábor Veress, Márta Szekeres, Erzsébet Illés, and Etelka Tombácz. 2015. "Magnetic Hyaluronate Hydrogels: Preparation and Characterization." *Journal of Magnetism and Magnetic Materials* 380: 175–180. doi:10.1016/j.jmmm.2014.10.139

Trabia, Sarah, Zakai Olsen, and Kwang Kim. 2017. "Searching for a New Ionomer for 3D Printable Ionic Polymer-Metal Composites: Aquivion® as a Candidate." *Smart Materials and Structures* 26 (October): 1–32. doi:10.1088/1361-665X/aa919f

Ullakko, K. 1996. "Magnetically Controlled Shape Memory Alloys: A New Class of Actuator Materials." *Journal of Materials Engineering and Performance* 5 (3): 405–409. doi:10.1007/BF02649344

Villar, Gabriel, Alexander D Graham, and Hagan Bayley. 2013. "A Tissue-Like Printed Material." *Science* 340 (6128). American Association for the Advancement of Science: 48–52. doi:10.1126/science.1229495

Wan, Xue, Fenghua Zhang, Yanju Liu, and Jinsong Leng. 2019. "CNT-Based Electro-Responsive Shape Memory Functionalized 3D Printed Nanocomposites for Liquid Sensors." *Carbon* 155: 77–87. doi:10.1016/j.carbon.2019.08.047

Wang, Jing, Zhongmin Xue, Gang Li, Yu Wang, Xuewei Fu, Wei-Hong Zhong, and Xiaoping Yang. 2018. "A UV-Curable Epoxy with 'Soft' Segments for 3D-Printable Shape-Memory Materials." *Journal of Materials Science* 53 (17): 12650–12661. doi:10.1007/s10853-018-2520-0

Wang, Shuai, Jia Min Lee, and Wai Yee Yeong. 2015. "Smart Hydrogels for 3D Bioprinting." *International Journal of Bioprinting* 1 (July): 3–14. doi:10.18063/IJB.2015.01.005

Wang, Xin, Man Jiang, Zuowan Zhou, Jihua Gou, and David Hui. 2017. "3D Printing of Polymer Matrix Composites: A Review and Prospective." *Composites Part B: Engineering* 110: 442–458. doi:10.1016/j.compositesb.2016.11.034

Wei, Hongqiu, Qiwei Zhang, Yongtao Yao, Liwu Liu, Yanju Liu, and Jinsong Leng. 2017. "Direct-Write Fabrication of 4D Active Shape-Changing Structures Based on a Shape Memory Polymer and Its Nanocomposite." *ACS Applied Materials & Interfaces* 9 (1). American Chemical Society: 876–883. doi:10.1021/acsami.6b12824

Wu, Hongzhi, Peng Chen, Chunze Yan, Chao Cai, and Yusheng Shi. 2019. "Four-Dimensional Printing of a Novel Acrylate-Based Shape Memory Polymer Using Digital Light Processing." *Materials & Design* 171: 107704. doi:https://doi.org/10.1016/j.matdes.2019.107704

Wu, Jing-Jun, Li-Mei Huang, Qian Zhao, and Tao Xie. 2018. "4D Printing: History and Recent Progress." *Chinese Journal of Polymer Science* 36 (5): 563–575. doi:10.1007/s10118-018-2089-8

Xie, Tao. 2011. "Recent Advances in Polymer Shape Memory. Polymer." *Polymer* 52 (October): 4985–5000. doi:10.1016/j.polymer.2011.08.003

Yang, Hui, Wan Ru Leow, Ting Wang, Juan Wang, Jiancan Yu, Ke He, Dianpeng Qi, Changjin Wan, and Xiaodong Chen. 2017. "3D Printed Photoresponsive Devices Based on Shape Memory Composites." *Advanced Materials* 29 (33): 1701627. doi:10.1002/adma.201701627

Yin, Guoxiao, Qingsong He, Xiangman Zhou, Yuwei Wu, Hongkai Li, and Min Yu. 2021. "Printing Ionic Polymer Metal Composite Actuators by Fused Deposition Modeling Technology." *International Journal of Smart and Nano Materials* 12 (2). Taylor & Francis: 218–231. doi:10.1080/19475411.2021.1914766

Yu, Kai, Alexander Ritchie, Yiqi Mao, Martin Dunn, and H. Qi. 2015. "Controlled Sequential Shape Changing Components by 3D Printing of Shape Memory Polymer Multimaterials." *Procedia IUTAM* 12 (December): 193–203. doi:10.1016/j.piutam.2014.12.021

Yu, Ran, Xin Yang, Ying Zhang, Xiaojuan Zhao, Xiao Wu, Tingting Zhao, Yulei Zhao, and Wei Huang. 2017. "Three-Dimensional Printing of Shape Memory Composites with Epoxy-Acrylate Hybrid Photopolymer." *ACS Applied Materials & Interfaces* 9 (2): 1820–1829. doi:10.1021/acsami.6b13531

Zafar, Muhammad Qasim, and Haiyan Zhao. 2020. "4D Printing: Future Insight in Additive Manufacturing." *Metals and Materials International* 26 (5): 564–585. doi:10.1007/s12540-019-00441-w

Zarek, Matt, Michael Layani, Ido Cooperstein, Ela Sachyani, Daniel Cohn, and Shlomo Magdassi. 2016. "3D Printing of Shape Memory Polymers for Flexible Electronic Devices." *Advanced Materials* 28 (22): 4449–4454. doi:10.1002/adma.201503132

Zarek, Matt, Nicola Mansour, Shir Shapira, and Daniel Cohn. 2017. "4D Printing of Shape Memory-Based Personalized Endoluminal Medical Devices." *Macromolecular Rapid Communications* 38 (2): 1600628. doi:10.1002/marc.201600628

Zeng, Chengjun, Liwu Liu, Wenfeng Bian, Yanju Liu, and Jinsong Leng. 2020. "4D Printed Electro-Induced Continuous Carbon Fiber Reinforced Shape Memory Polymer Composites with Excellent Bending Resistance." *Composites Part B: Engineering* 194: 108034. doi:10.1016/j.compositesb.2020.108034

Zeng, H, Piotr Wasylczyk, Camilla Parmeggiani, Daniele Martella, Matteo Burresi, and Diederik Wiersma. 2015. "Light-Fueled Microscopic Walkers." *Advanced Materials (Deerfield Beach, Fla.)* 27 (May): 3883–3887. doi:10.1002/adma.201501446

Zeng, Hao, Daniele Martella, Piotr Wasylczyk, Giacomo Cerretti, Jean-Christophe Gomez Lavocat, Chih-Hua Ho, Camilla Parmeggiani, and Diederik Sybolt Wiersma. 2014. "High-Resolution 3D Direct Laser Writing for Liquid-Crystalline Elastomer Microstructures." *Advanced Materials* 26 (15): 2319–2322. doi:10.1002/adma.201305008

Zhang, Fenghua, W Linlin, Zheng Zhichao, Yanju Liu, and J Leng. 2019. "Magnetic Programming of 4D Printed Shape Memory Composite Structures." *Composites Part A-Applied Science and Manufacturing* 125: 105571.

Zhang, Quan, Dong Yan, Kai Zhang, and Gengkai Hu. 2015. "Pattern Transformation of Heat-Shrinkable Polymer by Three-Dimensional (3D) Printing Technique." *Scientific Reports* 5: 8936. doi: 10.1038/srep08936

Zhang, Quan, Kai Zhang, and Gengkai Hu. 2016. "Smart Three-Dimensional Lightweight Structure Triggered from a Thin Composite Sheet via 3D Printing Technique." *Scientific Reports* 6 (1): 22431. doi: 10.1038/srep22431

Zhao, Wei, Fenghua Zhang, Jinsong Leng, and Yanju Liu. 2019. "Personalized 4D Printing of Bioinspired Tracheal Scaffold Concept Based on Magnetic Stimulated Shape Memory Composites." *Composites Science and Technology* 184: 107866. doi: 10.1016/j.compscitech.2019.107866

Zhibin, Wen, Keke Yang, and Jean-Marie Raquez. 2020. "A Review on Liquid Crystal Polymers in Free-Standing Reversible Shape Memory Materials." *Molecules* 25 (March): 1241. doi: 10.3390/molecules25051241

Zhou, Quan, Muhammad Hayat, Gang Chen, Song Cai, Xuanhui Qu, Huiping Tang, and Peng Cao. 2018. "Selective Electron Beam Melting of NiTi: Microstructure, Phase Transformation and Mechanical Properties." *Materials Science and Engineering: A* 744 (December): 290–298. doi: 10.1016/j.msea.2018.12.023

14 Performance of Smart Alloys in Manufacturing Processes during Subtractive and Additive Manufacturing

A Short Review on SMA and Metal Alloys

Suman Chatterjee
Department of Mechanical Engineering, National Institute of
Technology, Rourkela, India

Jinyang Xu
State Key Laboratory of Mechanical System and Vibration,
School of Mechanical Engineering, Shanghai Jiao Tong
University, Shanghai, PR China

T. V. Huynh
Faculty of Mechanical Engineering, Ho Chi Minh City
University of Technology and Education, Ho Chi Minh City,
Vietnam

Kumar Abhishek
Department of Mechanical Engineering, Institute of
Infrastructure Technology Research and Management,
Ahmedabad, India

Soni Kumari
Department of Mechanical Engineering, GLA University,
Mathura, Uttar Pradesh, India

DOI: 10.1201/9781003189404-14

Ajit Behera

Department of Metallurgical and Materials Engineering,
National Institute of Technology, Rourkela, India

CONTENTS

14.1   INTRODUCTION

With recent advancement in science and technology, the discovery and need for development of new materials and alloys takes place. Starting from easy-to-machine to hard-to-machine materials, they contribute to different engineering aspects. With introduction and evolution of industry 4.0 to 5G in the manufacturing and production sectors, everything becomes smart in recent times, such as smart cities, smart design, smart manufacturing, smart materials, or smart alloys. A report by Grand View Research in 2019–2020 stated that by the year 2025, smart material will make a market worth 98.2 billion dollars with a growth rate of 13.5% each year [1]. The extensive research activities have expanded the use of smart alloys in industries, medical science, robotics, aviation, and automobile and aerospace applications. To understand the physical and mechanical behavior of smart materials, researchers have conducted a rigorous study on these materials. Smart materials are materials that can withstand the changes applied to them and respond accordingly [2]. These smart materials can modify the material properties in a controllable and reversible manner [3]. These smart materials are also termed as responsive materials or active materials. These characteristics shown by the materials help to develop a sensor and actuator from the smart materials. There are several categories of smart materials, such as piezoelectrics materials, bi-component fibers, magnetostrictive materials, shape memory polymers, hydrogels, electroactive polymers, shape memory alloys, and graphene [4].

Smart alloys are part of smart materials, which can memorize the original shape and structure at its original shape. Technically, the smart alloy is also known as a shape memory alloy (SMA). The machinability of SMA is quite a challenge due to its shape memory effect (SME) [5]. Several research works have been performed to understand the quality behaviour of the SMA products during subtractive manufacturing and additive manufacturing process [6]. Subtractive manufacturing is a process designated for different meticulous material removal and machining processes. Subtractive manufacturing can be employed for any range of materials from soft materials to hard-to-machine materials, as well as for smart alloys that are given shape by eradicating materials through grinding, drilling, turning, boring, and cutting. In contrast to the subtractive manufacturing process that takes place by removing materials from a work material to give the desired shape, the additive manufacturing process develops an object through layer-by-layer deposition [7]. Additive manufacturing (AM) is a computer-controlled and assisted-manufacturing process to develop three-dimensional objects by depositing materials, usually by layers. There are various additive manufacturing processes, such as powder bed fusion, binder jetting, directed energy deposition, material extrusion, material jetting, sheet lamination, and vat polymerization [8]. The AM technologies are classified viz. as sintering, direct metal laser sintering (DMLS), stereolithography (SLA), direct metal laser melting (DMLM), and electron beam melting (EBM) [9].

The research work has been carried out to determine the product quality by physical and mechanical characterization of the developed products and additive manufacturing processes. In addition to this, Yuan et al. [10] have studied the different additive manufacturing processes used for the development of smart polymeric composites. The study performed by Yuan et al. [10] also provides information on the development of new fields of additive manufacturing. A similar type of study was carried by Lee et al. [11]. Lee et al. [11] have discussed the potential aspects of fiber composites in the fused-deposition modeling (FDM) process. The study also discusses the modification and development in the field of additive manufacturing processes and structural materials used in the process. Jiménez et al. [12] have discussed the recent advances in the field of additive manufacturing processes and provided brief notes on materials and metals used in the powder-based additive manufacturing process. The study discusses the advantages and complexities related to the powder-based additive manufacturing process. The study also discusses some testing methods, such as the visual inspection method, liquid penetration method, X-ray method, ultrasonic test, eddy current test, and magnetic particle test used for inspecting the additive manufactured parts. Haghdadi et al. [13] have studied the different challenges associated with AM of steels. The study discusses the different processing treatments required to enhance the quality of the additive manufactured product. Chen et al. [14] have studied the microstructural features and laser metal interaction during the direct energy deposition process during the fabrication of titanium-based alloys. Chen et al. [14] have used synchrotron X-ray imaging and in-situ process to address the key changes during additive manufacturing of Ti-based materials. It was observed that laser power is the most significant parameter for building efficiency, and the lack of fusion regions can be effectively reduced by the higher transverse speed of the process. Moghaddam et al. [15] have done a review analysis on difficulties associated with the fabrication of smart and high entropy materials. Kotadia et al. [16] have studied the laser powder bed fusion process for the fabrication of aluminum-based materials. The study discusses the difficulties and unfavorable microstructural characteristics under rapid directional solidification conditions for wrought Al-alloys. Yang et al. [17] have investigated the effect of parameters such as scanning speed and laser power during additive manufacturing and subtractive manufacturing of stainless steel. Physical and mechanical characterization has been carried out to understand the parametric effect on the additive and subtractive manufactured product. The study was also carried out to understand the effect of built-up direction on the mechanical strength of the material fabricated by additive and subtractive hybrid manufacturing processes. The study shows that the mechanical strength of the products in the H direction was maximum as compared to those built in the E or V direction. Zhang et al. [18] have investigated the effect of milling on the dimensional accuracy and surface quality of additive and subtractive hybrid manufacturing by a wire-deposition process. It was observed that machining allowance and surface waviness were minimized as compared to additive manufactured products. It is normally known that products manufactured by the direct energy deposition have some irregularities and stress risers, including hot stress, voids, hot cracks, and partially melted or unmelted powder due to each portion of additively developed parts suffering from rapid melting and solidification with a unique thermal history.

The present study focuses on studying the surface quality of the smart alloys during the subtractive and additive manufacturing process. Previous literature enlightens the effect of process parameters on the microstructure and mechanical properties of the products during subtractive and additive manufacturing. The present study will provide a brief introduction to each process.

14.2 MATERIAL PROCESSING OF SMART MATERIALS

Fabrication of smart alloys is one of the important and challenging tasks for researchers. Researchers and industries have invested a lot of research and efforts to understand the irregularities associated with the physical and mechanical properties of the fabricated parts developed using smart alloys. The smart alloys indefinitely change their dimension and properties under thermodynamic properties and external forces. So, it is essential to learn the behaviour of the developed products using subtractive and additive manufacturing. Wright and Chow [19] have investigated the machinability of nickel alloys during cutting operations. The study stated that the shear force and cutting-edge temperature in primary and secondary machining zone increase with speed. Shih et al. [20] have studied the difficulties that arise during the fabrication of TiNi shape memory alloys using the co-sputtering fabrication process. Pragana et al. [21] have fabricated the stainless steel 316L parts using additive manufacturing based on the laser bed fusion process. Unlike smart alloys, traditional hard-to-machine alloys like stainless steel 316L can produce fully dense products by laser powder bed fusion without post-processing by laser remelting as reported by Tan et al. [22] and Bayati et al. [23]. Bayati et al. [23] have reported that smart alloys fabricated using AM process offer micropores and microvoids. The process also offers rough surfaces due to fabrication techniques. To avoid and overcome such irregularities, Bayati et al. [23] have proposed remelting procedure during post-processing operation. The process offers improvement in surface roughness and irregular voids. The result does not offer much significant improvement in the mechanical properties, unlike the fatigue life of the products. Lobo et al. [24] have studied mechanical behaviour and surface integrity of shape memory alloys during the fabrication process. The study shows the relevance of the kinematic and thermal effect on the mechanical behavoiur of the developed SMAs.

14.3 SUBTRACTIVE MANUFACTURING OF SMART MATERIALS

The fabricating processes where removal of unwanted and undesired materials in a defined and controlled manner through milling, boring, shaping, drilling, or cutting to achieve the desired form of the product is known as subtractive manufacturing [25]. Material removable of smart alloys is a challenging task due to its physical and mechanical behaviour. The shape memory behaviour of a few smart alloys makes it more difficult during the subtractive manufacturing process. Researchers have invested a lot of resources in studying the subtractive manufacturing processes. The surface quality is one major concern while fabricating such materials. Figure 14.1 indicates the schematic representation.

Yu et al. [27] have used tomography data during the subtractive manufacturing process to produce the duplicate product, as shown in Figure 14.2. The product developed using subtractive manufacturing attracts researchers toward its use of the process. The major concern regarding the subtractive manufactured product is its dimensional accuracy and surface quality. To address this issue, Braian et al. [28] have studied the dimensional accuracy by standard deviation and mean accuracy of the products. The study shows the accuracy ranges between 1%–2% of its original products.

Wang et al. [29] have performed milling operation of smart alloy like NiTi to understand the work-hardening behaviour on shape memory effect and super-elasticity of the material. It was observed that work hardening affects the direction of deformation during the milling process. The study also reveals that at up to 200 m/min, the work hardening of the material decreases and increases after 200 m/min speed. The study

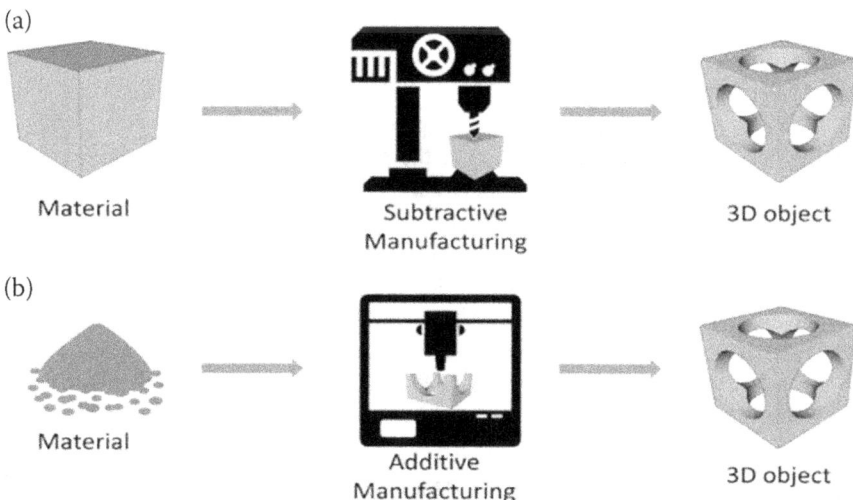

FIGURE 14.1 Schematic process of subtractive manufacturing and additive manufacturing [26].

FIGURE 14.2 Product developed using subtractive manufacturing is at right and original at left [27].

suggests the selection of speed attributes in the work hardening of the materials. Chaudhari et al. [30] have performed wire-cut electrical discharge machining on shape memory alloys to attain the desired surface quality of the final product. The study reveals that spark energy has an immense effect on the surface roughness of the product. Gupta and Laubsher [31] have studied the behaviour of titanium alloys during the subtractive manufacturing process. The study stated the challenges associated with the machining of titanium alloys during the machining process. The study observed that the surface finish of the materials was affected quite adversely under high speed and feed rate. The study also reveals the importance of lubrication during subtractive manufacturing processes. The study suggested that the implementation of cryogenic manufacturing under laser-assisted machining shows a high impact on the surface quality of the products. Daneshmand et al. [32] have performed machining of smart alloys like SMA using electro-discharge machining process using Al_2O_3 to increase surface quality and material removal rate during the subtractive process. The study further extended to understand the effect of tool rotation on performance during the machining process. The study established a statistical model to understand the parametric influence in the manufacturing process. Yin et al. [33] have conducted helical drilling of super alloys (Ni alloy) using femtosecond laser during the subtractive manufacturing process. The study was performed to understand the parametric behaviour of geometrical features and surface quality of drilled holes. The researchers state that with an increase in laser energy quality of drilled holes increases in the helical laser drilling process. The study shows the effect of high intensity of laser during machining of super alloys.

14.4 ADDITIVE MANUFACTURING OF SMART MATERIALS

Building a 3D object by printing using a layer-by-layer technology is called an additive manufacturing (AM) process. AM have huge application in various fields, such as medical science, automobile, aerospace, aviation, artifacts preservation, circuit printing, etc. [34]. There are various AM processes involved to contribute to these fields such as stereolithography [35], fuse-deposition melting [36], wire arc additive manufacturing [37], selective laser melting [38], laser sintering [39], direct-metal laser melting [40], and electron beam melting [41]. It can print a wide range of materials from plastics [42], metals [43], metal alloys [44], and concrete [45]. There is a wide aspect of possibilities that still need to be explored in AM processes. To study and explore the possibilities in this field, researchers and industries have invested a lot of resources to study the aforementioned [46]. It was observed that apart from the subtractive manufacturing process, additive manufacturing is a more environmentally friendly process [47]. The researchers state that less material is wasted in AM as compared to the SM process [48]. There are other challenges associated with AM that have been stated by researchers in this study.

Andani et al. [49] have fabricated smart shape memory alloys using the SLM process. The study was conducted to evaluate the pore morphology of mechanical properties of smart alloys. The study states that the modulus of elasticity is higher for dense fabricated materials as compared to materials having porosity. The study also shows that even though the material has porosity, it has reasonable SME. Pragana et al. [21] have studied the effect of process parameters on the selective laser-melting process during the fabrication of complex parts manufactured by stainless steel. The study presents a review of the present situation of the process. The study enlightens the effect of shielding gas and vector size during the SLM process. The study also suggests that in the SLM process, dense parts can be achieved by laser remelting of fabricated objects. Zhu et al. [50] have studied the effect of process parameters during the fabrication of smart alloys using L-PBF. The study suggests that the pore and mechanical strength are directly affected by hatch distance and laser density during the process. Tan et al. [22] have studied the effect of process parameters during the fabrication of Ti-rich TiNi using LPBF technology. The study was also focused to understand the effect of LPBF on the formation of inter-metallics. Walker et al. [51] have studied the effect of process parameters on structural and functional abilities of nickel-based smart alloys using SLM manufactured products. The study demonstrates the effect of laser density on the surface texture of the product. The research output suggests that at higher and lower laser densities the surface texture was greatly varied by waviness and discontinuous melts respectively. Both of these qualities were undesirable properties of any developed product.

14.5 PROPERTIES AND PERFORMANCE

It is very necessary to evaluate the mechanical and metallurgical properties of the fabricated by additive and subtractive manufacturing processes. The quality of the product is determined by various standards and specifications, such as mechanical properties, wear resistance, surface finish, porosity, etc. The present section discusses the recent standards and properties to analyze the fabricated products.

14.5.1 MECHANICAL PROPERTIES

Additive manufacturing is also known as additive layer manufacturing, which manufactures products at the highest degree of complexities, intricate shapes, and geometries. For the additive manufacturing process, the engineers associated with this process are specifically interested in functional optimization or design, rather than machinability. However, considering such potential aspects for design refinements and optimization, it is also required to ensure that the fabricated product using additive manufacturing can have similar mechanical properties, such as yield strength, fatigue, ultimate tensile strength, and hardness, to conventional fabrication processes such as milling, forming, forging, casting, etc. As researchers, it is essential to consider studying the mechanical properties of the fabricated products and to enhance the mechanical properties of the additive manufactured products as much as possible. To discuss this, the present section discusses some recent contributions related to the enhancement of mechanical and other properties of additive manufactured products.

Podgornik et al. [52] have investigated the mechanical strength, resistance to cracking, and resistance of the additive manufactured products using a laser-based powder bed fusion process. In the study, the effect of different build-up directions such as the inclination of 45°, 90°, and 0° (horizontal directions) was studied. The samples were post-processed inside the solution for aging and heat treatment of the fabricated samples. The aging heat treatment of the samples increases the mechanical strength and resistance to wear. The post-processing also helps in the improvement of the homogeneity of the microstructure of the additive manufactured products. The study also reports the enhancement of elastic properties, and this results in an improvement in resistance to fracture and initiation of cracks. A similar type of observation has been reported by Vishwakarma et al. [53], Tan et al. [54], and Bai et al. [55]. Here, it was reported that the aging of products (maraging steel) manufactured by a selective laser-melting process has reported a decrement in residual stress [52], increase in mechanical strength and hardness values [53–56], and also resulted in an increase in elongation properties from the products without aging heat treatment. To study the effect of post-heat treatment on other materials manufactured by the additive manufacturing process, research works are carried out. Karabulut et al. [56] have studied the effect of the heat treatment process on samples manufactured by selective laser melting using In718. The results show that mechanical behavior, wear behavior, and micro hardness improved as compared with wrought In718. The outcomes provide enough evidence that post-heat treatment is capable of altering microstructural aspects of as-built specimens.

14.5.2 Wear Resistance

The property that helps in resisting the material loss by some mechanical action is known as wear resistance. In this context, it can be said that materials can be tough and resistant to wear but not particularly hard. This is one of the properties that need to be studied for any fabricated products. To determine the quality and behavior of additively manufactured high entropy alloys and stainless steel, Yang et al. [57] have proposed scanning electron microscopy (SEM), X-ray diffraction (XRD), and wear test analysis of the fabricated products. The microstructural study using SEM shows the presence of interdendritic and dendrite microstructures over the manufactured layer. A comparative study of wear resistance analysis over the high entropy alloys and stainless steel has been carried out. They illustrate that high entropy alloys have steady coefficient friction and relatively less worn surfaces as compared to stainless steel. The study exhibits the wear resistance can be effectively used for analyzing the wear resistance to compare the developed products. The wear resistance has been carried out by several researchers to estimate the surface wear resistance of the products using different materials and different additive manufacturing processes. Wesemann et al. [58] have studied the wear behavior (wear resistance) of the samples fabricated using injection molding and subtractive and additive manufacturing processes. The comparative study was performed between the products developed by the injection molding process and products fabricated by a combination process of milling (subtractive manufacturing) and additive manufacturing. Two body-wears test were performed to analyze the wear behavior of the developed product. The study reports that occlusal devices and products manufactured by injection molding and subtractive and additive manufacturing demonstrate similar wear resistance after a simulated wear time of one year [58].

14.5.3 SURFACE FINISH

Overall, surface finish is one of the other properties that demonstrate or define the quality of the fabricated products. As discussed earlier in the previous chapters and sections, the necessity of the surface irregularities and factors affect the surface finish. In the present section, a discussion on the improvement of surface finish by subtractive manufacturing is discussed. Chan et al. [59] and Frazier [60] have studied the products developed by additive manufactured products, such as laser beam melting and electron beam melting, to analyze the relationship between surface finish and fatigue life of the additive manufactured products. The study reported that there is a relationship between the surface finish and fatigue performance of the product [59,60]. Frazier [60] has also reported in their study that feature definition and built-up rate are closely affecting the surface finish of the product. It was observed that the surface finish is mostly affected by the factors such as built-up rate, built-up direction [59], manufacturing process, and selection of process parameters. Wesemann et al. [58] have carried out the study to improve the surface quality of polymer composites using a milling process integrated with an additive manufacturing process. Kashouty et al. [61] have used subtractive manufacturing, like the end-milling process on samples prepared by selective laser melting process used for automotive sectors. The products end processed by this hybrid process offered better dimensional accuracy as compared to products directly manufactured by selective laser melting process.

14.5.4 POROSITY

One of the major concerns regarding the additive manufacturing process is the development of porosity in the fabricated parts. The presence of voids or a pore in the fabricated products is termed as porosity, or it can also be known as lack of material in the segment of fabricated products. This category of defect is most specifically seen in the selective laser melting process as compared to different types of additive manufacturing processes. This happened because of the selection of an inappropriate range of process parameters. To understand this, it can be noted that selective laser melting is a layer-by-layer deposition process, and each layer is also fabricated line by line [62]. In this process, developing a small product, fabricated at high scan speed and low laser power may result in insufficient melting powder at the laser bed, and may also result in spacing between each sintered line and adjacent lines. This can result in the formation of gaps, which results in the development of porosity in the final fabricated products. To address this issue and to find optimum fabricating conditions, researchers have performed rigorous studies. One of the easy ways suggested by Bland and Aboulkhair [62] is to increase the laser energy/power for the materials having low absorptivity and high reflectivity. However, this type of parametric setting cannot be considered for fabrication because of various limitations, for example, the range of the machine or machine capabilities. Also, to address the concern for fabrication is to minimize energy consumption. In the study, Bland and Aboulkhair [62] have discussed the two types of pores, the first based on morphology and the second based on pores. The present observation from the study shows many research opportunities to eliminate this process. The study is also helpful for the development of products needed to have porosity as one of the major concerns [63–65].

14.6   CONCLUSION AND ASPECTS OF RESEARCH

The present study focuses on the approaches established in the subtractive and additive manufacturing processes. The following conclusions have been drawn within the constraints of this in our vitro study.

- The study suggests additive manufacturing (AM) helps in the manufacturing of more precise materials.
- AM is an environmentally friendly fabrication process involving less material waste during manufacturing as compared to subtractive manufacturing processes.
- Subtractive manufacturing can be performed on any range of materials and products as compared to the additive manufacturing process.
- Mechanical properties of the subtractive manufactured products are quite superior as compared to additive manufactured products.
- Post-processing heat treatment processes can be more helpful in reducing voids and increasing the mechanical properties of any AM product.

Future work can be extended by a combination of both processes, which can be adopted firmly to provide a good range of products. The literature survey shows that nickel-based smart alloy has been studied. Future work can be extended to explore the parametric effect during the fabrication of other smart materials.

REFERENCES

1. https://www.grandviewresearch.com/press-release/global-smart-materials-market
2. Mohamed, A.S.Y., 2017. Smart materials innovative technologies in architecture; towards innovative design paradigm. *Energy Procedia*, 115, pp. 139–154.

3. Khoo, Z.X., Teoh, J.E.M., Liu, Y., Chua, C.K., Yang, S., An, J., Leong, K.F. and Yeong, W.Y., 2015. 3D printing of smart materials: A review on recent progresses in 4D printing. *Virtual and Physical Prototyping*, 10(3), pp. 103–122.
4. Spaggiari, A., Castagnetti, D., Golinelli, N., Dragoni, E. and Scirè Mammano, G., 2019. Smart materials: Properties, design and mechatronic applications. Proceedings of the Institution of Mechanical Engineers, Part l: Journal of Materials: Design and Applications, 233(4), pp. 734–762.
5. Alidoosti, A., Ghafari-Nazari, A., Moztarzadeh, F., Jalali, N., Moztarzadeh, S. and Mozafari, M., 2013. Electrical discharge machining characteristics of nickel–titanium shape memory alloy based on full factorial design. *Journal of Intelligent Material Systems and Structures*, 24(13), pp. 1546–1556.
6. Li, J., Chen, Z. and Yuan, T., 2020, September. Optimized Design of a Miniaturized Irregular Spherical Resonator with Enhanced Subtractive/Additive Manufacturing Process Compatibility. In 2020 IEEE MTT-S International Wireless Symposium (IWS) (pp. 1–3). IEEE.
7. Chalvin, M., Campocasso, S., Hugel, V. and Baizeau, T., 2020. Layer-by-layer generation of optimized joint trajectory for multi-axis robotized additive manufacturing of parts of revolution. *Robotics and Computer-Integrated Manufacturing*, 65, p. 101960.
8. Dutta, B. and Froes, F.H., 2014. Additive manufacturing of titanium alloys. *Advanced Materials & Processes*, 172(2), pp. 18–23.
9. Murr, L.E., Martinez, E., Amato, K.N., Gaytan, S.M., Hernandez, J., Ramirez, D.A., Shindo, P.W., Medina, F. and Wicker, R.B., 2012. Fabrication of metal and alloy components by additive manufacturing: Examples of 3D materials science. *Journal of Materials Research and Technology*, 1(1), pp. 42–54.
10. Yuan, S., Li, S., Zhu, J. and Tang, Y., 2021. Additive manufacturing of polymeric composites from material processing to structural design. *Composites Part B: Engineering*, 219, p. 108903.
11. Lee, C.H., Padzil, F.N.B.M., Lee, S.H., Ainun, Z.M.A.A. and Abdullah, L.C., 2021. Potential for natural fiber reinforcement in PLA polymer filaments for fused deposition modeling (FDM) additive manufacturing: A review. *Polymers*, 13(9), p. 1407.
12. Jiménez, A., Bidare, P., Hassanin, H., Tarlochan, F., Dimov, S. and Essa, K., 2021. Powder-based laser hybrid additive manufacturing of metals: A review. *The International Journal of Advanced Manufacturing Technology*, 114, pp. 63–96.
13. Haghdadi, N., Laleh, M., Moyle, M. and Primig, S., 2021. Additive manufacturing of steels: A review of achievements and challenges. *Journal of Materials Science*, 56(1), pp. 64–107.
14. Chen, Y., Clark, S.J., Sinclair, L., Leung, C.L.A., Marussi, S., Connolley, T., Atwood, R.C., Baxter, G.J., Jones, M.A., Todd, I. and Lee, P.D., 2021. Synchrotron X-ray imaging of directed energy deposition additive manufacturing of titanium alloy Ti-6242. *Additive Manufacturing*, 41, p. 101969.
15. Moghaddam, A.O., Shaburova, N.A., Samodurova, M.N., Abdollahzadeh, A. and Trofimov, E.A., 2021. Additive manufacturing of high entropy alloys: A practical review. *Journal of Materials Science & Technology*, 77, pp. 131–162.
16. Kotadia, H.R., Gibbons, G., Das, A. and Howes, P.D., 2021. A review of laser powder bed fusion additive manufacturing of aluminium alloys: Microstructure and properties. *Additive Manufacturing*, 46, p. 102155.
17. Yang, Y., Gong, Y., Qu, S., Yin, G., Liang, C. and Li, P., 2021. Additive and subtractive hybrid manufacturing (ASHM) of 316L stainless steel: Single-track specimens, microstructure, and mechanical properties. *JOM*, 73, pp. 759–769.
18. Zhang, S., Zhang, Y., Gao, M., Wang, F., Li, Q. and Zeng, X., 2019. Effects of milling thickness on wire deposition accuracy of hybrid additive/subtractive manufacturing. *Science and Technology of Welding and Joining*, 24(5), pp. 375–381.

19. Wright, P.K. and Chow, J.G., 1982. Deformation characteristics of nickel alloys during machining. *Journal of Engineering Materials and Technology*, 104(2), pp. 85–93.

20. Shih, C.L., Lai, B.K., Kahn, H., Phillips, S.M. and Heuer, A.H., 2001. A robust co-sputtering fabrication procedure for TiNi shape memory alloys for MEMS. *Journal of Microelectromechanical Systems*, 10(1), pp. 69–79.

21. Pragana, J.P., Pombinha, P., Duarte, V.R., Rodrigues, T.A., Oliveira, J.P., Bragança, I.M., Santos, T.G., Miranda, R.M., Coutinho, L. and Silva, C.M., 2020. Influence of processing parameters on the density of 316L stainless steel parts manufactured through laser powder bed fusion. *Proceedings of the Institution of Mechanical Engineers, Part B: Journal of Engineering Manufacture*, 234(9), pp. 1246–1257.

22. Tan, C., Li, S., Essa, K., Jamshidi, P., Zhou, K., Ma, W. and Attallah, M.M., 2019. Laser powder bed fusion of Ti-rich TiNi lattice structures: Process optimisation, geometrical integrity, and phase transformations. *International Journal of Machine Tools and Manufacture*, 141, pp. 19–29.

23. Bayati, P., Safaei, K., Nematollahi, M., Jahadakbar, A., Yadollahi, A., Mahtabi, M. and Elahinia, M., 2021. Toward understanding the effect of remelting on the additively manufactured NiTi. *The International Journal of Advanced Manufacturing Technology*, 112(1), pp. 347–360.

24. Lobo, P.S., Almeida, J. and Guerreiro, L., 2015. Shape memory alloys behaviour: A review. *Procedia Engineering*, 114, pp. 776–783.

25. Paris, H. and Mandil, G., 2018. The development of a strategy for direct part reuse using additive and subtractive manufacturing technologies. *Additive Manufacturing*, 22, pp. 687–699.

26. https://bitfab.io/blog/additive-manufacturing/

27. Yu, J., Lynn, R., Tucker, T., Saldana, C. and Kurfess, T., 2017. Model-free subtractive manufacturing from computed tomography data. *Manufacturing Letters*, 13, pp. 44–47.

28. Tao, Y., Yin, Q. and Li, P., 2021. An Additive manufacturing method using large-scale wood inspired by laminated object manufacturing and plywood technology. *Polymers*, 13(1), pp. 144.

29. Wang, G., Liu, Z., Niu, J., Huang, W. and Xu, Q., 2019. Work hardening influencing on shape memory effect of NiTi alloy by varying milling speeds. *Smart Materials and Structures*, 28(10), p. 105034.

30. Chaudhari, R., Vora, J.J., Patel, V., López de Lacalle, L.N. and Parikh, D.M., 2020. Surface analysis of wire-electrical-discharge-machining-processed shape-memory alloys. *Materials*, 13(3), p. 530.

31. Gupta, K. and Laubscher, R.F., 2017. Sustainable machining of titanium alloys: A critical review. *Proceedings of the Institution of Mechanical Engineers, Part B: Journal of Engineering Manufacture*, 231(14), pp. 2543–2560.

32. Daneshmand, S., Monfared, V. and Neyestanak, A.A.L., 2017. Effect of tool rotational and Al_2O_3 powder in electro discharge machining characteristics of NiTi-60 shape memory alloy. *Silicon*, 9(2), pp. 273–283.

33. Yin, C.P., Wu, Z.P., Dong, Y.W., You, Y.C. and Liao, T., 2019. Femtosecond laser helical drilling of nickel-base single-crystal super-alloy: Effect of machining parameters on geometrical characteristics of micro-holes. *Advances in Production Engineering & Management*, 14(4).

34. Jiménez, M., Romero, L., Domínguez, I.A., Espinosa, M.D.M. and Domínguez, M., 2019. Additive manufacturing technologies: An overview about 3D printing methods and future prospects. *Complexity*, 2019, pp. 1–30.

35. Zhang, X., Jiang, X.N. and Sun, C., 1999. Micro-stereolithography of polymeric and ceramic microstructures. *Sensors and Actuators A: Physical*, 77(2), pp. 149–156.

36. Melocchi, A., Parietti, F., Maroni, A., Foppoli, A., Gazzaniga, A. and Zema, L., 2016. Hot-melt extruded filaments based on pharmaceutical grade polymers for 3D printing by fused deposition modeling. *International Journal of Pharmaceutics*, 509(1–2), pp. 255–263.
37. Williams, S.W., Martina, F., Addison, A.C., Ding, J., Pardal, G. and Colegrove, P., 2016. Wire+ arc additive manufacturing. *Materials Science and Technology*, 32(7), pp. 641–647.
38. Yadroitsev, I., Bertrand, P. and Smurov, I., 2007. Parametric analysis of the selective laser melting process. *Applied Surface Science*, 253(19), pp. 8064–8069.
39. Simchi, A., 2006. Direct laser sintering of metal powders: Mechanism, kinetics and microstructural features. *Materials Science and Engineering: A*, 428(1–2), pp. 148–158.
40. Keshavarzkermani, A., Sadowski, M. and Ladani, L., 2018. Direct metal laser melting of Inconel 718: Process impact on grain formation and orientation. *Journal of Alloys and Compounds*, 736, pp. 297–305.
41. Murr, L.E., Martinez, E., Gaytan, S.M., Ramirez, D.A., Machado, B.I., Shindo, P.W., Martinez, J.L., Medina, F., Wooten, J., Ciscel, D. and Ackelid, U., 2011. Microstructural architecture, microstructures, and mechanical properties for a nickel-base superalloy fabricated by electron beam melting. *Metallurgical and Materials Transactions A*, 42(11), pp. 3491–3508.
42. Wu, T., Jahan, S.A., Zhang, Y., Zhang, J., Elmounayri, H. and Tovar, A., 2017. Design optimization of plastic injection tooling for additive manufacturing. *Procedia Manufacturing*, 10, pp. 923–934.
43. Atzeni, E. and Salmi, A., 2012. Economics of additive manufacturing for end-usable metal parts. *The International Journal of Advanced Manufacturing Technology*, 62(9–12), pp. 1147–1155.
44. Johnson, L., Mahmoudi, M., Zhang, B., Seede, R., Huang, X., Maier, J.T., Maier, H.J., Karaman, I., Elwany, A. and Arróyave, R., 2019. Assessing printability maps in additive manufacturing of metal alloys. *Acta Materialia*, 176, pp. 199–210.
45. Lee, D., 2018. Investigation of laser ablation on acrylonitrile butadiene styrene plastic used for 3D printing. *Journal of Welding and Joining*, 36(1), pp. 50–56.
46. Camacho, D.D., Clayton, P., O'Brien, W.J., Seepersad, C., Juenger, M., Ferron, R. and Salamone, S., 2018. Applications of additive manufacturing in the construction industry–A forward-looking review. *Automation in Construction*, 89, pp. 110–119.
47. Kumar, S. and Czekanski, A., 2018. Roadmap to sustainable plastic additive manufacturing. *Materials Today Communications*, 15, pp. 109–113.
48. Paris, H., Mokhtarian, H., Coatanéa, E., Museau, M. and Ituarte, I.F., 2016. Comparative environmental impacts of additive and subtractive manufacturing technologies. *CIRP Annals*, 65(1), pp. 29–32.
49. Andani, M.T., Saedi, S., Turabi, A.S., Karamooz, M.R., Haberland, C., Karaca, H.E. and Elahinia, M., 2017. Mechanical and shape memory properties of porous Ni50. 1Ti49. 9 alloys manufactured by selective laser melting. *Journal of the Mechanical Behavior of Biomedical Materials*, 68, pp. 224–231.
50. Zhu, J.N., Borisov, E., Liang, X., Farber, E., Hermans, M.J.M. and Popovich, V.A., 2021. Predictive analytical modelling and experimental validation of processing maps in additive manufacturing of nitinol alloys. *Additive Manufacturing*, 38, p. 101802.
51. Walker, J.M., Haberland, C., Taheri Andani, M., Karaca, H.E., Dean, D. and Elahinia, M., 2016. Process development and characterization of additively manufactured nickel–titanium shape memory parts. *Journal of Intelligent Material Systems and Structures*, 27(19), pp. 2653–2660.
52. Podgornik, B., Šinko, M. and Godec, M., 2021. Dependence of the wear resistance of additive-manufactured maraging steel on the build direction and heat treatment. *Additive Manufacturing*, 46, p. 102123.

53. Vishwakarma, J., Chattopadhyay, K. and Srinivas, N.S., 2020. Effect of build orientation on microstructure and tensile behaviour of selectively laser melted M300 maraging steel. *Materials Science and Engineering: A*, 798, p. 140130.

54. Tan, C., Zhou, K., Kuang, M., Ma, W. and Kuang, T., 2018. Microstructural characterization and properties of selective laser melted maraging steel with different build directions. *Science and Technology of Advanced Materials*, 19(1), pp. 746–758.

55. Bai, Y., Yang, Y., Wang, D. and Zhang, M., 2017. Influence mechanism of parameters process and mechanical properties evolution mechanism of maraging steel 300 by selective laser melting. *Materials Science and Engineering: A*, 703, pp. 116–123.

56. Karabulut, Y., Tascioglu, E. and Kaynak, Y., 2021. Heat treatment temperature-induced microstructure, microhardness and wear resistance of Inconel 718 produced by selective laser melting additive manufacturing. *Optik*, 227, p. 163907.

57. Yang, S., Liu, Z. and Pi, J., 2020. Microstructure and wear behavior of the AlCrFeCoNi high-entropy alloy fabricated by additive manufacturing. *Materials Letters*, 261, p. 127004.

58. Wesemann, C., Spies, B.C., Sterzenbach, G., Beuer, F., Kohal, R., Wemken, G., Krügel, M. and Pieralli, S., 2021. Polymers for conventional, subtractive, and additive manufacturing of occlusal devices differ in hardness and flexural properties but not in wear resistance. *Dental Materials*, 37(3), pp. 432–442.

59. Chan, K.S., Koike, M., Mason, R.L. and Okabe, T., 2013. Fatigue life of titanium alloys fabricated by additive layer manufacturing techniques for dental implants. *Metallurgical and Materials Transactions A*, 44(2), pp. 1010–1022.

60. Frazier, W.E., 2014. Metal additive manufacturing: A review. *Journal of Materials Engineering and Performance*, 23(6), pp. 1917–1928.

61. El Kashouty, M., Rennie, A. and Ghazy, M., 2015, November. Assessing additive and subtractive manufacturing technologies for the production of tools in the automotive industry. In The 23rd CAPE Conference: Manufacturing Research and its Applications in the 21st Century. Lancaster University, UK, (pp. 1–8).

62. Bland, S. and Aboulkhair, N.T., 2015. Reducing porosity in additive manufacturing. *Metal Powder Report*, 70(2), pp. 79–81.

63. Ghods, S., Schur, R., Schultz, E., Pahuja, R., Montelione, A., Wisdom, C., Arola, D. and Ramulu, M., 2021. Powder reuse and its contribution to porosity in additive manufacturing of Ti6Al4V. *Materialia*, 15, p. 100992.

64. Gupta, V., Alam, F., Verma, P., Kannan, A.M. and Kumar, S., 2021. Additive manufacturing enabled, microarchitected, hierarchically porous polylactic-acid/ Lithium iron phosphate/carbon nanotube nanocomposite electrodes for high performance Li-Ion batteries. *Journal of Power Sources*, 494, p. 229625.

65. Paxton, N.C., Dinoro, J., Ren, J., Ross, M.T., Daley, R., Zhou, R., Bazaka, K., Thompson, R.G., Yue, Z., Beirne, S. and Harkin, D.G., 2021. Additive manufacturing enables personalised porous high-density polyethylene surgical implant manufacturing with improved tissue and vascular ingrowth. *Applied Materials Today*, 22, p. 100965.

15 Manufacturing of 3D Print Biocompatible Shape Memory Alloys

Raj Manik and Ajit Behera
Department of Metallurgical and Materials Engineering,
National Institute of Technology, Rourkela, India

CONTENTS

DOI: 10.1201/9781003189404-15

15.1   INTRODUCTION

The shape memory alloy (SMA) have a unique property of recovering large in-elastic strain by heating or unloading, depending on the application of a stimulating agent. There are various types of shape memory alloys based on their composition, like copper-based SMA (such as CuZn, CuAl, etc.) [1,2], nickel-based SMA (such as NiTi, NiAl, etc.) [3,4], iron-based SMA (such as FeMnSi, FeNiC, etc.) [5,6]. The most common type of commercially adopted SMA includes NiTi alloy. This alloy is most widely used due to its better functional stability, biocompatibility, damping characteristics, and high resistance to oxidation. The major disadvantage of NiTi includes poor workability that results in high tool wear, which prevents it from making complex structures by conventional methods of manufacturing. Similarly, it is difficult to fabricate polycrystalline Cu-based SMAs due to their brittle nature. To overcome this difficulty, alternative methods of manufacturing were discovered, and, with the advent of additive manufacturing, several limitations of SMA are corrected. The additive manufacturing method allows the complex shapes to be drawn directly from CAD prototyping and avoids the use of tooling. There are various modes of manufacturing of SMAs by AM that includes binder jetting, powder bed fusion, material extrusion, layer fusion techniques, and many other advanced techniques [7–10]. The main advantages of using AM for SMA include higher deposition rates, multiple powder feeders, large build envelope, a wider range of powders, repair and feature addition, less support structure, heated build chamber (which helps in stress relief), 3D nesting of parts build in support chamber, good accuracy, ability to build complex structures, and many others. However, the major disadvantage by AM includes lack of fusion, stress and distortion, delamination, microcrack, porosity voids and defects, CNC software limitations in case of powder bed fusion, fewer material options in case of electron beam melting, slower cooling time, high cost of powder materials in the case of electron beam deposition, and many others. The mechanical properties of materials obtained from AM are considered to have cons due to the presence of defects [11].

The most widely used method of manufacturing of SMA is selective laser melting, which starts by slicing the 3D CAD file data into layers, usually from 20 to 100 μm thick, creating a 2D image of each layer; this file format is the industry standard stl file used on most layer-based 3D printing or stereolithography technologies. Thereafter, the file is being loaded inside a file preparation software package, which further assigns parameters, values, and physical supports that allow the file to be interpreted and built by different types of AM machines. The other most commonly used technique is electron beam melting, which involves placing the raw material (metal powder or wire) under a vacuum and fused together from heating by an electron beam. These modern AM techniques have made it very easy to manufacture complex parts of biocompatible NiTi SMA that was considered tough in earlier times.

15.2 ADDITIVE MANUFACTURING TECHNIQUES

15.2.1 History of Additive Manufacturing

The history of additive manufacturing dates back to the 19th century. With the emergence of photography in the early 1800s, inventors began to inspect how to extend the process to a third dimension and recreate physical objects. Currently, AM techniques are similar to that of two ideas originating from that century-photo sculpture in the 1860s and topography in the 1890s. A photo sculpture is the reproduction of persons, animals, and things in 3D form by taking a series of photos in the round and using them as synchronized photo projections to create a sculpture. Topography deals with the artistic representation of a particular thing. The 18th and 19th centuries were a boon for topographical surveying and topographic maps. These two fields are not only responsible for the rapid increase of AM techniques but also made great advancements in computers, computer-aided design (CAD), computer numerical control (CNC) machining, and lasers [12,13]. One of the major problems responsible for hindering CNC success is the lack of an exact scheme for describing 3D parts. PADL (part and assembly description language) allows engineers to fully define their parts and gives them the option to assign tolerances to their parts. The first demonstration of the powder bed fusion ASTM standard process of AM was done for a molding process for forming 3D articles in layers using heat and powdered feedstock [14].

The initial commercialization of additive manufacturing started with the patenting of the idea related to stereolithography (SLA). The process involves solidifying a liquid photopolymer by using a scanning laser and further making it in a layered manner. A highly crosslinked polymeric chain was formed due to the high energy of the laser via a chemical reaction. The rapid prototyping in the world of additive manufacturing started with Hull's process, which involved vat polymerization. 3D Systems was a company founded by Hull, and its major breakthrough was to develop standard triangulation language (STL) file format. Its function was to display the surface geometry of 3D objects. The use of STL format is mostly prevalent in additive manufacturing (AM) and computer-aided manufacturing (CAM) [15]. The next breakthrough in AM process came in the year 1986 by Carl Deckard and Joseph Beaman. The invention was selected laser sintering, which was a commercial method of powder bed fusion. The apparatus and method for forming integral objects from laminations were developed in the year 1987. This marked the development of sheet lamination. There was severe expansion in the market of additive manufacturing from 1988 to 1993 that includes building heavy pieces of machinery for the advancement of additive manufacturing. Fused deposition modeling was developed in 1989, which was the first metal extrusion technique used in the field of AM. In 1994, it was mostly used for the process of electron beam melting, which is one of the most used methods for metal 3D printing. In the late 1990s, various advanced applications of SLA technology were used for rapid prototyping of metals. The industry of additive manufacturing is expected to grow tremendously in the 21st century, with various companies adopting the technology to produce a final product that includes sectors of automobiles, iron and steel industry, and many others [16,17].

15.2.2 Modern Methods of Additive Manufacturing

3D printing and additive manufacturing bring more development in the field of technologies, such as information, computing, robotics, and materials. Rapid prototyping with plastics and polymers launched the use of STL models and layer-wise deposition. Weld cladding demonstrated metal buildup of coatings and shapes, while laser welding and cladding provided a better understanding of the metallurgy of rapidly solidified AM metal deposits. The powder metallurgy industry continues to lead the way in new and improved methods of powder production. The origin technologies of AM metal provide a rich source of information with engineering studies and applications relevant to AM metal processing, such as metal powder and production facility safety [18,19]. ASTM International categorizes the AM processes into seven main categories, according to the adhesion and bonding method. These categories are: (i) VAT photopolymerization, (ii) material jetting, (iii) material extrusion, (iv) powder bed fusion, (v) binder jetting, (vi) direct energy deposition, (vii) sheet lamination, and (viii) wire arc additive manufacturing.

15.2.2.1 VAT Photopolymerization

In the Vat polymerization method, the model is constructed layer by layer with the use of a vat of liquid photopolymer resin. To harden the resin, ultraviolet (UV) light is used. Simultaneously, a platform is used to move the object downward after each new layer is cured. As the process uses liquid to form objects, there is no structural support from the material during the build phase, unlike powder-based methods, where support is given from the unbound material. In this case, support structures will often need to be added. Resins are cured using a process of photopolymerization or UV light, where the light is directed across the surface of the resin with the use of motor-controlled mirrors. Where the resin comes in contact with the light, it cures or hardens. Advantages of the VAT photopolymerization method are high level of accuracy and good finish, relatively quick process, and large build areas. Disadvantages of the VAT photopolymerization method are the relatively high expense, lengthy post-processing time, limited material use of photo-resins, and often requires support structures and post-curing for parts to be strong enough for structural use [20].

15.2.2.2 Material Jetting

Material jetting (MJ) is one of the fastest and most accurate 3D printing technologies. Here, object creation is similar to that of a two-dimensional inkjet printer. Using either a continuous or drop on demand (DOD) approach, the material is jetted onto a build platform where it solidifies and the model is built layer by layer. Material is deposited from a nozzle that moves horizontally across the build platform. There is a lot of variation in machines' complexity and in their methods of controlling the deposition of material. The material layers are then hardened using ultraviolet (UV) light. As material must be deposited in drops; the number of materials available to use is limited. The most suitable and commonly used materials are polymers and waxes because of their viscous nature and ability to form drops. Advantages of the MJ method are build speed and outstanding dimensional accuracy, ideal for making aesthetical prototypes, and the process allows for multiple material parts and colors under one process. Disadvantages of the MJ method are that it is the most expensive 3D printing technology, MJ parts are structurally weak, materials are limited, and only polymers and waxes can be used [21].

15.2.2.3 Material Extrusion

Metal extrusion in additive manufacturing is a new process, very similar to that of the plastic-based FDM process; here, the filament is first heated, and then it is drawn through a nozzle and deposited layer by layer. The filament used here is an amalgamation of thermoplastic material and metallic particles. For a given layer, here the nozzle moves in both the x- and y-axis. To make room for new layers, the build platform lowers. After completion of the above part, it is placed into a sintering furnace to burn out the remaining plastic and sinter the metal particles together. Recently, material extrusion based additive manufacturing method has developed to create metal parts. Otherwise, it is generally used for plastics and polymers only. Because of its fast build times and significantly reduced cost as compared to laser sintering, it proved to be an extremely capable prototyping technology. Advantages of metal extrusion method are very low-cost process, great for prototypes, and high precision with fine layers. Disadvantages of the metal extrusion method are limited production potential, high binder content that makes sintering difficult, significant part shrinkage in the furnace, required supports, and parts are lower in density [22].

15.2.2.4 Powder Bed Fusion

The principle of selective laser sintering is being used for powder bed fusion. The fusion of layer occurs due to a laser beam that is into a bed of powder, the cross-sectional area of the sliced part model, and a scan path of overlapping weld beads define it. The successive layers of powder form the part that occurs by dropping the initial part, and a powder bed and a roller or blade spread a new layer of powder to allow the fusion to successively coat it. An important point is that the thickness, which is being formed by the powder layer, is greater than the thickness formed by fused deposits. The penetration is comparatively larger when compared with the depth-of-deposit layer thickness as this has the capacity to penetrate three or more layers in-depth to more fully fuse the deposit. This reason validates the formation of fully dense materials by using 3D models designed by software. The materials that are being used for this purpose include steel, nickel, titanium, cobalt chrome molybdenum (CoCrMo), metal matrix composite materials, and other specialty metals. The merits of using powder-based techniques include ready-made prototypes, simplicity of using STL files, complex structures that can be easily framed using this technique, increased accuracy, and moreover, it also facilitates multiple instances in a single build cycle. In spite of all the merits, there are certain drawbacks of using powder bed fusion; these include limitation in size of the parts to be made, the high capital cost of powder material, the chance of distortion and cracking, the slow rate of deposition, porosity voids, and defects [23].

15.2.2.5   Binder Jetting

Two material classes are involved in the process of binder jetting process that includes powder-based material and a binder. The role of the binder is to act as an adhesive between the powder layers. The state of the binder and the build material are always dissimilar as the binder is always in liquid form and the build material is in powder form. The working of the binder-jetting process involves movement of the print head along the two perpendicular axes of the machine that deposits alternating layers of the build material and the binding material. After each layer, the object being printed is lowered onto its build platform. Due to the method of binding, the material characteristics are not always suitable for structural parts, and despite the relative speed of printing, additional post-processing can add significant time to the overall process. As with other powder-based manufacturing methods, the object being printed is self-supported within the powder bed and is removed from the unbound powder once completed. The process is being used for the bonding of stainless steel, iron, and tungsten. The merits of using this process involve making structures of large size, less production time as the speed is comparatively higher than the other processes, and the ability to manufacture a wide variety of materials. Certain demerits possess certain constraints on the application; they include the greater cost of the process, and there is no technical ceramics support [24].

15.2.2.6   Direct Energy Deposition

Direct energy deposition is also known as LENS or DMD. The metal 3D solid model is prepared by fusing metal filler in a computerized motion control system. In the case of DED, a powder-delivery nozzle is used for delivering metal powder in a molten pool or focal zone that is dissimilar when compared to the sintering bed of powder in PBF. The fusing of deposits onto the substrate is done after the laser/ powder delivery head is traversed, followed by the melt pool. Since the microstructure is fully evolved from a molten state, the liquid phase sintering is not an ideal method for DED processing. In the case of DED, there is no specific requirement of remelting by the subsequent layers to achieve the full density. Instead, to achieve full densification, fusion mixing is done within the molten pool. Advantages of the direct-energy deposition method are repair and feature addition, multiple powder feeders, large build envelope, wider range of powders, and higher deposition rates. Disadvantages of the direct-energy deposition method are stress and distortion, porosity, voids and defects, cnc software limitations, and less accurate, less complex shapes [25].

15.2.2.7   Sheet Lamination

Sheet lamination is one of the additive manufacturing (AM) techniques. Here, thin sheets of material are usually supplied via a system of feed rollers, and then the sheets are bonded together layer by layer to form a single piece that is then cut into a 3D object. Examples of sheet-lamination techniques are laminated object manufacturing (LOM) and ultrasonic consolidation (UC). This technique can use a variety of materials, such as paper, polymer, and metal, but here, each material needs a different method to bind the sheets together. In the case of paper sheets, heat and pressure are used for binding. For certain polymers, the same application of heat and pressure is used to melt the sheets together. Metal sheets are bound together with the help of ultrasonic vibrations. Advantages of the sheet-lamination technique are: it is a fast and low-cost way to 3D print nonfunctional prototypes, its designs mainly consist of easily handled materials, and it is used to make composite materials. The disadvantage of the sheet-lamination technique is low accuracy [26].

15.2.2.8 Wire Arc Additive Manufacturing

Wire arc additive manufacturing (WAAM) is a fusion manufacturing process in which the heat energy of an electric arc is employed for melting the electrodes and depositing material layers for wall formation or for simultaneously cladding two materials to form a composite structure. This method does not employ laser or electron beams; instead, an electric arc is used to melt the metal and fabricate it layer by layer. One of the important leverages of using wire arc additive manufacturing is that it is made for creating complex shapes in a stipulated amount of time. The deposition rate of using the WAAM technique is quite high, ranging from 1–10 kg/h, and it also provides a good resolution of close to 1 mm. The two fundamental pieces of equipment for the WAAM technique are arc welding set-up and 3D axes manipulator. There is high demand for WAAM in aerospace, automobile, nuclear, molds, and dies industries that clearly demonstrate compatibility and reflect comprehensiveness. This technique for producing at different stages requires manual intervention, and therefore, we can conclude that this method is not an automatic one. The geometrical requirements cannot be achieved by using this method, as further finishing is a base requirement to have a good surface finish. The materials that can be printed using the WAAM techniques are steel, aluminum, and titanium [27].

15.3 FABRICATION PROCEDURE OF SMA BY AM TECHNIQUES

The major drawback of Ni-based SMA is that it suffers from poor workability that further leads to high tool wear. Hence, it is not easy to use the material for framing complex geometries using the conventional mode of fabrication. Similarly, Cu-based SMA is difficult to fabricate because of the brittleness of the material. These types of flaws that are present in SMA can be removed by additive manufacturing, which avoids the need for tooling and thereby facilitates the development of SMA structure directly from CAD drawings. The additive manufacturing techniques of shape-memory alloy fabrication are tested for different alloy compositions. Although there is a requirement for future progress, the technology has definitely made a mark to make feasible various kinds of SMA engineering designs that were previously unexpected using conventional technologies [28]. There are various factors that influence the fabrication of SMA, which include initial powder characteristics, AM process parameters, and heat treatment. There was also evidence from various researchers that the functional and mechanical properties of various additively manufactured nickel-based alloys were influenced by selective laser melting (SLM) parameters. The technology of additive manufacturing was not only favorable for the fabrication of nickel-based alloy but also for Cu-based, Co-Ni, and NiMnGa shape memory alloys. The amount of work done in the field of fabrication of SMAs other than NiTi is limited. Therefore, the immediate focus in this chapter is to summarize the fabrication technique used for various SMAs, including NiTi. This will help future researchers to dive deeper in the search of the application of non-nickel SMAs as fabrication techniques would not be a hurdle. Most of the additive manufacturing techniques used to fabricate SMAs start with a CAD model followed by powder bed fusion and flow-based methods that are generally in use to fabricate SMAs.

15.3.1 Fabrication of Nitinol by Additive Manufacturing

There are broadly two methods to manufacture nitinol, namely powder-based technique and flow-based technique. Some widely used methods of the powder-based technique include selective laser sintering (SLS), direct metal sintering (DMLS), selective laser melting (SLM), and laser CUSING. There are various kinds of flow-based or direct deposition techniques that include laser engineered net shaping (LENS), direct light fabrication (DLF), laser consolidation (LC), laser cladding, and shape deposition manufacturing (SDM).

15.3.1.1 Powder-Based Technique for Fabrication of Nitinol

It is essential in the powder-based technique that the required CAD model must be divided into layers with desired thickness by defining the process parameters. With the help of a roller, the powder layer is deposited with a similar thickness as that of the CAD layer, which was divided initially. Furthermore, with the available information about the geometry of divided CAD layers, the laser beam melts the powder layer. The layer-by-layer deposition occurs after solidification, as the building piston drops down by the assigned thickness. The above procedure is repeated until the final structure is completed. Finally, the loose powder needs to be removed. To minimize the possibility of oxidation, the amount of oxygen must be kept in check by filling the chamber with argon. When fabricating NiTi, the substrate must be preheated. If the substrate is not pre-heated, then there are high chances for developing residual stress due to the temperature gradient between the initial layer and the build platform. Therefore, it is essential to reduce the amount of residual stress as it may lead to the separation of sample from the substrate [29].

15.3.1.2 Flow-Based Technique for Fabrication of Nitinol

In the case of powder-based technique, the required CAD model must be divided into layers with desired thickness by defining the process parameters. Furthermore, there is an injection of powder in a controlled manner inside a molten pool that is simultaneously scanned by a laser. There are two ways of supplying powder to the platform. Both the methods are based on the technique that either the table (or platform) moves or the focusing lens and nozzle moves on the basis of sliced CAD data. After the fabrication of the previous layer is completed, the nozzle and focusing lens move by a thickness layer according to the data given. This procedure is continued until the final product is fabricated. The ratio of Ni/Ti needs to be selected based on the functional properties of the final product that needs to be fabricated. A higher level of Ti leads to higher transformation temperatures and shape-memory properties in the final product. A higher Ni content leads to developing a superelastic behavior [30].

15.3.2 Fabrication of Copper-Based Alloy by Additive Manufacturing

The main reason for developing Cu-based shape-memory alloys by adding Al, Ni, and Zn is due to the fact that they have good shape-memory properties and low cost. There are two families of Cu-based SMAs developed so far. The Cu-based SMAs are the second most popular shape-memory alloys because they are used in prominent applications like medical, aerospace, automotive, and many others. The widely used method of additive manufacturing for manufacturing Cu-based alloy is selective laser melting. Due to the high thermal conductivity and low absorption energy of fiber laser, it is pretty much tougher to fabricate parts made of pure copper. However, researchers tested that the fabrication of 99.5% Cu could be done by electron beam melting. High-density Cu parts could be easily manufactured by the selective laser melting process [31].

15.3.2.1 Fabrication of CuAlNi by Selective Laser Melting (SLM)

The use of selective laser melting for the preparation of CuAlNi SMA is found to be a promising technique. While fabricating a material using selective laser melting, the metallic powder particles are fused together in a non-equilibrium state by using a high-energy laser to develop a refined microstructure. In the fabrication of CuAlNi, the powder atomization is generally combined with laser Nd-YAG in an inert gas atmosphere with a relative density of more than 92% and exhibited an average grain size of 32 mm. CuAlNi is a good shape-memory alloy as it can operate at a higher temperature. The rapid solidification method could be employed to observe a better shape memory effect in this alloy. Therefore, it is easier to fabricate CuAlNi by the method of selective laser melting. The chemical composition of alloys generally used for fabrication is Cu-82%, Ni-14%, Al-4%. CuAlNi alloy was mainly fabricated by using Yb-fiber laser. The optimum process parameter for fabrication of this kind of alloy requires laser power of 500 W, the scan speed of 800 mm/s, an energy density of 62.5 J/mm^3 [32].

15.3.2.2 Fabrication of CuAlNiMn by Selective Laser Melting

Fabrication of CuAlNiMn by SLM allows significant grain refinement that is possible due to rapid solidification techniques. The fabrication of parts that are of complex geometries can be manufactured by selective laser melting. Hence, it is possible to manufacture any complex structure layer by layer by CAD data. Due to the high cooling rates and the small size of the melt pool, the molten spots generally solidify. The precipitation of metastable phases occurs due to excessive cooling rates. Selective laser melting is one of the broad manufacturing processes by which processing parameters the properties of the tool to be produced as well as the properties of the material can be varied. The use of proper design and laser energy helps in reducing the final density of the product to close to 20%. The microstructure of SMA depends on the phase, distribution, and also size. Therefore, selective laser melting can offer net shape processing, even of complex parts. The alloy was prepared using a master ingot with nominal composition Cu-11.85Al-3.2Ni-3Mn (wt%) in an induction furnace with high purity elements (more than 99.5%). These ingots were used to prepare powders by nitrogen gas atomization in an atomization chamber. The parameters used with an atomization pressure of 0.5 MPa, nozzle diameter of 6 mm, the gas-metal ratio of 1.93, and superheating of 90 °C [33].

15.3.3 FABRICATION OF NIMNGA ALLOY BY ADDITIVE MANUFACTURING

NiMnGaFe ferromagnetic shape-memory alloys are fabricated mostly by selective laser melting. The average particle size is around 17.6 µm, which is generally produced by milling of melt-spun ribbons. The mechanical milling of melt-spun ribbons in a vibration mill leads to the formation of powders. Certain traces of powder were heat-treated in an argon atmosphere for a time duration close to one hour. Furthermore, those powders were employed for additive manufacturing using the selective laser melting technique. The process parameters used for the preparation of Ni-Mn-Ga include laser beam of power 40 W, beam speed up to 250 mm/s, and line spacing close to 50 µm. There were two types of melting techniques used in SLM; in the first case, melting was done along the plate direction, and in the second case, by using an oscillating laser patch of amplitude close to 100 µm. Methods like X-ray diffraction, electron microscopy, and vibrating sample magnetometry are used to investigate the microstructure and phase transformation. The materials obtained are chemically homogenous and also showcase typical layered microstructure. The magnetic response shown by NiMnGa is better than quaternary alloys, and the martensite formation takes place at a higher temperature in the case of NiMnGa than its corresponding quaternary alloys. The selective laser melting technique allows NiMnGa to be fabricated with homogeneity [34].

15.4 NANOSCALED FABRICATIONS OF SMA THROUGH AM

Nanoscale production is enabled by the development of nanoscale thickness during the fabrication of SMA thin films. Focusing on the nanoscale fabrication, it is possible to develop an efficient actuation system, such as useful forces and displacements embedded in small-scale robotics. In these cases, the complex shape in nanostructure is possible in case of shape-memory alloys by retaining the shape-memory effect and pseudoelasticity behavior. Due to nanoscaled fabrication by AM techniques, greater flexibility is achieved than those of conventional fabrication processes [35].

15.5 CHARACTERISTICS OF SMA OBTAINED FROM AM

The properties of shape-memory alloy made by additive manufacturing generally depend on microstructure, macro-defects, texture, precipitate, dislocation, and twins. The anisotropy in performance is caused due to texture formed by columnar grain, which is formed by SLM-fabricated NiTi. Therefore, optimizing the process parameters is essential to get a shape memory that is composed of all essential properties.

15.5.1 Mechanical Properties of Additive Manufactured SMA

In the case of using LENS (laser engineered net shaping), the nitinol with equiatomic content shows martensitic phase after being solution treated, whereas in the case of Ni-rich alloy, it shows austenite and martensite phase upon aging for a longer duration. The nitinol alloy generally shows a similar kind of compressive response, irrespective of the change of composition when subjected to compressive loading and unloading. When the NiTi is fabricated by selective electron-beam melting, the printed NiTi samples had to undergo compression as well as tension. It is interesting to note that the compressive curve underwent a four-stage compressive deformation. The elastic deformation of austenite occurs in stage I, whereas, in the case of stage II, the elastic deformation of the transformed martensite and some slip of the yet-to-be-transformed austenite may occur, which generally depends on the orientation of crystals. There are chances for stage II and stage III of the four-stage compressive deformation to overlap. It is noted that stage II and stage III may well overlap. The final stage of fracture occurs when the value plastic deformation reaches the value of critical stress for plastic deformation of martensite [36]. Furthermore, when the load is removed, there is a phase change from martensite to austenite. There are also possibilities for recovering further strain by heating to a temperature greater than austenite finish temperature without any stress value. Moreover, the extent of plastic deformation in both phases of plastic deformation upon loading is predicted by the amount of irreversible strain.

15.5.2 Fracture Analysis of Additive Manufactured SMA

The fracture analysis of the additively manufactured SMA is generally done by SEM and EDS analysis. It was generally observed that the fractured surface contained clusters of dimples and clevage-like regions. It is also evident from the microstructure that there is the presence of voids and pores. Furthermore, at a higher magnification, transgranular fracture surfaces with pockets of dimples and fine microcracks are observed. Therefore, this testifies the presence of a mixed-mode fracture mechanism. There is also the presence of impurities in form of black regions. Moreover, identification of elements present as black regions were done using EDS analysis. Mostly, elements like C, Si, Al, O, and Ni were generally present as impurities in NiTi SMA [37]. In certain cases, the presence of a region of unmelted powder was also reported. To investigate the structural properties of additively manufactured SMA, the morphology of fractured surfaces at the unmelted regions was studied. Fracture toughness is a property describing the ability of a material containing a crack to resist fracture. In the case of both SLM-built and EBM-built SMA, anisotropy is observed in fracture toughness. Furthermore, the pathway of movement of crack was greatly influenced by the anisotropy in fracture toughness. The propagation of cracks is generally along the columnar grains in vertically oriented NiTi sample, while the propagation is through the columnar grains in the case of horizontally oriented samples. The value of fracture toughness is higher in the case of EBM-built SMA as compared to SLM-built SMA due to the difference in the microstructure [38].

15.5.3 DISLOCATIONS PRESENT IN ADDITIVE MANUFACTURED SMA

Dislocations are generally visible with the help of TEM (transmission electron microscope) images. The dislocations inside SLM-fabricated NiTi alloy are highly dense and twisted kinds of dislocations. The average density of dislocation was observed to be in the range of 10^{11} cm^2. Furthermore, there is an observation of straight dislocations and long flat kinks when viewed with higher magnification. In certain cases, small spiral-shaped dislocations are also observed. In the case of metals with a high concentration of vacancies, there are helical dislocations; this type of dislocations are generally observed in quench steels and aluminum alloys. The thermal motion of dislocation, along with the movement of bending dislocation in opposite direction, leads to the formation of helical dislocations. When AM manufactured NiTi shows helical dislocations, this is a clear indication of a higher concentration of vacancies. The dislocation formed in AM manufactured NiTi forms wavy or bowed morphologies as the dislocations tend to align themselves to minimize their energy [39]. The dislocation reaction happening during the formation of SMA by additive manufacturing can be known by investigating more on the bending dislocations. Methods like double beam analysis are used to investigate bending dislocations.

15.5.4 PRECIPITATES FORMED IN ADDITIVE MANUFACTURED SMA

The study of formations of precipitates was done by analyzing the microstructure of nitinol with different Ni contents. The analysis of EDS, XRD showcases a total of three different phases: $NiTi_2$ (black), $NiTi$ (grey), and Ni_4Ti_3 (white). When the sample of NiTi fabricated from additive manufacturing was viewed from the top, then precipitates were observed. $NiTi_2$ precipitates were found near the surface of the sample, as well as on the center. The precipitates of Ni_4Ti_3 are lenticular shaped, and the density of this kind of precipitates increases with the increasing amount of Ni content. The size of the Ni_2Ti_2 precipitate found near the surface was larger in size when compared to the precipitates located away from the surface. Moreover, the area surrounding the surface was also filled with Ti-rich precipitates. The region near the surface had the highest amount of congestion of precipitates as compared to the other areas. Therefore, in contrast, there was an insignificant amount of precipitates found in areas away from the substrate. The number of precipitates formed by the conventional method is comparatively less when compared to AM manufactured SMA. Furthermore, to study the incoherent characteristics of precipitates, the formation of precipitates was studied with aging time. The longer aging time increases the amount and size of Ni_4Ti_3 precipitate [40]. There were different sets of Ni_4Ti_3 precipitates that clearly indicate that it was formed incoherently. The coherency of the precipitate phase can be explained by the variation of difference in phase transformation.

15.5.5 PHASE TRANSFORMATION TEMPERATURE OF ADDITIVE MANUFACTURED SMA

The solid-state phase transformations of NiTi SMA are based on three character-istics temperature A_s austenite start, A_f austenite finish, and Rs R-phase start. The hysteresis loop in the rebuilding process is projected by these temperatures. The fastest transition moment is the transition from austenite and R-phase, which also represents peak transition temperature. The two-step martensitic transformation of B2→R→B19′ during cooling and one-step B19′→B2 reverse martensitic trans-formation during heating. The R-phase transformation temperature (i.e., B2→R) is significantly separated from martensite transformation (i.e., R→B19′). The DSC curve generally helps in analyzing the relation between temperature and heat flow for both endothermic as well as exothermic conditions. The R-phase is generally visible before complete solidification. The R-phase is generally visible with a small hysteresis, which is calculated by the formula T=TA−TR and the value of 'T' ranges in 4 °C–5 °C [41]. The two-stage transformation is formed due to the pre-sence of precipitates in the grain boundaries. The chemical composition of the alloy influences the transformation temperature by increasing the Ni content by 1% re-sults in a decrease in transformation temperature. The increase in nickel content generally influences the starting temperature of martensite (M_s). Moreover, the austenite final temperature can be raised due to thermal treatments. The annealing has an impact on the transition temperatures especially the austenitic transformation.

15.6 MERITS AND DEMERITS OF AM

The additive manufacturing technique has a great number of advantages that ex-plain its increased usage in the current time. The traditional route of manufacturing complex parts was a painstaking job but with the advent of additive manufacturing, this problem was greatly resolved as it cost less to print the entire structure that could be drawn via CAD software. Furthermore, if anything needs to be changed then it could be done directly modifying those changes in the original CAD file. The number of parts to be manufactured also decreases as the moving parts can be directly printed into the metal with the help of additive manufacturing. It also re-duces the time as engineers could directly proceed for property testing instead of waiting for the prototype model by printing the STL file [42]. There is an almost negligible constraint to the variety of things that could be manufactured by AM. Moreover, wastage of materials is also reduced as whatever structure is required that is only printed. A wide range of software for the powder bed fusion method can do generating STL files. There are a good range of things that can be done by editing software including flixing, editing, slicing, and preparation for 3D printing. Magnetically driven mirrors using galvanometers can be used for laser scanning optics. The multiple instances of the same part can be made all at once and that is the prime opportunity provided by the powder bed method. Moreover, it allows it to be built simultaneously in a similar time frame. Therefore, to optimize the positioning of various parts within the build volume software is provided. In addition, multiple instances of different parts may be built at the same time. There are also developments in recent time that includes enhancing the speed of processing by heating the powder. Regarding critical application, inert gases of high purity are used for reactive metals. This helps in cooling after the building process is completed. In the case of direct energy deposition, in order to allow a change in material composition during deposition multiple powder or wire feeders are used which delivers different powder to the molten pool. Therefore, it helps in building a functionally graded deposit of metal. The deposition of different features using different materials is possible due to the switching between powder feeders [43]. The control of surface chemistry is possible due to the ability to change the powder delivery gas. The increased deposition rates provided by CNC machining may not require additional accuracy as to achieve the final dimensions high precision parts are required. The original accuracy of the deposition is of less importance if important surfaces and dimensions are machined. Due to the mass availability of a varied range of certified powder and weld wire, the usage of the DED-L technique is most preferable when compared to PBF-L. Repair, remanufacturing, refurbishment, or enhancement of existing components is made easier by DED-L than by powder bed systems as the surface preparation, measurement, repositioning, deposition, finishing and in-process inspection may all occur in one setup, in sequence, on one machine [44].

On the hind side, there are certain demerits that are yet to be overcome. The speed at which the material is manufactured is comparatively slower when it comes to the traditional methods. As the method needs a considerable amount of time, therefore the cost of production also increases. Furthermore, setting the process parameters is also a taxing job, and material after getting manufactured needs post-processing. The mechanical properties of the material obtained from AM routes are prone to defects compared to the traditional routes. The defects arising from additive manufacturing AM include lack of fusion, porosity, voids, cracking, oxidation, discoloration, distortion, irregular surface profile, or surface stair stepping. The most common defect is lack of fusion that occurs when the deposit is not fully melted and it fuses into the substrate. Lack of penetration results in delamination of the part, it's a failure to fuse deeply into the part. There is a good chance for lack of fusion to get unnoticed if a part fails in testing or service. Moreover, poorly bonding locations are hard to be detected by radiographic inspection. A cold lap is a term sometimes used to describe a lack of fusion defect in which coalescence of material is prevented by an oxide layer or thin film. There is a probability of cracking if the boundary has got poorly fused powder particles. The AM build structure should be modified if there is a failure during destructive testing, functional testing, or while in service. The fracture initiating points can be detected by the use of a scanning electron microscope under higher magnification and this inspecting technique is termed fractography. The type of defect that occurs in a region built with insufficient support structure is called slumping and it is a dimensional defect. The general features for slumping defects include small design features, thin walls, downward facing surfaces (also referred to as down skin surfaces), or overhangs. Shrinkage and distortion can occur because of localized melting and solidification. The wrapping and distortions are created due to shrinkage-induced stresses. Proper design, material and parameter selection, and process development can reduce or eliminate the risk of these types of defects [45]. Furthermore, the presence of irregular surface conditions highlights the occurrence of poorly designed, supported, or oriented parts. A poorly developed or controlled process may result in stair-stepping, balling, lack of fusion, surface-breaking delamination, undercutting, holes, porosity, or voids. Porosity is a kind of defect that occurs in additive manufactured material, it arises due to gases and it is evolved in melted and fused material. The shape is often spherical or oblong. The cause of cracking in a material manufactured by AM technique can be due to thermal, mechanical, and metallurgical conditions. In case of cracks occurring in a welded reactor vessel can be detected by dye penetrant testing. The AM manufactured components can be susceptible to the same types of cracking mechanisms present in welded or weld clad structures. The types of cracking present in AM components include crater cracks, hot cracks, hot tearing, and cold cracking.

15.7 SUMMARY

The manufacturing of biocompatible SMAs can be done by additive manufacturing techniques preferentially using the selective laser melting that is widely used because it provides the leverage to be used for a large range of materials, the ability to tune properties during the processing of the parts, increased functionality, relatively low cost, and the production of near-net-shaped components. Electron beam melting is also used for the fabrication of Cu-based SMAs and it is considered effective because EBM has the capability of processing brittle materials that generally cannot be processed by SLM. Various influencing parameters as well as the advantages and limitations of SMA fabrications has successfully discussed here.

REFERENCES

1. Kenneth Kanayo Alaneme, Eloho Anita Okotete, Reconciling viability and cost-effective shape memory alloy options-A review of copper and iron based shape memory metallic systems, *Engineering Science and Technology, an International Journal*, Volume 19, Issue 3, 2016, Pages 1582–1592, 10.1016/j.jestch.2016.05.010
2. R.V.N. Melnik, A.J. Roberts, K.A. Thomas, Computing dynamics of copper-based SMA via centre manifold reduction of 3D models, *Computational Materials Science*, Volume 18, Issues 3–4, 2000, Pages 255–268, 10.1016/S0927-0256(00)00104-X
3. Sato Kinji, Goto Hideaki, Tomita Nobuhisa, The shape memory heat treatment and environmental temperature for improvement of forming limit on Ti-Ni based shape memory alloy, Editor(s): W.B. Lee, *Advances in Engineering Plasticity and Its Applications*, Elsevier, 1993, Pages 1117–1125, 10.1016/B978-0-444-89991-0.50153-0
4. Y.C. Lin, Ling Li, Dao-Guang He, Ming-Song Chen, Guo-Qiang Liu, Effects of pre-treatments on mechanical properties and fracture mechanism of a nickel-based superalloy, *Materials Science and Engineering: A*, Volume 679, 2017, Pages 401–409, 10.1016/j.msea.2016.10.058

5. M.R. Izadi, E. Ghafoori, M. Shahverdi, M. Motavalli, S. Maalek, Development of an iron-based shape memory alloy (Fe-SMA) strengthening system for steel plates, *Engineering Structures*, Volume 174, 2018, Pages 433–446, 10.1016/j.engstruct.2018.07.073
6. Mohammadreza Izadi, Masoud Motavalli, Elyas Ghafoori, Iron-based shape memory alloy (Fe-SMA) for fatigue strengthening of cracked steel bridge connections, *Construction and Building Materials*, Volume 227, 2019, 116800, 10.1016/j.conbuildmat.2019.116800
7. N. Sabahi, W. Chen, C.H. Wang et al., A Review on additive manufacturing of shape-memory materials for biomedical applications. *JOM*, Volume 72, 2020, Pages 1229–1253, 10.1007/s11837-020-04013-x
8. Kirstie R. Ryan, Michael P. Down, Craig E. Banks, Future of additive manufacturing: Overview of 4D and 3D printed smart and advanced materials and their applications, *Chemical Engineering Journal*, Volume 403, 2021, 126162, 10.1016/j.cej.2020.126162
9. M Bhuvanesh Kumar, P. Sathiya, Methods and materials for additive manufacturing: A critical review on advancements and challenges, *Thin-Walled Structures*, Volume 159, 2021, 107228, 10.1016/j.tws.2020.107228
10. M. Zafar, H. Zhao, 4D printing: Future insight in additive manufacturing. *Metals and Materials International*, Volume 26, 2020, Pages 564–585, 10.1007/s12540-019-00441-w
11. M.R. Karamooz Ravari, S. Nasr Esfahani, M. Taheri Andani, M. Kadkhodaei, A. Ghaei, H. Karaca, M. Elahinia, On the effects of geometry, defects, and material asymmetry on the mechanical response of shape memory alloy cellular lattice structures, *Smart Materials and Structures*, Volume 25, Issue 2, 2016, 025008, 10.1088/0964-1726/25/2/025008
12. Mriganka Roy, Olga Wodo, Data-driven modeling of thermal history in additive manufacturing, *Additive Manufacturing*, Volume 32, 2020, 101017, 10.1016/j.addma.2019.101017
13. Y. Huang, M.C. Leu, J. Mazumder, A. Donmez, Additive manufacturing: Current state, future potential, gaps and needs, and recommendations, *ASME. J. Manuf. Sci. Eng.*, Volume 137, Issue 1, February 1, 2015, 014001, 10.1115/1.4028725
14. J. Xiao, N. Anwer, A. Durupt et al., Information exchange standards for design, tolerancing and additive manufacturing: A research review. *Int J Interact Des Manuf* 12, 2018, Pages 495–504, 10.1007/s12008-017-0401-4
15. Giovanni Moroni, Wahyudin P. Syam, Stefano Petrò, Functionality-based part orientation for additive manufacturing, *Procedia CIRP*, Volume 36, 2015, Pages 217–222, 10.1016/j.procir.2015.01.015
16. I. Campbell, D. Bourell, I. Gibson, Additive manufacturing: Rapid prototyping comes of age, *Rapid Prototyping Journal*, Volume 18, Issue 4, 2012, Pages 255–258, 10.1108/13552541211231563
17. Syed A.M. Tofail, Elias P. Koumoulos, Amit Bandyopadhyay, Susmita Bose, Lisa O'Donoghue, Costas Charitidis, Additive manufacturing: Scientific and technological challenges, market uptake and opportunities, *Materials Today*, Volume 21, Issue 1, 2018, Pages 22–37, 10.1016/j.mattod.2017.07.001
18. Alexander V. Manzhirov, Advances in the theory of surface growth with applications to additive manufacturing technologies, *Procedia Engineering*, Volume 173, 2017, Pages 11–16, 10.1016/j.proeng.2016.12.008
19. Ch. Achillas, D. Aidonis, E. Iakovou, M. Thymianidis, D. Tzetzis, A methodological framework for the inclusion of modern additive manufacturing into the production portfolio of a focused factory, *Journal of Manufacturing Systems*, Volume 37, Part 1, 2015, Pages 328–339, 10.1016/j.jmsy.2014.07.014

20. Ali Davoudinejad, Lucia C. Diaz-Perez, Danilo Quagliotti, David Bue Pedersen, José A. Albajez-García, José A. Yagüe-Fabra, Guido Tosello, Additive manufacturing with vat polymerization method for precision polymer micro components production, *Procedia CIRP*, Volume 75, 2018, Pages 98–102, 10.1016/j.procir.2018.04.049

21. Yee Ling Yap, Chengcheng Wang, Swee Leong Sing, Vishwesh Dikshit, Wai Yee Yeong, Jun Wei, Material jetting additive manufacturing: An experimental study using designed metrological benchmarks, *Precision Engineering*, Volume 50, 2017, Pages 275–285, 10.1016/j.precisioneng.2017.05.015

22. J. Blindheim, T. Welo, M. Steinert, First demonstration of a new additive manufacturing process based on metal extrusion and solid-state bonding, *Int J Adv Manuf Technol*, Volume 105, 2019, Pages 2523–2530, 10.1007/s00170-019-04385-8

23. J.P. Oliveira, A.D. LaLonde, J. Ma, Processing parameters in laser powder bed fusion metal additive manufacturing, *Materials & Design*, Volume 193, 2020, 108762, 10.1016/j.matdes.2020.108762

24. P.K. Gokuldoss, S. Kolla, J. Eckert, Additive manufacturing processes: Selective laser melting, electron beam melting and binder jetting—Selection guidelines, *Materials*, Volume 10, 2017, Page 672, 10.3390/ma10060672

25. Tarun Bhardwaj, Mukul Shukla, C.P. Paul, K.S. Bindra, Direct energy deposition – Laser additive manufacturing of titanium-molybdenum alloy: Parametric studies, microstructure and mechanical properties, *Journal of Alloys and Compounds*, Volume 787, 2019, Pages 1238–1248, 10.1016/j.jallcom.2019.02.121

26. Prahar M. Bhatt, Ariyan M. Kabir, Max Peralta, Hugh A. Bruck, Satyandra K. Gupta, A robotic cell for performing sheet lamination-based additive manufacturing, *Additive Manufacturing*, Volume 27, 2019, Pages 278–289, 10.1016/j.addma.2019.02.002

27. Bintao Wu, Zengxi Pan, Donghong Ding, Dominic Cuiuri, Huijun Li, Jing Xu, John Norrish, A review of the wire arc additive manufacturing of metals: Properties, defects and quality improvement, *Journal of Manufacturing Processes*, Volume 35, 2018, Pages 127–139, 10.1016/j.jmapro.2018.08.001

28. M. Mehrpouya, A. Gisario, A. Rahimzadeh et al., A prediction model for finding the optimal laser parameters in additive manufacturing of NiTi shape memory alloy, *Int J Adv Manuf Technol*, Volume 105, 2019, Pages 4691–4699, 10.1007/s00170-019-04596-z

29. G.S. Altug-Peduk, S. Dilibal, O. Harrysson et al., Characterization of Ni–Ti alloy powders for use in additive manufacturing, *Russ. J. Non-ferrous Metals*, Volume 59, 2018, Pages 433–439, 10.3103/S106782121804003X

30. Ali N. Alagha, Shahadat Hussain, Wael Zaki, Additive manufacturing of shape memory alloys: A review with emphasis on powder bed systems, *Materials & Design*, Volume 204, 2021, 109654, 10.1016/j.matdes.2021.109654

31. Wenlong Lu, Wenzheng Zhai, Jian Wang, Xiaojun Liu, Liping Zhou, Ahmed Mohamed Mahmoud Ibrahim, Xiaochun Li, Dong Lin, Y. Morris Wang, Additive manufacturing of isotropic-grained, high-strength and high-ductility copper alloys, *Additive Manufacturing*, Volume 38, 2021, 101751, 10.1016/j.addma.2020.101751

32. T. Gustmann, A. Neves, U. Kühn, P. Gargarella, C.S. Kiminami, C. Bolfarini, J. Eckert, S. Pauly, Influence of processing parameters on the fabrication of a Cu-Al-Ni-Mn shape-memory alloy by selective laser melting, *Additive Manufacturing*, Volume 11, 2016, Pages 23–31, ISSN 2214-8604, 10.1016/j.addma.2016.04.003

33. S. Sekar, S. Brown, A. Cockburn, P. Iyamperumal Anand, C.P. Paul, W. O'Neill, Investigating the various properties of cold sprayed CuAlNi shape memory alloys developed by post annealing process, *Proceedings of the Institution of Mechanical Engineers, Part B: Journal of Engineering Manufacture*, Volume 235, Issue 4, 2021, Pages 663–672, 10.1177/0954405420967867

34. Matthew P. Caputo, Ami E. Berkowitz, Andrew Armstrong, Peter Müllner, C. Virgil Solomon, 4D printing of net shape parts made from Ni-Mn-Ga magnetic shape-memory alloys, *Additive Manufacturing*, Volume 21, 2018, Pages 579–588, 10.1016/j.addma.2018.03.028

35. A. Mitchell, U. Lafont, M. Hołyńska, C. Semprimoschnig, Additive manufacturing – A review of 4D printing and future applications, *Additive Manufacturing*, Volume 24, 2018, Pages 606–626, 10.1016/j.addma.2018.10.038

36. Q. Portella, M. Chemkhi, D. Retraint, Influence of surface mechanical attrition treatment (SMAT) post-treatment on microstructural, mechanical and tensile behaviour of additive manufactured AISI 316L, *Materials Characterization*, Volume 167, 2020, 110463, 10.1016/j.matchar.2020.110463

37. Mohammad H. Elahinia, Mahdi Hashemi, Majid Tabesh, Sarit B. Bhaduri, Manufacturing and processing of NiTi implants: A review, *Progress in Materials Science*, Volume 57, Issue 5, 2012, Pages 911–946, 10.1016/j.pmatsci.2011.11.001

38. Ch Srinivasa Rakesh, A. Raja, Priyanka Nadig, R. Jayaganthan, N.J. Vasa, Influence of working environment and built orientation on the tensile properties of selective laser melted AlSi10Mg alloy, *Materials Science and Engineering: A*, Volume 750, 2019, Pages 141–151, 10.1016/j.msea.2019.01.103

39. S. Shiva, N. Yadaiah, I.A. Palani, C.P. Paul, K.S. Bindra, Thermo mechanical analyses and characterizations of TiNiCu shape memory alloy structures developed by laser additive manufacturing, *Journal of Manufacturing Processes*, Volume 48, 2019, Pages 98–109, 10.1016/j.jmapro.2019.11.003

40. Beth A. Bimber, Reginald F. Hamilton, Jayme Keist, Todd A. Palmer, Anisotropic microstructure and superelasticity of additive manufactured NiTi alloy bulk builds using laser directed energy deposition, *Materials Science and Engineering: A*, Volume 674, 2016, Pages 125–134, 10.1016/j.msea.2016.07.059

41. Jun Wang, Zengxi Pan, Yangfan Wang, Long Wang, Lihong Su, Dominic Cuiuri, Yuhong Zhao, Huijun Li, Evolution of crystallographic orientation, precipitation, phase transformation and mechanical properties realized by enhancing deposition current for dual-wire arc additive manufactured Ni-rich NiTi alloy, *Additive Manufacturing*, Volume 34, 2020, 101240, 10.1016/j.addma.2020.101240

42. S.H. Huang, P. Liu, A. Mokasdar et al., Additive manufacturing and its societal impact: A literature review, *Int J Adv Manuf Technol*, Volume 67, 2013, Pages 1191–1203, 10.1007/s00170-012-4558-5

43. I. Gibson, D. Rosen, B. Stucker, Directed energy deposition processes, *Additive Manufacturing Technologies*. Springer, New York, 2015, pp. 498, 10.1007/978-1-4939-2113-3_10

44. A.D. Iams, M.Z. Gao, A. Shetty, T.A. Palmer, Influence of particle size on powder rheology and effects on mass flow during directed energy deposition additive manufacturing, *Powder Technology*, Volume 396, Part A, 2022, Pages 316–326, 10.1016/j.powtec.2021.10.059

45. Swee Leong Sing, Jia An, Wai Yee Yeong, Florencia Edith Wiria, Laser and electron-beam powder-bed additive manufacturing of metallic implants: A review on processes, *Materials and Designs*, 2015, Volume34, Issue3, March 2016, Pages 369–385, 10.1002/jor.23075

16 Fused Deposition Modeling (FDM) and Nano-Fillers Impact on Shape Memory Properties of 3D Printed Thermoplastic Polyurethane (TPU) Filament

Jigar Patdiya and
Balasubramanian Kandasubramanian
Rapid Prototyping Laboratory, Department of Metallurgical and Materials Engineering, Defence Institute of Advanced Technology (DU), Ministry of Defence, Pune, India

CONTENTS

DOI: 10.1201/9781003189404-16

16.1 INTRODUCTION

The technique of rapid prototyping, generally termed as additive manufacturing or 3D printing, has taken a fixed place in laboratory and industrial developments. The rapid spread of 3D printing was observed in recent years in vital regions, particularly in the instantaneous design, prototypes, evolution, as well as manufacture geometry, as customer needs were unfeasible by virtue of previously accessible methods (Belka and Bączek 2021). 3D printing captures tremendous advantages that apply to numerous commercial market applications, such as in biomedical, aerospace, electronics, textile, mechanical device, defence, retail, and many others (Rastogi and Kandasubramanian 2019).

Recently, 3D printing had been experimented with, producing magnificent breakthroughs in technology and application. 3D printing originates a fourth level of resolution as it incorporates stimuli-responsive materials for dynamic structure fabrication, known as 4D printing. Additive manufacturing serves as the backbone in the fourth-dimension printing technique, contributing to the elaborate shape-morphing parts when triggered by stimuli medium like thermal actuation, ohmic actuation, magnetic, light, water, or natural (biological) provocation (Patadiya et al. 2021; Subash and Kandasubramanian 2020).

Third-dimension fabrication applied for printing Centimetre (cm) level to micrometre (μm) level of wide range in various input materials such as liquid-based, powder-based, or solid wire-based. Among all the processes, the most achievable, affordable, and easy-to-use technology was fused deposition modeling (FDM) (Montero, Roundy, and Odell 2001). Besides inert materials like natural PLA and ABS, FDM were also utilized for shape memory polyurethane (SMPU), shape memory polylactic acid (SPLA), and polycaprolactone (PCL) functioning as shape alteration. Shape memory polyurethane employed to FDM for stimuli-responsive object printing due to its enormous strain, flexibility, low cost, high shape recovery ratio (~98%), and utilization as foam, coating elastomer, adhesive, and rubber. Polyurethanes are constructed with segments of rigid and soft molecules containing polyether or polyester group, diisocyanates, and diols engendering urethane linkage of $-NHCOO-$, its relative proportion governs crosslinking as well as thermal stability. Soft segments aim for shape fixation, and hard segments for shape recovery to permanent geometry (Lendlein and Kelch 2002). Though SMPU inherits thermal and mechanical performance, reinforcement can induce other distinct features, such as multi stimuli,` to be availed for various applications in aerospace, medicine, soft robotics and actuator, space with the non-contact mode of actuation, automotive, and also in fashionwear (textiles) for body-temperature sensing (Lendlein and Kelch 2002).

Different fillers and their interaction with 3D-printed filament materials significantly impact the final properties listed below. Foreign materials in the primary matrix cause indirect heating or activation stimuli because of several advantages, like local and uniform heating, remote control, and convenience (Patdiya and Kandasubramanian 2021).

1. Carbon-based filters like carbon black (CB) acts as an energy shifter by absorbing electromagnetic radiation, generating local joule heating, and reducing resistance. CB filler is spherical and well distributed in matrix, resulting in increased recovery time due to the softening effect (E. Wang et al. 2019; Yu, Ge, and Qi 2014), while CNT strengthens the mechanical and electrical properties better than the CB due to its high aspect ratio. CNT also improves thermal stability and mechanical strength on curing with UV radiations at the end. CNT builds strong interphase interaction in a polymer matrix.

2. Metal and metal oxides are other widely used fillers for multi-stimulation in SMPCs. In general, metal and metal oxide have super magnetic and electrical conductivity; for example, Fe, Fe_2O_3, Ni, Cu, and some metals are biodegradable and biocompatible, like Au, Ag. Fe_3O_4 is added to electric, and magnetic energy conversion occurs, decreasing break elongation and recovery in the free recovery cycle (Xie et al. 2020).

3. SiO_2, glass, and SiC have a natural 3D irregular network structure that drastically modifies shape memory behavior and mechanical characteristics. SiO_2 nanoparticles in polymer base material matrix observed that a high aspect ratio of nanoparticles leads to a well-dispersed arrangement and non-crystalline nature of filler strengthen toughness, modulus (up to three times), and shape fixation and shape recovery approximate 100% even after the tenth cycle (E. Wang et al. 2019; Liu et al. 2018).

This chapter is deliberate to point out current achievements in the evolution of functional 3D printing filament material of shape memory polyurethane (SMPU) and filler materials in SMPU; the additive-processing effect on shape memory properties were proposed. Time-piloted alteration of thermoplastic polyurethane was utilized to formulate composite from additives like MWCNT, nano clay, activated charcoal, and polypyrrol to investigate material properties change and shape memory properties of the pre-printing wire. These incorporated reinforcements were tested for FTIR to verify the characteristic vibrations, TGA to witness thermal stability. At the same time, for shape memory properties characterization of extrudate filaments, a spring force assembly was designed and fabricated, followed by the involute theory of angular measurement for observing percentage fixation of structure and percentage recovery of structure; it has more incredible application in the medical sector (as for stimuli-responsive stitching), biomimetic application (as soft tendril mimetics). Further experiments are extended for scrutinizing the consequence of 3D printing on geometry-memorizing parameters of SMPU, utilizing by spring force assembly followed by involute theory to utilize SMPU as a timely morphing 3D printing structure.

16.2 EXPERIMENTAL METHODOLOGY

16.2.1 MATERIALS

Economically available transparent SMPU (shape memory polyurethane) evincing a phase-transition temperature of 55°C and filament-printing temperature range was 195°C–210°C, obtained from KYORAKU Co. Ltd., Japan. Industrial-grade powder of activated charcoal with ≤ 1 wt.% solubility in water was bought from Merck Specialities Pvt. Ltd., India. Powder of polypyrrole (H-$(C_4H_2NH)_n$-H) with a 10–50 S/cm conductivity was purchased from Sigma Aldrich, India. Nanoclay, whose powder surface has been modified and comprised of aminopropyltriethoxysilane in 0.5 wt.% to 5 wt.%, octadecyl amine in 15 wt.% to 35 wt.%, was acquired from Sigma Aldrich, India. Particle size of a −300 mesh and a purity of 99% graphite powder was obtained from Alfa Aesar, U.S. Multi-walled carbon nanotube (MWCNT) embodied −OH functionalities on its outer surface and radial dimensions in the range from 10 nm to 20 nm on its external. In contrast, longitudinal dimensions of 10 μm to 30 μm) were procured. Reinforcement fillers applied for SMPU-composite were incorporated without any additional surface modification, i.e., used as received. Spring-of-force applying assembly was purchased from the local spring industry with 17 mm and 1.5 mm outer diameter and wire diameter, respectively, free length mark as 50 mm, and the number of active coils was six. The assembly was fabricated using wood and polytetrafluoroethylene (PTFE), commonly known as Teflon, as it requires to work as high-temperature water.

16.2.2 METHODS

16.2.2.1 Composite Processing

Homogenizing fillers in polymer matrix fabricate the composite of SMPU initiated with compounding in a mini single screw extruder (LabTech engineering corporation, model no. LTE20–40) with an output capacity of 20 kg/h, diameter of screw was 20 mm, and ratio of L/D was 40:1. Extruder temperature was maintained at 190°C, 190°C, 195°C and 195°C in the feeding zone, melting zone, melt-conveying zone, and die, respectively. A fillers material, polypyrrole, MWCNT, charcoal, and nanoclay in weight percentages of 2 with SMPU pellets, was streamed into an extruder at a pace equivalent to 18 revolution/min, retard enough to furnish mixing (Figure 16.1). The resulting output was in uniform diameter filament, which was collected for preparing the sample via 3D printing.

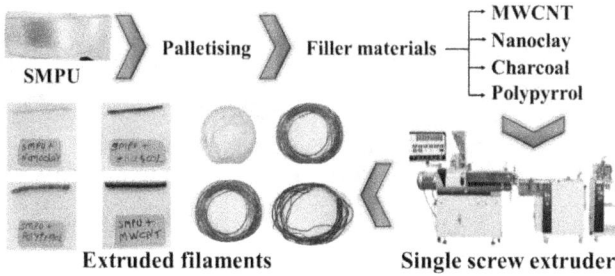

FIGURE 16.1 Procedure of producing SMPU composite filaments of different fillers using a single screw extruder.

16.2.2.2 3D Printing

The strip sample of SMPU with the dimension 135 × 3 × 1.75 mm (L × W × T) was fabricated using fused deposition modeling (Fab × pro). Before this sample design in Solidworks software, successively design was imported to Creo software for slicing and coversion into G-code and M-Code, as per the standard procedure of 3D printing. SMPU filament was kept at 60°C for 2 h prior to extrusion in a hot air oven to eliminate any adhered moisture or volatile particles. Pre-printing aids, for instance, bed temperature (50°C), nozzle temperature (210°C), z-axis position (safe distance between nozzle and platform (2 mm)), nozzle speed (20 mm/s), layer resolution (200 μm), infill density (100%) was optimized before commanding the printer for dispensing to manufacture desire geometry, as portrayed in Figure 16.2.

16.2.2.3 Force Assembly Fabrication

As for shape, as mentioned earlier, memory properties evolute by ratio of shape fixation and ratio shape recovery of the material. To fulfill one of the research aims,

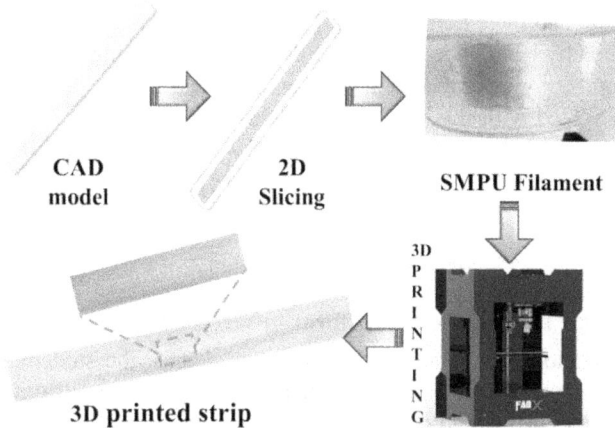

FIGURE 16.2 3D printing approach to fabricate SMPU strip with the help of CAD model and slicer.

(a)

(b)

(c)

(d)

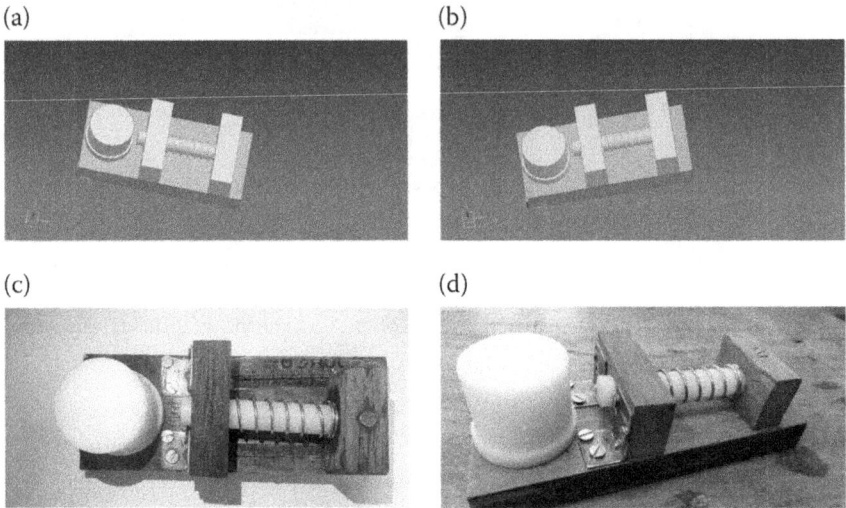

FIGURE 16.3 The spring-force assembly (a) CAD model isometric view (b) CAD model top view (c) fabricated assembly top view (d) fabricated isometric assembly view.

check shape memory properties of extrudate filaments of SMPU composite and simple SMPU and utilize for further study of analyzing 3D printing effect on SMPU. A shape fixation and recovery are obtained by programming components, including heating, loading, constrain cooling, and unloading for shape fixity while reheating the shape's recovery. A spring-force assembly was designed and fabricated to propose a compression load on bending pre-printing wire and post-printing strip samples during high-temperature loading and constrain cooling. The CAD model of the assembly was designed by NX essential design software and fabricated with mechanical components like drilling, grinding, etc., as represented in Figure 16.3. A wire and strip wound on PTFE cylinder (white vertical cylinder) of diameter 40mm, and the end was fixed by the compression load produced when the spring is pushed by hand. The actual load during programming was a load required to hold both ends of wire and strip in a close circle, which is equal to applied spring force and calculated with the help of (Equation 16.1).

$$\text{Applied force}(F) = \text{Spring displacement}(x) \times \text{Spring stiffness}(K) \quad (16.1)$$

16.2.3 Apparatus for Characterization

16.2.3.1 ATR-FTIR (Fourier Transform Infrared Spectroscopy)

This characterization technique was performed to deduce the stretching and vibrational bonding present among polymer chains. The spectra (transmittance vs. wavenumber) of neat SMPU and composited composition (charcoal + SMPU, nanoclay + SMPU, MWCNT + SMPU, and polypyrrol + SMPU) were obtained from ATR-FTIR spectrometer in range 4000 cm^{-1} to 500 cm^{-1} supplied by (Bruker).

16.2.3.2   TGA (Thermogravimetric Analysis)

This device involve thermic uniformity of filaments based on weight loss witnessed during a temperature rise at a controlled rate. TGA set-up obtained from Perkin Elmer was employed, and samples, both pure and composite, were provided with progressive thermal energy commencing from 30°C, with the final temperature being 800°C. The nitrogen gas supply suspends unwanted reactions with oxygen with rate of 20 mL/min, while the rate of temperature rising was kept at 20°C/min; these environmental testing conditions were maintained for the samples.

16.2.4   SHAPE MEMORY PROPERTIES

16.2.4.1   Compression Test for Spring of Force Assembly

The compression test performs for knowing the stiffness of the spring – compression force on SMPU and SMPU composite using spring force as per equation 16.1. Compression tests are performed using the universal testing machine (UTM) purchasing from Instron. UTM has a capacity of ± 100 kN axial loading and axial stoke of 150mm. Initially, cold molding was done on both the end of the spring for flatness and straightforward loading, as shown in Figure 16.4. Testing was conducted with 10 kg load cell at the rate of 0.5 mm/min to compression of 20 mm as 80% displacement of total length.

16.2.4.2   Shape Memory Properties Evolution

SM characteristics data of filament and strip samples were acquired at 60°C temperature (exceeding Tg of specimens) in DI water to transmit actuation energy from a stimulus (thermal) uniformly across the surfaces with consequential faster actuation. To perceive shape transformation, constructed entities were energized and stabilized at 60°C water for 2 min, followed by angular deformation to 360°C (programming angle-θp) under stress by a spring assembly. Patterns were expeditiously relocated to room temperature (RT), below Tg, water to fix the sample

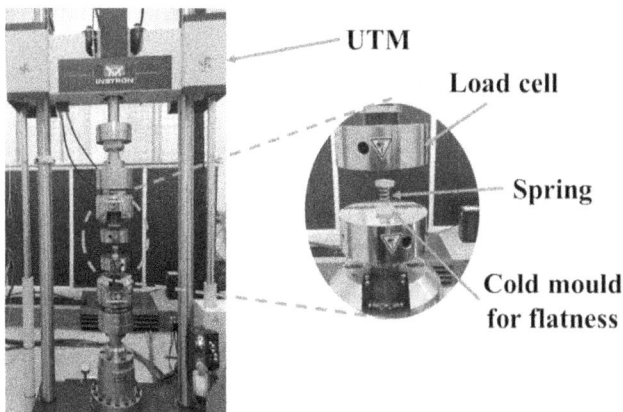

FIGURE 16.4   Representation of UTM (universal testing machine) employed a load cell of 10 kg, spring sample, and cold molding for flatness.

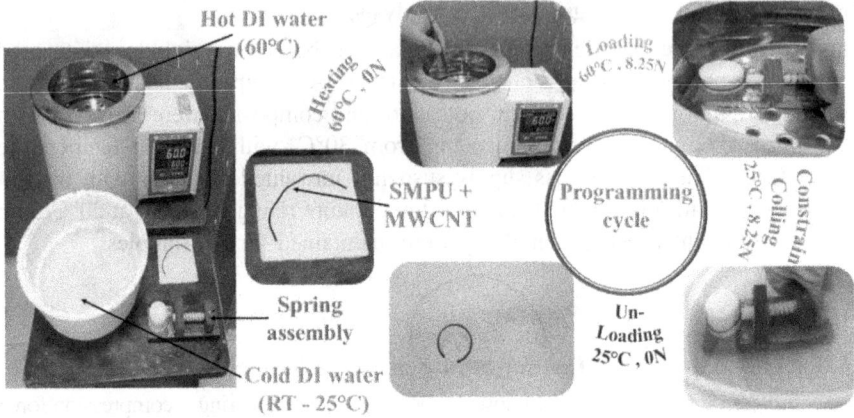

FIGURE 16.5 Representation of programming cycle for SMPU and SMPU composite starting with heating to 60°, then force apply as 8.25N followed by forcefully cooling and unloading.

stressed profile, followed by its release. Composite designs were preserved in the same environment (RT) for 10 min, and θ_t (Figure 16.5) was calculated owing to un-actuated angular recovery. Strain release in terms of angle pertaining to 10 min was calculated, and corresponding shape fixity was determined. Lastly, entities were subjected to 60°C hydrothermal actuations to determine recovery angle- θr, thereby determining, shape recovery and shape fixity (Equations 16.2 & 16.3). Figure 16.6 portrayed an example diagram for various angles: programming angle, fixation angle, and angle after recovery, used in (Equations 16.2 and 16.3).

$$\text{Percentage fixation of structure} = R_f = \frac{360° - \theta_t}{360°} \times 100\% \qquad (16.2)$$

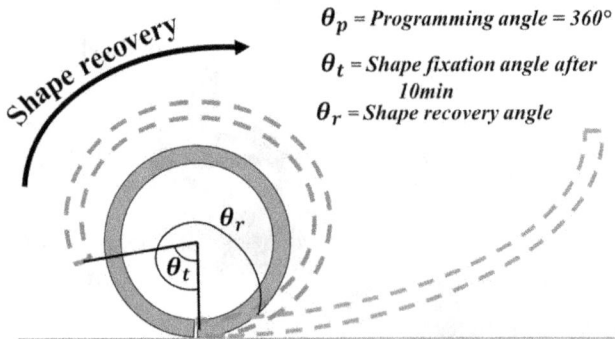

θ_p = Programming angle = 360°

θ_t = Shape fixation angle after 10min

θ_r = Shape recovery angle

FIGURE 16.6 Illustration of shape bending to the programming angle (θp = 360°) along with shape fixation angle after 10 min (θ_t) and shape recovery angle (θ_r) during recovery process when actuated by stimuli.

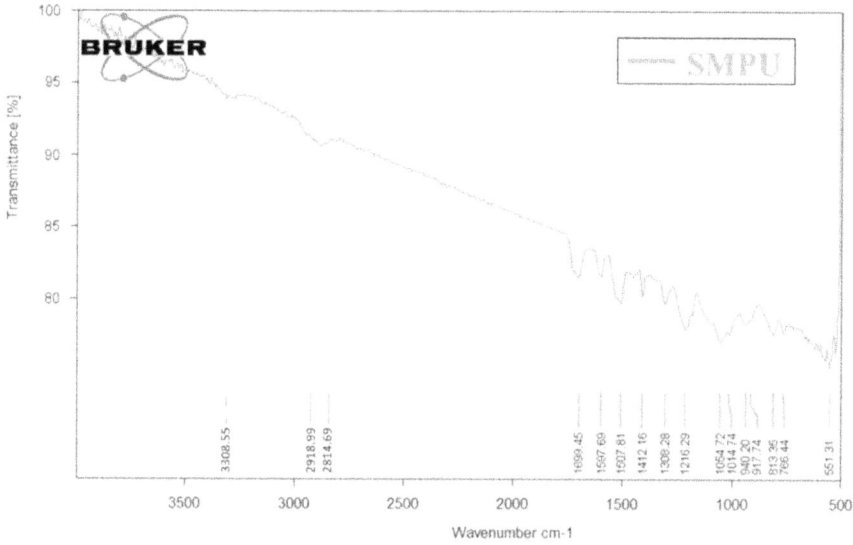

FIGURE 16.7 FTIR graphical representation of pure SMPU and its associate peaks.

$$\text{Percentage recovery of structure} = R_r = \frac{\theta_r}{360°} \times 100\% \qquad (16.3)$$

16.3 RESULTS AND DISCUSSIONS

16.3.1 FTIR INVESTIGATION

The result peaks of pure SMPU depict in Figure 16.7 and its composites determined by ATR-FTIR containing the reinforcement of 2 wt.%. Spectra of neat SMPU displayed the presence of 3309 cm^{-1} and 3360 cm^{-1} assigned to N-H bond stretching, 1307 cm^{-1}, 1519 cm^{-1}, and 1411 cm^{-1} for bending of N-H. Furthermore, for C-H bonding, peaks were noted at 2920 cm^{-1} and 2869 cm^{-1} (Chen, Xia, and Ni 2018; Khan et al. 2015; Yanilmaz et al. 2012; Vargas et al. 2018; Fonseca et al. 2013; Cervantes-Uc et al. 2009). Linkage in the urethane group (-NHCOO-) was identified at 1706 cm^{-1} which signified C=O bonding defines the existence of dipole-dipole interaction and hydrogen-bonding with –NH group from adjacent chains (B. Yang et al. 2006). However, validation of ether linkages in SMPU observed by absorption peaks at 1105 cm^{-1} and 1213 cm^{-1} (Vargas et al. 2018). The existence of 3360 cm^{-1} and 1411 cm^{-1} spectra also determine –OH group and C-C bond stretching, individually, 1054 cm^{-1} for stretching of C-O-C linkages, also peak observed at 1213 cm^{-1} for stretching of C-O bonds in SMPU (Chen, Xia, and Ni 2018; Yanilmaz et al. 2012). Reinforcements containing functional groups were added to matrix, which probably instigated interactions with the matrix polymer chains, thereby tailoring the intensification of infrared transmittance peaks and shifting in wave number. Those characteristic peaks also evidenced hydrogen

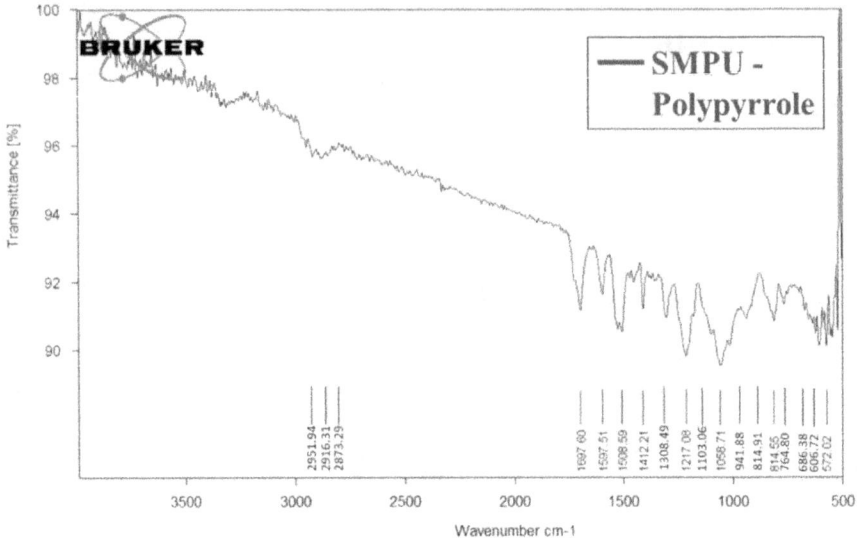

FIGURE 16.8 FTIR peaks of Polypyrrole composite.

bonding (while the magnitude of shifting defines its strength) and possibly Van der Waals forces (Yanilmaz et al. 2012; B. Yang et al. 2006).

Consequently, slight shifting was noticed in the absorption peaks of composite of polypyrrole + SMPU (Figure 16.8) for example $1307 \rightarrow 1309$ cm^{-1}, $1706 \rightarrow 1710$ cm^{-1}, $2920 \rightarrow 2916$ cm^{-1}, and $3309 \rightarrow 3301$ cm^{-1}. Peaks of polypyrrole ring observed at 1376 cm^{-1}, 1054 cm^{-1}, 942 cm^{-1}, and 764 cm^{-1} in which 1054 cm^{-1} and 764 cm^{-1} portrayed in-plane and out-of-plane stretching of C-H bond vibrations (H. Wang et al. 2015; Saeed et al. 2020). Most likely, 1411 cm^{-1}, 1457 cm$^{-1,}$ and 1515 cm^{-1} infrared absorption indicate characteristic of C-C bond stretching, C-N bond symmetrically, and C=C bond anti-symmetrically in the pyrrole ring, respectively (H. Wang et al. 2015). These results were also verified by FT-IR results for polypyrrole powder (pellets in conjunction with KBr) as per literature, which probably interpreted to peaks 1549 cm^{-1} & 1469 cm^{-1} $\rightarrow$ stretching of C=C-C bond characteristics of π-electrons delocalization; 1290 cm$^{-1} \rightarrow$ stretching of C-N bond; 1191 cm^{-1} $\rightarrow$ in-plane bending of C-H bond; 1038 cm^{-1} $\rightarrow$ bending of C-H or in-plane N-H vibration; 781cm^{-1} $\rightarrow$ out-of-plane bending of C-H bond; 964 cm^{-1} & 677 cm^{-1} $\rightarrow$ C-C bond of pyrrole ring out-of-plane vibration (Sattar, Kausar, and Siddiq 2016; Merlini et al. 2017). For abovesaid interaction between reinforcement and matrix during composite formation, shifting in the peak absorption was witnessed.

For another reinforcement, i.e., nanoclay, FT-IR graph (pure nanoclay) revealed broad peaks fluctuating from 1000 cm^{-1} to 1040 cm^{-1} associated with the characteristic of Si-O bond stretching in structure. Additionally, it also represented vibrations of –OH stretching at 3633 cm^{-1} in octahedral and tetrahedral position in the layered structure, 916 cm^{-1} for presence of bond Al-Al-OH, 797 cm^{-1} for silica, 521 cm^{-1} derived from Al-O-Si bond, and 469 cm^{-1} owing to bond Si-O-Si. Furthermore, bending and stretching spectra at 1613 cm^{-1} and 3425 cm^{-1}

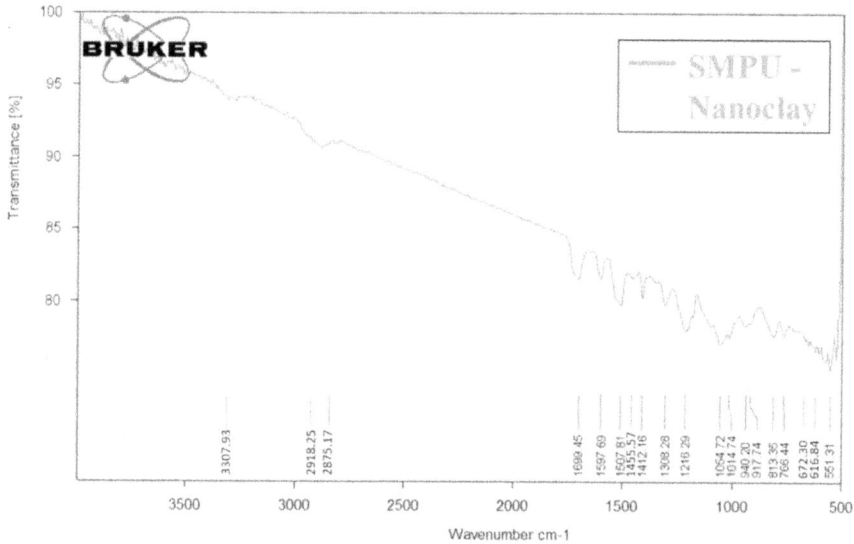

FIGURE 16.9 FTIR peaks graph of SMPU filled with Nanoclay.

respectively can be evidence for its water (H-O-H) existence (bending and stretching respectively) (Zacaroni et al. 2015; Ashhari and Sarabi 2017). Nevertheless, composite NC-SM-PU (Figure 16.9) probably manifested shifted absorptions at 3413 cm^{-1}, 815 cm^{-1}, and 546 cm^{-1} wave numbers attributable to stretching of water, silica structure, and Al-O-Si bond. In addition, it can also be presented that those peak intensities were perhaps decreasing as a consequence of 32 newly formed nanoclay linkages with SMPU chains, except for Al-O-Si and silica, which demonstrated an increase.

FTIR spectra obtained for charcoal-SMPU composite displayed overlapping (with SMPU) characteristic stretching peak at 3417 cm^{-1} for –OH functional group, may be derived from aliphatic or aromatic alcohol, 2916 cm^{-1} & shoulder spectra at 2950–2960 cm^{-1} for stretch of C-H, 1710 cm^{-1} for stretching of C=O group, and 762 cm^{-1}, 800 cm^{-1} could be contributed to C-H bond out of plane in aromatic structure bending (Rajak et al. 2018; Zhao et al. 2019) (Figure 16.10). At the same time, other peaks of stretching vibrations associated with C=C, C-O, etc., manifested small interaction of fillers with a matrix.

Vibrations probably characterized for -OH stretching in enol (C=C-OH) –OH bond, was observed stretching at 1598 cm$^{-1,}$ and also an enhanced intensity at 3413 cm^{-1} concerning SMPU could also represent hydroxyl functionalization in MWCNT (Figure 16.11) (D. Yang et al. 2008). However, as with the above-described composites, other carbon bonds evinced overlapping absorptions and exhibited a slight shift in the absorption of incoming energy by bonds that channelize it into bending, stretching, and other vibrations.

Available FTIR results on neat material and composites displayed similarity (though shifting in peak vibration absorption in composites was witnessed) between the graphs with only modulations in intensities of peaks, which indicate a minor

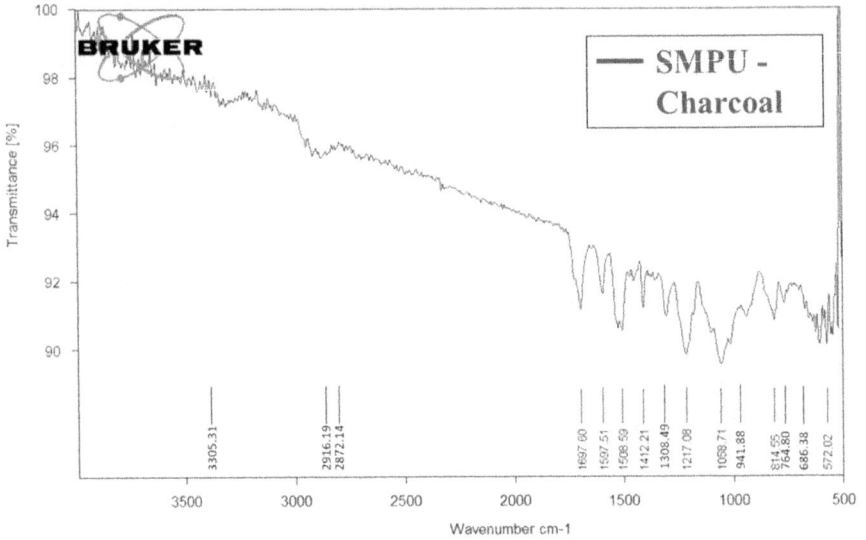

FIGURE 16.10 FTIR transmission peaks for activated charcoal + SMPU composite.

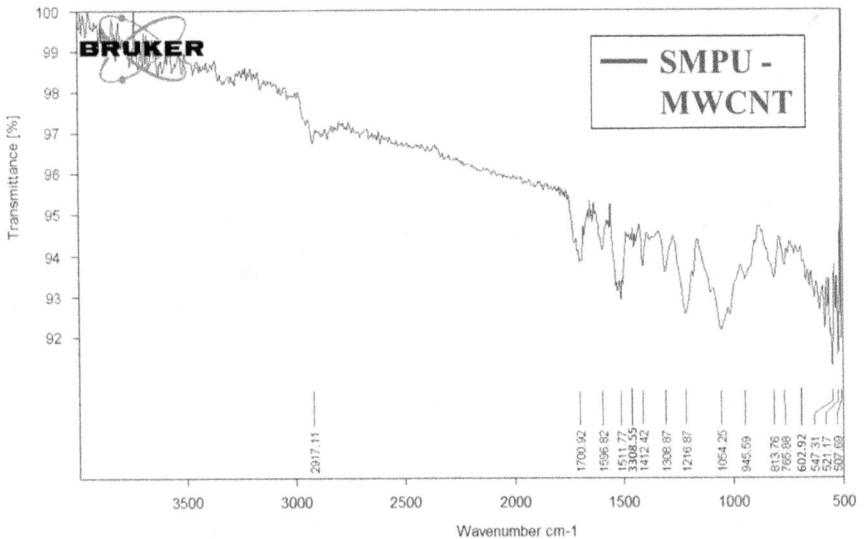

FIGURE 16.11 Depiction of characterizing peak of -OH functionalize MWCNT filled SMPU composite.

influence of reinforcement on the SMPU matrix. Another possibility for this similitude can be attributed to the overshadowing of reinforcement peaks by composites due to their absorption strength (Zhang et al. 2016). Combined FTIR distinctive peaks and effects of respective reinforcements were brief in Table 16.1.

TABLE 16.1

Outline of FTIR Absorption Wavelengths according to Inter-Component Bonds

Bonds	Wavenumber (Cm^{-1})	References
N-H	3309, 3360, 1307, 1518, 1411	(Chen, Xia, and Ni 2018; Khan et al. 2015; Yanilmaz et al. 2012; Vargas et al.
C-H	2920, 2950, 2869, 1191, 781	2018; Fonseca et al. 2013; Cervantes-Uc et al. 2009)(Sattar, Kausar, and Siddiq 2016; Merlini et al. 2017)
C=O in urethane linkage (NHCOO-)	1706	(B. Yang et al. 2006)
Ether linkage in SMPU	1105, 1213	(Vargas et al. 2018)
O-H	3360	(Yanilmaz et al. 2012; Chen, Xia, and Ni
C-C	1411	2018)
C-O-C	1054	
C-O	1214	
Polypyrrole ring	1376, 942, 764	(H. Wang et al. 2015; Saeed et al. 2020;
C-N	1457, 1290	Sattar, Kausar, and Siddiq 2016; Merlini
C=C	1515	et al. 2017)
C=C-C	1549, 1469	(Sattar, Kausar, and Siddiq 2016; Merlini
C-C in pyrrole ring	964, 677	et al. 2017)
Si-O	1000–1040	(Zacaroni et al. 2015; Ashhari and Sarabi
O-H in octahedral & tetrahedral position in clay	3633	2017)
Al-O-Si	521	
Al-Al-OH	916	
Silica	797	
Si-O-Si	469	
H-O-H	3425, 1613	
C=C in benzene	1565, 1582–1588, 1606	(Tian et al. 2017; Anadão, Sato, and
C=O	1722	Wiebeck 2018; Kim et al. 2018)

16.3.2 TGA ANALYSIS

A TGA tool captured degradation demeanour of neat SMPU and its composites having a filler weight % of 2, which delineated their thermal stability. TGA results displayed those reinforcements had declined the rate of degradation by changing the offset temperature to higher magnitudes. Decomposition temperature at onset in prepared samples was determined with the thermogram for similarity and conflicts the effects of additives in a composite. In neat SMPU, DT was approximately 270.3°C; however, DT for MWCNT and polypyrrole composite witnessed increments in onset to 296.2°C and 290.9°C, respectively, while charcoal and nanoclay

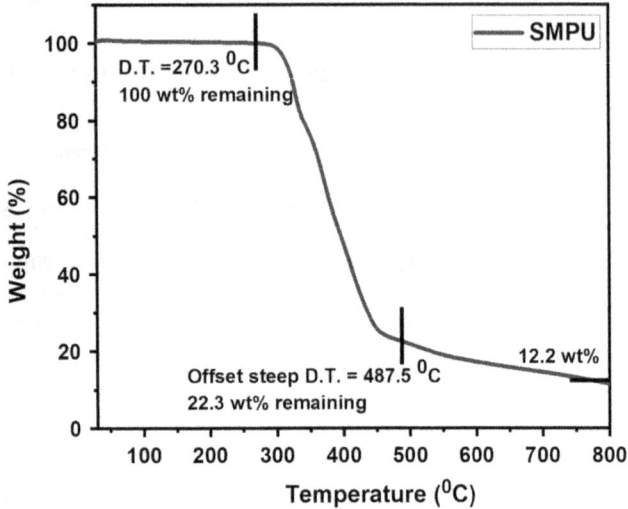

FIGURE 16.12 TGA weight loss and degradation representation for SMPU.

had been relocated to 280.9°C and 299.8°C respectively. SMPU steep degradation was continual at ~487.5°C and preserved 12.2% of its original weight; however, composites had retarded degradation, thus ameliorating offset temperature to close to 563.2°C (Nanoclay + SMPU), 535.1°C (MWCNT + SMPU), 558.9°C (charcoal + SMPU), and 521.2°C (polypyrrole + SMPU) for employed reinforcements (Figure 16.12, 16.13, & 16.14).

Furthermore, composites proclaimed decrease in 40% weight loss at offset more likely on account of the impediment engendered due to interaction with the matrix during compounding. Degradation of neat SMPU displayed two inflection regions, one being at ~487.5°C from 270.3°C, which witnessed a meteoric transformation in weight stability with elevating temperature owing to hard segment breakdown. The second being a gradual change from 511°C, which may be attributed to the onset of soft segment breakdown in SMPU (Pielichowska, Bieda, and Szatkowski 2016). Corresponding to SMPU, polypyrrole + SMPU had also portrayed a swift decrement beyond 290.9°C, which stabilized at ~521.2°C and a gradual inflection region commencing from 578°C (improved for neat matrix) but preserved reduce the amount, 10.9% of its weight at 730°C in contrast to SMPU (12.2 wt.%). Steady change after ~578°C can be ascribed to the combined effect of soft segment degradation and pyrrole ring of polypyrrole, while an abrupt decrease was accredited to hard segments of SMPU (He et al. 2018). Nevertheless, residual weights of charcoal + SMPU and MWCNT + SMPU composites were 12.6 wt.% and 11.6 wt.% at a maximum temperature of 730°C with thereby validating the enhancement in thermal stability pertaining to carbon material along with buildout of interaction between matrix-reinforcement and barrier effect. However, NC-SM-PU had augmented weight loss in the composite to ~90% at 730°C. TGA analysis revealed amelioration in temperature stability of MWCNT and polypyrrole composite at onset might be accredited to hydrogen bonding from –OH group in MWCNT and

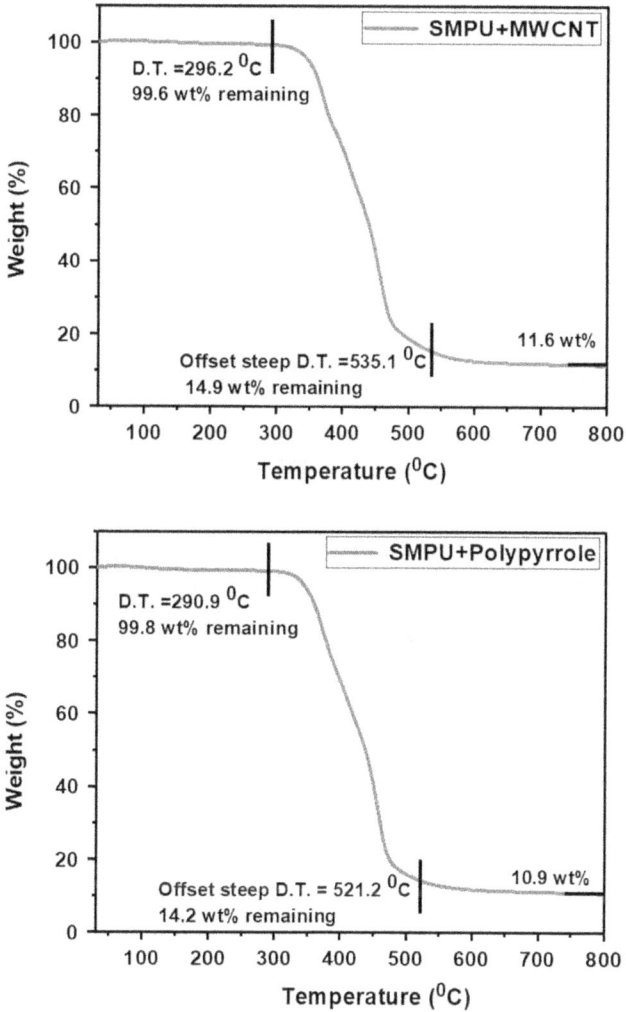

FIGURE 16.13 Weight loss and thermal stability depiction of SMPU + MWCNT and SMPU + polypyrrole.

–NH group in pyrrole. However, all composites had delayed the weight loss by degradation compared to pure SMPU due to interference with neat polymer chains (barrier effect), thus, compatible with the 3D printing temperature range for designing an entity.

16.3.3 Spring Force Calculation

A compression test was performed to estimate spring stiffness for the utilization in force provided into the samples of composite filaments and pure SMPU in filament

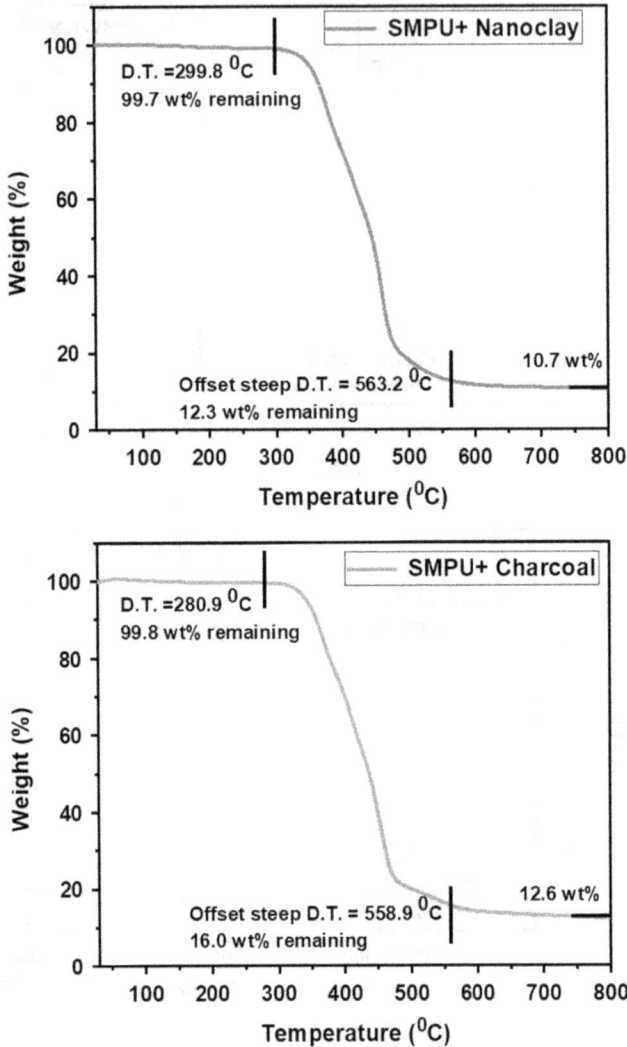

FIGURE 16.14 Nanoclay and charcoal composite displaying the thermal stability and weight Loss in TGA graph.

form and 3D strip form during loading of the programming cycle. As per Figure 16.15, spring displacement was 11mm, and spring stiffness was 0.75 N/mm observed. Spring force was calculated using (Equation 16.1),

$$\text{Applied force}(F) = \text{Spring displacement (x)} \times \text{Spring stiffness}(K) = 11 \times 0.75$$

$$= 8.25 \text{ N}$$

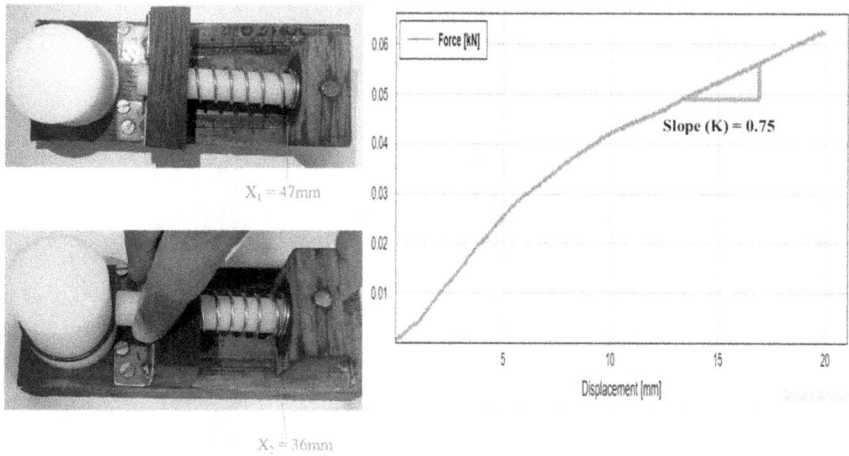

FIGURE 16.15 Portray of spring displacement, Initial position X_1, final position X_2, and spring stiffness result of slope of force vs. displacement graph.

16.3.4 Shape Memory Properties (SMP)

16.3.4.1 SMP Performance of SMPU Composite Filaments

Shape memory testing at temperatures above Tg was determined to understand the shape recovery and fixity of the printed entity. The process begins when thermal energy at 60°C softens the materials, and the programmed shape was affixed with the assistance of inherent hard segments by restricting chain mobility at the lower temperature of 25°C. Strain recovery at 25°C associated for 10 min was captured for determining time-dependent shape fixity. However, when the cooled entity was exposed to thermal energy, soft segments aids printed design to retrieve its prior geometry, bestowing shape recovery to the material. When discovery shapes the extrudate part's memory properties using equations 16.2 and 16.3, a critical problem occurs in finding θ_t and θ_r due to its opening shape from the circle. Involute curve theory was implemented to resolve this bottleneck issue in which an involute profile is obtained when a thread is opening from the circle. When a thread is wounding in circular geometry and opening, tracing the endpoint renders one curve known as the involute curve. When shape memory composite filaments bend in a circle and are provoked by stimuli, the author found that it follows the same theory and involute path as shown in Figure 16.16.

As represented in Figure 16.16, extrudate filaments obey the path of the involute curve, so it is possible to discover the shape memory angle of stressed structure and recovered structured when stimuli experiences medium change. Involute curves provided any point of the curve on the base circle when the normal of the point was drawn. With the help of a standard line of endpoints of fixed and recover shape easy drawing, give the point on the base circle, which is further utilized for measuring θ_t and θ_r (Figure 16.17).

For the experiment, two samples of composite filament were taken as per dimension and evolute θ_t and θ_r succeeding by R_f and R_r for both. Arithmetic means

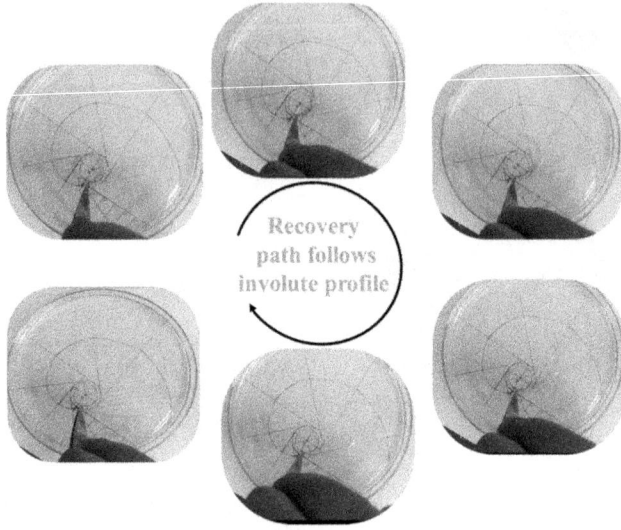

FIGURE 16.16 Display of filament recovery path which follows involute curve line.

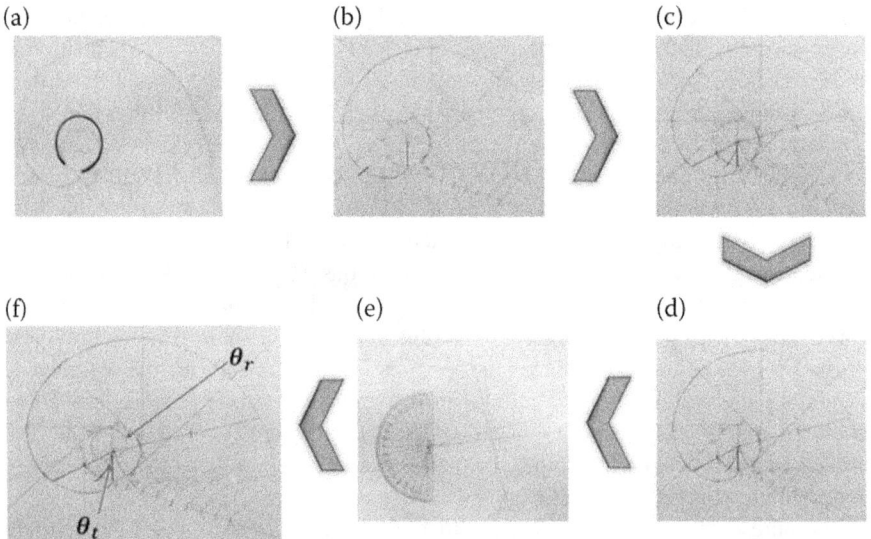

FIGURE 16.17 Calculation of recovery angle and fixity angle. (a) fixation wire on the involute path, (b) marking of endpoints of stressed sample and recovery sample, (c) marked points join with a center of base circle and arc made from the center of the base circle as a mid-point of the line taking as a fixed point, the (d) middle point of base circle connect with an intersecting point on the circumference, (e) measuring of angle with the help of protector, and (f) shape fixity angle and shape recovery angle.

TABLE 16.2
Shape Memory Property of Polymer and Its Composites Having 2 wt.% Reinforcement Actuated in the Heated Water for Three Shape Memory Cycles

Shape Memory Cycle	Shape Memory Properties	SMPU	MWCNT + SMPU	NC + SMPU	Ch + SMPU	Ppy + SMPU
1st	$\theta_t - 1$	5°	86°	32°	116°	139°
2nd	$\theta_t - 2$	8°	78°	31°	127°	113°
1st	$\theta_r - 1$	346°	285°	304°	306°	299°
2nd	$\theta_r - 2$	339°	293°	307°	315°	308°
1st	$R_f - 1$	98.6%	76.1%	91.1%	67.8%	61.4%
2nd	$R_f - 2$	97.8%	78.3%	91.1%	64.7%	68.6%
1st	$R_r - 1$	96.1%	79.2%	84.4%	85.0%	83.1%
2nd	$R_r - 2$	94.2%	81.4%	85.3%	87.5%	85.6%
Average	$\mathbf{R_f}$	**98.2%**	**77.2%**	**91.1%**	**66.2%**	**65.0%**
Average	$\mathbf{R_r}$	**95.15%**	**80.3%**	**84.8%**	**86.2%**	**84.35%**

θ_p = programming angle at 60°C = 360°, θ_t = shape fixation angle at 25°C after 10 min of cooling, θ_r = recovery angle at 60°C, R_f = percentage fixation of structure, and R_r = percentage recovery of structure, SMPU = shape memory polyurethane, MWCNT = multi-walled carbon nanotube, NC = nanoclay, Ppy = polypyrrole, and Ch = activated charcoal.

of all values were reported as an average value to minimize error. Two shape memory cycles were performed for each material (SMPU and SMPU composite), and results were summarized in Table 16.2, which displayed that those reinforcements (2wt.%) had maintained recovery more significantly than 80%. Shape recovery decreases to nea SMPU can be attributed to reinforcements with soft segments, contributing to shaping recovery. Meanwhile, adding reinforcements had minimal effect on hard segments. Thereby, shape fixity was greater than 95% for pure SMPU and SMPU + nanoclay composite, while decrement was observed for a composite of MWCNT, polypyrrole, and charcoal.

16.3.4.2 3D Printing Effect on SMP of SMPU

Thus, these experiments predict that neat matrix and composite were capable of displaying actuation via contact and non-contact mode, which can be utilized for various medial, sensing, and other engineering applications. The filament is exploited for the application of 4D printing technology. Literature surveys demonstrate that the 3D printing parameter is highly affected for properties, mechanical, optical, and shape memory. This study further expands to describe the effect of processing (3D printing) on SMPU shape memory properties, which was carried out with the same experiment as filament, a 3D strip of dimension 135 × 3 × 1.75 was printed, succeeding by shape memory analysis using spring force assembly and involute theory. Table 16.3 gives contrast shape memory properties of

TABLE 16.3

Comparative Summary of Pre-printing and Post-printing Geometry Memorising Parameters of SMPU

Geometry Memorising Cycle	Geometry Memorizing Parameters	Pre-printing SMPU	Post-printing SMPU
1st	$\theta_t - 1$	5°	8°
2nd	$\theta_t - 2$	8°	8°
1st	$\theta_r - 1$	346°	354°
2nd	$\theta_r - 2$	339°	328°
1st	$R_f - 1$	98.6%	97.8%
2nd	$R_f - 2$	97.8%	97.8%
1st	$R_r - 1$	96.1%	98.3%
2nd	$R_r - 2$	94.2%	91.6%
Average	**R_f**	**98.2%**	**97.8%**
Average	**R_r**	**95.15%**	**94.7%**

SMPU in pre-printing and post-printing conditions. Two sample trials were performed, and arithmetic means take as an average value for accuracy. Results of Table 16.3 display shape recovery and fixity close to pre-printing values as 94.7% and 97.8% correspondingly for post-printing, while pre-printing values are 95.15% for recovery of geometry and 98.2% for stressed structure. Conclusion of effected processing on shape memory polymer, especially on SMPU, was not much affected on shape transforming so that it will operate for 4D printing, but processing parameters need to choose carefully as per end application necessity.

16.4 CONCLUSION AND FUTURE SCOPE

16.4.1 FORTHCOMING PROSPECTIVE AND DIFFICULTIES

Fourth-dimension fabrication (4D printing) and shape memory polymers are expeditiously researched and an enormous area for exploring studies. The current topic is suffering from challenges listed below and some scope in the future for exploring ideas.

1. Extrusion with different parameters and observed effect of parameters on properties, especially on shape memory performance. Further, analyze the 3D processing effect on composite, same as SMPU material studied here.
2. Modeling of recovery material and its validation with experiments for recognized 4D printing modeling.
3. Reprogramming – Investigated materials mainly show one-way actuation that demands reprogramming after completing a cycle, not desire in plenty of applications.
4. The mechanism for actuation – Stimuli-responsive materials change shape when triggered by the environment. Actual utilization of material beyond

the laboratory necessitates actuation mechanisms such as control thermal environment in the colossal area, electrical supply or magnetic field generation, light actuation arrangements, etc., and handling and maintaining the surrounding at desire parameters.

5. The shape memorizing ability of material is not stable. After every cycle, it varies or degrades. It is a massive challenge to stabilize shape memory properties for end-application execution. Also, composite material is better for shape-changing, but compositional degradation and microstructural change limit its practical implementation.

6. Upcoming analysis can focus on 2W SMEs, such as liquid crystalline elastomers in a laboratory and practical application.

7. Development of characterization technique for 4D-printed parts – Researchers examine shape recovery and shape fixity on uncomplicated geometry like 3D strip; however, application of 4D printing cover aerospace, medical, soft actuator, robotics, and electronics, in which actual complicated structural change and its shape memory properties, stimuli effect, and composition effect need to optimize. Programming and artificial intelligence can resolve this by knowing the volumetric change of 4D-printed material with 3D scanning.

16.4.2 CONCLUSION

The incorporation of material's intelligence with additive manufacturing has fascinated the globe as a consequence of its magical time-dependant provocation comparable to the fantasy magic wand (the spell for 4D printing is stimuli). Present work was first challenged by the processing of composite in which dispersion of fillers in the matrix, rough surface during extruding owing to inappropriate temperature, porosity in wires on account of rotating speed of screw or zonal temperatures, and output diameter of wire, which requires due consideration of winding speed. Other challenges after processing could be printing parameters like the temperature of the bed or nozzle, infill density, speed of nozzle printing, etc., which required optimization for smooth printing of software-designed entity. The present work delineates SMPU printing with FDM under the optimized machine and environmental conditions. FTIR results confirmed the presence of polyurethane characteristic bonds inclusive of the ether-based matrix. It also delineated characteristic peaks of reinforcements though they displayed overlapping with a neat matrix with a tuned intensity of transmittance. Similarly, TGA results described stability and degradability of composites in which MWCNT+SM-PU and Polypyrrole+SM-PU bestowed a modest increase in onset temperature while NC-SMPU offered the lowest resistance. Nevertheless, all reinforcements had exhibited increased residual weight at 700°C, indicating intervention in chain scission. However, determination of transformation in geometry spring force assembly was a design and fabricated further analysis with involute path theory, body temperature region. In addition, transform capability was also examined at 60°C, and all composites endowed fixity > 90% owing to rigid molecular chains responsible for sustaining temporary shape while recovery > 80% for SMPU and nanoclay composite, another composite adversely performs with only > 65% shape fixity. The shape alternation effect also measured after processing on SMPU points out that no significant change occurs after 3D printing. Thus, formulated filaments after extruding and composites after 3D printing can have wide-range applications in biomedical, sensing, aerospace, automotive, and other engineering applications with gradient printing for multi-actuation.

REFERENCES

Anadão, P., L. F. Sato, and H. Wiebeck. 2018. "Study of the Influence of Graphite Content on Polysulfone-Graphite Composite Membrane Properties." *Journal of Thermal Analysis and Calorimetry* 134 (3): 1647–1656. 10.1007/s10973-018-7700-2

Ashhari, Shabnam, and Ali Asghar Sarabi. 2017. "Effects of Organically Modified Nanoclay Particles on the Mechanical Properties of Aliphatic Polyurethane/Clay Nanocomposite Coatings." *Polymer Composites* 38 (6): 1167–1174. 10.1002/pc.23680

Belka, Mariusz, and Tomasz Bączek. 2021. "Additive Manufacturing and Related Technologies – The Source of Chemically Active Materials in Separation Science." *TrAC Trends in Analytical Chemistry* 142 (September): 116322. 10.1016/j.trac.2021.116322

Cervantes-Uc, J.M., J.I. Moo Espinosa, J.V. Cauich-Rodríguez, A. Ávila-Ortega, H. Vázquez-Torres, A. Marcos-Fernández, and J. San Román. 2009. "TGA/FTIR Studies of Segmented Aliphatic Polyurethanes and Their Nanocomposites Prepared with Commercial Montmorillonites." *Polymer Degradation and Stability* 94 (10): 1666–1677. 10.1016/j.polymdegradstab.2009.06.022

Chen, Hairong, Hong Xia, and Qing-Qing Ni. 2018. "Study on Material Performances of Lead Zirconate Titanate/Shape Memory Polyurethane Composites Combining Shape Memory and Piezoelectric Effect." *Composites Part A: Applied Science and Manufacturing* 110 (July): 183–189. 10.1016/j.compositesa.2018.04.029

Fonseca, M.A., B. Abreu, F.A.M.M. Gonçalves, A.G.M. Ferreira, R.A.S. Moreira, and M.S.A. Oliveira. 2013. "Shape Memory Polyurethanes Reinforced with Carbon Nanotubes." *Composite Structures* 99 (May): 105–111. 10.1016/j.compstruct.2012.11.029

He, Jiahui, Haian Xie, Jie Hong, Zhaodongfang Gao, Zhikang Liu, and Chuanxi Xiong. 2018. "Self-Suspended Polypyrrole with Liquid Crystal Property." *Journal of Polymer Research* 25 (2): 56. 10.1007/s10965-018-1462-1

Khan, Tabrez A., Momina Nazir, Equbal A. Khan, and Ufana Riaz. 2015. "Multiwalled Carbon Nanotube–Polyurethane (MWCNT/PU) Composite Adsorbent for Safranin T and Pb(II) Removal from Aqueous Solution: Batch and Fixed-Bed Studies." *Journal of Molecular Liquids* 212 (December): 467–479. 10.1016/j.molliq.2015.09.036

Kim, Jong-Min, Jeong-Hyeon Kim, Jae-Hyeok Ahn, Jeong-Dae Kim, Sungkyun Park, Kang Hyun Park, and Jae-Myung Lee. 2018. "Synthesis of Nanoparticle-Enhanced Polyurethane Foams and Evaluation of Mechanical Characteristics." *Composites Part B: Engineering* 136 (March): 28–38. 10.1016/j.compositesb.2017.10.025

Lendlein, Andreas, and Steffen Kelch. 2002. "Shape-Memory Effect From Permanent Shape." *Angewandte Chemie (International Ed. in English)* 41: 2034–2057.

Liu, Ruiyuan, Xiao Kuang, Jianan Deng, Yi Cheng Wang, Aurelia C. Wang, Wenbo Ding, Ying Chih Lai, et al. 2018. "Shape Memory Polymers for Body Motion Energy Harvesting and Self-Powered Mechanosensing." *Advanced Materials* 30 (8): 1–8. 10.1 002/adma.201705195

Merlini, Claudia, Guilherme M.O. Barra, Matthäus D.P.P. da Cunha, Sílvia D.A.S. Ramôa, Bluma G. Soares, and Alessandro Pegoretti. 2017. "Electrically Conductive Composites of Polyurethane Derived from Castor Oil with Polypyrrole-Coated Peach Palm Fibers." *Polymer Composites* 38 (10): 2146–2155. 10.1002/pc.23790

Montero, Michael, Shad Roundy, and Dan Odell. 2001. "Material Characterization of Fused Deposition Modeling (FDM) ABS by Designed Experiments." *Proceedings of Rapid Prototyping & Manufacturing Conference*, 1–21. http://ode11.com/publications/sme_rp_2001.pdf.

Patadiya, Jigar, Adwait Gawande, Ganapati Joshi, and Balasubramanian Kandasubramanian. 2021. "Additive Manufacturing of Shape Memory Polymer Composites for Futuristic Technology." *Industrial & Engineering Chemistry Research* 60 (44): 15885–15912. 10.1021/acs.iecr.1c03083

Patadiya, Jigar, and Balasubramanian Kandasubramanian. 2021. "Progress in 4D Printing of Stimuli Responsive Materials." *Polymer-Plastics Technology and Materials*: 1–39. 10.1080/25740881.2021.1934016

Pielichowska, Kinga, Jakub Bieda, and Piotr Szatkowski. 2016. "Polyurethane/Graphite Nano-Platelet Composites for Thermal Energy Storage." *Renewable Energy* 91 (June): 456–465. 10.1016/j.renene.2016.01.076

Rajak, V. K., Sunil Kumar, N. V. Thombre, and Ajay Mandal. 2018. "Synthesis of Activated Charcoal from Saw-Dust and Characterization for Adsorptive Separation of Oil from Oil-in-Water Emulsion." *Chemical Engineering Communications* 205 (7): 897–913. 10.1080/00986445.2017.1423288

Rastogi, Prasansha, and Balasubramanian Kandasubramanian. 2019. "Breakthrough in the Printing Tactics for Stimuli-Responsive Materials: 4D Printing." *Chemical Engineering Journal* 366 (February): 264–304. 10.1016/j.cej.2019.02.085

Saeed, Mohsin Hassan, Shuaifeng Zhang, Yaping Cao, Le Zhou, Junmei Hu, Imran Muhammad, Jiumei Xiao, Lanying Zhang, and Huai Yang. 2020. "Recent Advances in The Polymer Dispersed Liquid Crystal Composite and Its Applications." *Molecules* 25 (23): 5510. 10.3390/molecules25235510

Sattar, Rabia, Ayesha Kausar, and Muhammad Siddiq. 2016. "Shape Memory and Physical Properties of Composites of Polyethylene Oxide/Poly(Propylene Glycol)- Block -Poly (Ethylene Glycol)- Block -Poly(Propylene Glycol)/2,4-Toluene Diisocyanate, Polypyrrole, and Modified MWCNT." *Polymer-Plastics Technology and Engineering* 55 (11): 1099–1114. 10.1080/03602559.2015.1132436

Subash, Alsha, and Balasubramanian Kandasubramanian. 2020. "4D Printing of Shape Memory Polymers." *European Polymer Journal* 134 (April): 109771. 10.1016/j.eurpolymj.2020.109771

Tian, Konghu, Zheng Su, Hua Wang, Xingyou Tian, Weiqi Huang, and Chao Xiao. 2017. "N-Doped Reduced Graphene Oxide/Waterborne Polyurethane Composites Prepared by in Situ Chemical Reduction of Graphene Oxide." *Composites Part A: Applied Science and Manufacturing* 94 (March): 41–49. 10.1016/j.compositesa.2016.11.020

Vargas, Patricia Cristine, Claudia Merlini, Sílvia Daniela Araújo da Silva Ramôa, Rafael Arenhart, Guilherme Mariz de Oliveira Barra, and Bluma Guenther Soares. 2018. "Conductive Composites Based on Polyurethane and Nanostructured Conductive Filler of Montmorillonite/Polypyrrole for Electromagnetic Shielding Applications." *Materials Research* 21 (5): 1–10, 10.1590/1980-5373-mr-2018-0014

Wang, Enliang, Yubing Dong, MD Zahidul Islam, Laiming Yu, Fuyao Liu, Shuaijie Chen, Xiaoming Qi, et al. 2019. "Effect of Graphene Oxide-Carbon Nanotube Hybrid Filler on the Mechanical Property and Thermal Response Speed of Shape Memory Epoxy Composites." *Composites Science and Technology* 169 (January): 209–216. 10.1016/j.compscitech.2018.11.022

Wang, Haihua, Yun Liu, Guiqiang Fei, and Jing Lan. 2015. "Preparation, Morphology, and Conductivity of Waterborne, Nanostructured, Cationic Polyurethane/Polypyrrole Conductive Coatings." *Journal of Applied Polymer Science* 132 (6). 10.1002/app.41445

Xie, Weilong, Ruoran Ouyang, Haoyu Wang, and Changren Zhou. 2020. "Construction and Biocompatibility of Three-Dimensional Composite Polyurethane Scaffolds in Liquid Crystal State." *ACS Biomaterials Science & Engineering* 6 (4): 2312–2322. 10.1021/acsbiomaterials.9b01838

Yang, B., W.M. Huang, C. Li, and L. Li. 2006. "Effects of Moisture on the Thermomechanical Properties of a Polyurethane Shape Memory Polymer." *Polymer* 47 (4): 1348–1356. 10.1016/j.polymer.2005.12.051

Yang, Dong, Guiquan Guo, Jianhua Hu, Changchun Wang, and Donglin Jiang. 2008. "Hydrothermal Treatment to Prepare Hydroxyl Group Modified Multi-Walled Carbon Nanotubes." *J. Mater. Chem.* 18 (3): 350–354. 10.1039/B713467C

Yanilmaz, Meltem, Fatma Kalaoglu, Hale Karakas, and A. Sezai Sarac. 2012. "Preparation and Characterization of Electrospun Polyurethane-Polypyrrole Nanofibers and Films." *Journal of Applied Polymer Science* 125 (5): 4100–4108. 10.1002/app.36386

Yu, Kai, Qi Ge, and H. Jerry Qi. 2014. "Effects of Stretch Induced Softening to the Free Recovery Behavior of Shape Memory Polymer Composites." *Polymer* 55 (23): 5938–5947. 10.1016/j.polymer.2014.06.050

Zacaroni, Lidiany Mendonça, Zuy Maria Magriotis, Maria das Graças Cardoso, Wilder Douglas Santiago, João Guilherme Mendonça, Sara S. Vieira, and David Lee Nelson. 2015. "Natural Clay and Commercial Activated Charcoal: Properties and Application for the Removal of Copper from Cachaça." *Food Control* 47 (January): 536–544. 10.1016/j.foodcont.2014.07.035

Zhang, Zhiyi, Huan Zhang, Ye Li, Haixiang Jia, Jinquan Shou, Kai Wen, Pengguang Bai, Yi Xue, Lizhong Bai, and Yaqing Liu. 2016. "Synthesis and Properties of Polyurethane/Graphite Elastomer by in Situ Technique." *Polymer Composites* 37 (5): 1318–1322. 10.1002/pc.23298

Zhao, Yufeng, Chul-Woong Cho, Longzhe Cui, Wei Wei, Junxiong Cai, Guiping Wu, and Yeoung-Sang Yun. 2019. "Adsorptive Removal of Endocrine-Disrupting Compounds and a Pharmaceutical Using Activated Charcoal from Aqueous Solution: Kinetics, Equilibrium, and Mechanism Studies." *Environmental Science and Pollution Research* 26 (33): 33897–33905. 10.1007/s11356-018-2617-7

Index

For Product Safety Concerns and Information please contact our EU
representative GPSR@taylorandfrancis.com
Taylor & Francis Verlag GmbH, Kaufingerstraße 24, 80331 München, Germany